World Cities
and Urban Form

How can the world's major cities and their urban regions be made more sustainable?
Is the planning and design of urban form a possible solution?

The book brings together theory and practice in the current debate about polycentric
development at the urban regional and city scale. At the regional scale, new
research and theory shows the forms metropolitan regions might take to enhance
sustainability. At the city scale, case studies based on the latest research and practice
from Europe, Asia and North America, show how both planning and flagship design
can propel cities into world class status, and also improve sustainability.

World Cities and Urban Form explores the tension between polycentric development
and urban fragmentation, not just physically, but also in a wider cultural, social and
economic context. Drawing on examples of cities that for one reason or another
have entered the world's consciousness, the book presents new information and
ideas that may resonate widely across the sector, and that will help to guide the
world's major cities towards more sustainable urban forms.

Mike Jenks is Professor Emeritus and Founder Director of the Oxford Institute for
Sustainable Development (OISD) at Oxford Brookes University, UK.

Daniel Kozak is Lecturer at the Faculty of Architecture, University of Buenos
Aires, Argentina, and PhD researcher at OISD, Oxford Brookes University, UK.

Pattaranan Takkanon is Lecturer at the Faculty of Architecture, Kasetsart
University, Bangkok, Thailand.

World Cities
and Urban Form

Fragmented, polycentric, sustainable?

Edited by
Mike Jenks, Daniel Kozak & Pattaranan Takkanon

LONDON AND NEW YORK

First published 2008
by Routledge
2 Park Square, Milton Park, Abingdon, Oxon, OX14 4RN

Simultaneously published in the USA and Canada
by Routledge
270 Madison Ave, New York NY 10016

Routledge is an imprint of the Taylor & Francis Group, an informa business

Transferred to Digital Printing 2009

Publisher's note
This book has been prepared from a camera-ready copy supplied by the editors

British Library Cataloguing in Publication Data
A catalogue record for this book is available from the British Library

Library of Congress Cataloging in Publication Data
World cities and urban form : fragmented, polycentric, sustainable? / edited by Mike Jenks,
Daniel Kozak and Pattaranan Takkanon.
p. cm.
Includes bibliographical references and index.
1. City planning--Case studies. 2. Cities and towns--Growth--Case studies. 3.
Metropolitan areas--Case studies. 4. Urban policy--Case studies. 5. Sustainable
development--Case studies. I. Jenks, M. (Michael) II. Kozak, Daniel. III. Takkanon,
Pattaranan.
HT166.W676 2008
307.1'216--dc22   2008003042

ISBN 10: 0–415–45184–1 (hbk)
ISBN 10: 0–415–45186–8 (pbk)
ISBN 13: 978–0–415–45184–0 (hbk)
ISBN 13: 978–0–415–45186–4 (pbk)

Contents

Section One Theoretical Approaches in a Global Context

Section Two Polycentric Regions and Cities: Perspectives from Europe, Asia and North America

Section Three Aspects of Fragmentation and Polycentrism

Contributors

Jaume Carné
Architect, Barcelona Regional,
Barcelona, Spain

Wanpen Charoentrakulpeeti
Lecturer, Department of Urban and
Regional Planning, King Mongkut's
Institute of Technology, Ladkrabang,
Bangkok, Thailand

Ben Derudder
Professor of Human Geography,
Department of Geography, Ghent
University, Ghent, Belgium; Research
Fellow of GaWC

Wael Salah Fahmi
Associate Professor of Urbanism,
Department of Architecture, Helwan
University, Cairo, Egypt

Oto Hudec
Associate Professor and Associate
Dean, Faculty of Economics, Technical
University of Kosice, Kosice, Slovak
Republic

Aleksander Ivančić
PhD, Mechanical Engineer, Barcelona
Regional, Barcelona, Spain

Mike Jenks
Professor Emeritus and Founder
Director, Oxford Institute for
Sustainable Development (OISD),
Oxford Brookes University,
Oxford, UK

Kiyonobu Kaido
Professor, Faculty of Urban Science,
Meijo University, Kani City, Japan

Nuttinee Karnchanaporn
Lecturer, School of Architecture and
Design, King Mongkut's University
of Technology Thonburi, Bangkok,
Thailand

Apiradee Kasemsook
Assistant Professor, Faculty of
Architecture, Silpakorn University,
Bangkok, Thailand

Tetsuo Kidokoro
Associate Professor, International
Development and Regional Planning
Unit, The University of Tokyo,
Tokyo, Japan

Daniel Kozak
Lecturer, Faculty of Architecture,
University of Buenos Aires, Argentina;
PhD researcher, Oxford Institute for
Sustainable Development, Oxford
Brookes University, Oxford, UK

Jeahyun Kwon
Doctoral candidate, Postgraduate
Course in Urban Science, Faculty of
Urban Science, Meijo University, Kani
City, Japan

Shih-wei Lo
Professor, Department of Architecture,
Tunghai University, Taichung,
Taiwan, ROC

Olli Maijala
Senior Advisor, Ministry of the
Environment, Land Use Department,
Helsinki, Finland

Peter Marcuse
Professor Emeritus of Urban Planning
and Special Lecturer, Division of
Urban Planning, Graduate School
of Architecture, Planning and
Preservation, Columbia University,
New York, USA

Vijay Neekhra
Doctoral candidate, International
Development and Regional Planning
Unit, The University of Tokyo,
Tokyo, Japan

Takashi Onishi
Professor, International Development
and Regional Planning Unit, The
University of Tokyo, Tokyo, Japan

Oana-Liliana Pavel
Doctoral candidate, Department of
Urban Planning, University of Paris–
Sorbonne, Paris, France

Zhu Qian
PhD candidate, Department of
Landscape Architecture and Urban
Planning, College of Architecture,
Texas A & M University, College
Station, USA

Darko Radović
Professor, Centre for Sustainable
Urban Regeneration (cSUR), The
University of Tokyo, Tokyo, Japan, and
the University of Melbourne, Australia

Judy Rogers
Lecturer, Landscape Architecture
Program, RMIT University,
Melbourne, Australia

Rauno Sairinen
Professor of Environmental Policy,
'Forestry, Environment and
Society' – Centre of Excellence,
University of Joensuu, Joensuu,
Finland

Sidh Sintusingha
Lecturer, Faculty of Architecture,
Building and Planning, the University
of Melbourne, Melbourne, Australia

May So
Intern Architect, Henriquez Partners
Architects, Vancouver, Canada

Pattaranan Takkanon
PhD graduate, University of
Queensland, Australia, and Lecturer,
Faculty of Architecture, Kasetsart
University, Bangkok, Thailand

Nataša Urbančíková
Associate Professor and Head of
Department of Regional Sciences and
Management, Faculty of Economics,
Technical University of Kosice, Kosice,
Slovak Republic

Frank Witlox
Professor of Economic Geography,
Department of Geography, Ghent
University, Ghent, Belgium

Jiaping Wu
Research Fellow, School for Social
and Policy Research, Charles Darwin
University, Australia

Perry Pei-Ju Yang
Assistant Professor, Department of
Architecture, School of Design and
Environment, National University of
Singapore, Singapore

Willi Zimmermann
Associate Professor, Asian Institute of
Technology, Bangkok, Thailand

Acknowledgements

Our thanks go to all those involved in the writing and production of this book. We are grateful to all the contributors for their chapters and for their patience during the long time it has taken for this book to be published. We are indebted to Margaret Ackrill for her pertinent comments and sub-editorial work throughout this book. Our gratitude also goes to Khatiya Chatpetch for the numerous illustrations she prepared in chapters throughout the book, and to Roberto Rocco, Devisari Tunas, Tuca Vieira and Sebastián Szyd for the use of their photographs to introduce sections two and three and in Chapter 15.

The tasks required in preparing this book in a camera-ready form would not have been possible without the support of the Dutch Ministry of Housing, Spatial Planning and the Environment (VROM), and our thanks go particularly to Hans Verspoor for his help. Our appreciation goes to colleagues of the International Urban Planning and Environment Association (IUPEA), to colleagues at the Faculty of Architecture, Kasetsart University in Bangkok who were responsible for the organisation of the UPE7 Symposium in January 2007, to Rod Burgess, Department of Planning, Oxford Brookes University, and to colleagues in the Oxford Institute for Sustainable Development at Oxford Brookes University, UK.

Our deepest gratitude goes to our families and friends for their support during the preparation of the book. In particular we would like to give our warmest thanks to Margaret Jenks, Jessica Borenstein, Lt. General Yongyoot Takkanon, Rampai Takkanon and Jiranat Takkanon for all their support and forbearance during the intensive work in bringing this book to fruition.

Introduction

Mike Jenks, Daniel Kozak and Pattaranan Takkanon

Introduction:
World Cities and Urban Form

Introduction

The tipping point from a predominantly rural to an urban world population is now predicted to happen in 2008 (United Nations, 2006). Although just a fraction of a percentage, the change is a highly emotive one. It has meant that cities have received more attention in the media and have become embedded in the public consciousness. The emotional impact comes from the fact that it is a world, rather than a regional change. While in 1950, Europe, Northern America and Oceania were largely urbanised, the rest of the world had predominantly rural populations. At that date the world's urban population was 732 million (approximately 30% of the total population). By 2030 it is estimated that some 60% of the population will be urban, representing a total of 4,912 million people, an increase of about seven times the numbers that were living in cities in the 1950s. Asia and Africa are the regions where explosive growth is occurring, and where cites and urban agglomerations of 10 million or more people are becoming ever more common. While in 1950 there were two mega-cities in the world – New York and Tokyo with populations of 10 million or more – by 2005 there were 20, and the number is growing (United Nations, 2006). Cities have been getting larger and the number of million plus cities now stands at 471 worldwide (Brinkhoff, 2007). Urban populations swell with the influx of people from rural areas with dreams of a better lifestyle and standard of living. Aspirations that are sharpened through globalisation, the ever-present influence of transnational corporations, of global brands, of the media, of communications through the sharp rise in mobile phone networks and the Internet, and of increasing inflows and outflows of foreign direct investment (e.g. UNCTAD, 2006; Smith, 2003). Throughout the world, major cities have the trappings of the global economy. Not only do the citizens aspire to the best that appears on offer globally, but also those who govern the cities have the ambition to make a mark on the world – to have their cities become 'world class'.

Fig. (opposite page). London – a global city, a world city, and also a 'world class' city embedded in the world's conciousness through its iconic (and cliched) cultural images
Source: Mike Jenks

Where better to start? This book has an ambitious aim, to link together a number of key concepts in the contemporary urban debate. By taking the idea of the world city and recognising the scale of urbanisation, and then raising the issue of the form of these urban agglomerations, leads to a number of questions.

World cities

First is the question of world cities. There is a vast theoretical corpus of work, mainly built up over the past fifteen years, about global networks and interconnected mega-

3

cities and regions. Many of these large cities are now classified as world cities, and some as global cities, and the definitions of them are precise and rigorous (see Chapter 1). The bases for these classifications usually relate to the extent to which such cities play a role in the world economy (e.g. Sassen, 2001), or their hierarchy as major service centres or centres for multinational enterprises (e.g. Friedmann, 1986), or through the extent and nature of the communication exchange between them (e.g. Hall and Pain, 2006). The resulting classifications, such as that by the Globalisation and World Cities Study Group and Network (Beaverstock *et al.*, 1999) provide a solid foundation for research and comparison. But whether this precision should lead to the exclusion of other terms is by no means consensually accepted. Frequently used terms, such as 'world class city', that are subject to many meanings and interpretations, do bring with them a qualitative dimension that most of the abovementioned studies lack. The question of urban space and form, in particular, has not been sufficiently examined in world city research. Many cities are part of the world's consciousness, either for the quality of their environment or architecture, for the quality of life they offer, or the experience of visiting them. It is interesting to note that the four cities that top the world city list – London, Paris, New York and Tokyo – are 39th, 33rd, 48th and 35th respectively on a quality of living survey, an interesting observation even if not methodologically comparable (Mercer, 2007). Indeed, if issues of sustainability were measured then it is probable that another entirely different ranking would occur. There is a further aspect that needs attention, and that is the spatial analyses and internal transformation of world and other large cities subject to the pressures of globalisation and growth. So the question arises as to whether there are other dimensions to world cities that could usefully be considered.

Polycentrism

A second consideration, then, is about urban form. Are there any aspects of world cities, mega-cities or global city regions that have some common currency in terms of their physical form? It is widely acknowledged that the form of cities is changing, and among the descriptions of these physical-spatial changes two concepts have come to the fore: polycentrism and urban fragmentation. At the largest scale the form of cities has clearly extended well beyond the European model of the contained compact city with its monocentric form. It is these forms that have generally been associated with claims to urban sustainability (e.g. Jenks *et al.*, 1996; Jenks and Burgess, 2000). But this is a particular form, as many cities have grown around clusters of small towns or villages and retain a measure of polycentrism, some, such as in Asia, have always been polycentric in form, and others have outgrown a simple single centre developing new centralities in suburbs and at their edge. A similar process occurs at the regional level, where towns and cities have formed polycentric urban regions. Along with growth, metropolitan regions and cities tend to move from monocentric forms towards polycentric structures. Usually this is an evolutionary process, one that happens over time, driven by external forces such as globalisation and population migration. But in some instances this process has not only been accepted but also incorporated into planning policy objectives to control and connect the centralities with efficient public transport to form a more integrated whole (Faludi *et al.*, 2002). Nevertheless, cities are clearly breaking up into new centralities, and despite some attempts at integration, polycentric forms often intensify fragmentation rather than reverse it.

Fragmentation

The fragmentation inherent in polycentric urban forms is the third of the concepts that demands some attention. The concept of fragmentation in urban studies has been used and interpreted in many various ways (see Chapter 15). When it is used to describe the 'dividing', 'splintering' or 'partitioning' of contemporary metropolises (Fainstein *et al.*, 1992; Graham and Marvin, 2001; Marcuse and van Kempen, 2002) it is widely understood as a negative phenomenon. It also seems to be generally agreed that currently this phenomenon is common to most large cities worldwide, even in the planned polycentric metropolitan regions. The characteristics of urban fragmentation are well-known and widely discussed (e.g. Burgess, 2005; Ghent Urban Studies Team, 2002). In the largest cities, new centralities can be of sufficient size and complexity to act as mini-cities in themselves, and usually are poorly connected and car-dependant. Spatial disparities between rich and poor are often clearly geographically delineated, and at the extremes characterised by gated 'communities', ironically disconnected to the realities of genuine local communities. Gentrification through urban regeneration and the privatisation of public spaces can act to further fragment cities and the new centralities. Even the introduction of mass rapid transit, which in theory should integrate, can lead to fragmentation simply by its cost and unaffordability to the poorer sectors of the urban population. Thus fragmentation occurs either through the impacts of globalisation, neo-liberal economic policies and *laissez-faire* planning, or through interventions with the best of intentions, but with unintended consequences. Generally urban fragmentation is a force that exacerbates problems of urban unsustainability.

Sustainability

This leads to a key question at the heart of this book. If world cities, metropolitan regions and mega-cities are developing polycentric urban forms, and these forms are tending to become fragmented – moving away from traditional 'sustainable' urban forms – what can be done about it? How can a measure of sustainability be achieved? However, it has to be admitted that cities are not, and probably will never be, sustainable – their ecological footprint will always extend well beyond their capacity to be self-sustaining, although when the metropolitan or wider region is taken into account then sustainability becomes more of a possibility. So the more realistic question is how can world cities and mega-cities be made more sustainable than they are at present, or at least, less unsustainable?

The tension between polycentric forms and urban fragmentation provides a key. Clearly the 'traditional' monocentric compact city has many sustainability advantages – these are the well known high density, mixed use, transport efficient and socially and economically diverse forms. But rapid urban growth, sheer scale and size make a 'compact' solution impractical. However, it could be argued that the emerging polycentric forms could be more manageable and possible to plan and design in a more sustainable way. If these centralities followed concepts of containment, of sustainable built form and design, of forms that encouraged walking and cycling, that managed public transport well, they might be more sustainable than if their growth was unplanned and unconstrained. In other words they could become a series of 'compact cities'. But that alone would not solve the problems when, as is often the case, these centralities were surrounded by a network of highways, and disconnected to each other and the city as whole. The need is to ensure that there are efficient, and affordable public transportation modes that link the whole city and regional network, and that the centralities of these networks are socially inclusive.

The debate about world cities has largely concentrated on globalisation, business and communication networks. Linked to world cities are the growth of mega-cities and very large cities, and the debate there has often been at the level of strategic and urban planning, and large scale issues of urban form. But there are gaps in the debate, which include the finer grain understanding of the impacts of these forces and large scale physical forms. They include issues of quality of life, quality of the urban experience and sustainability that embraces physical form, as well as the environment and social and economic sustainability. The aim of this book is to explore some of these complexities and tensions, and to provide some insights into ways in which to resolve them.

Structure of the book

The first section – *Theoretical Approaches in a Global Context* – sets the scene with an overview and analysis of the key concepts contained in the book's title. Derudder and Witlox provide definitions and the conceptual underpinning of the terms global city, world city and global city-region, and argue that great care is needed when using terms such as 'world class' which has little scientific basis or use as an analytical tool. The impact of globalisation on urban form is then discussed by Marcuse, who identifies a wide range of trends in the form of cities and built form, noting that however powerful the forces of globalisation, cities are still subject our influence. A further strand to the debate is offered by Radović who argues that quality is a key missing element, and demonstrates that it is of vital importance to include the concept 'urbanity' with respect to any cities that might be classified as 'world class'. Rogers turns our attention to the sustainable city uncovering some of the discourses that underlie the concept, and questions who it is for, noting that one of the missing aspects is that of social equity. Finally Jenks and Kozak link urban form to sustainability, concentrating on polycentric forms and the concept of urban fragmentation – looking towards the potential of urban form to enhance the sustainability of world cities.

The second section – *Polycentric Regions and Cities: perspectives from Europe, Asia and North America* – considers polycentric urban forms through a number of case studies. In the European context, the Helsinki Metropolitan Region, and the mega region of Bratislava/Vienna are considered from different points of view. Maijala and Sairinen show the impact of urban consolidation policies in a context of urban sprawl, regional fragmentation and polycentrism, and what can be done to promote more consolidated and sustainable urban forms. Hudec and Urbančiková consider how a city in a formerly communist country can compete and co-operate with the economically more powerful city of Vienna, considering the importance of human and social capital, and of competitiveness. Finally, the exemplary model of Barcelona is analysed by Carne and Ivančić. They show how the city has evolved into a model of urbanity and sustainable urban form over the past 30 years, and draw on this experience to indicate what will develop in the future.

Three perspectives on Asian regions are offered. Sintusingha analyses cases in cities in three countries – Melbourne, Bangkok and Hanoi – considering the tension between 'world class' and global aspirations, and the local contexts of sprawling suburban development. He argues that an open-ended diversification may be a better path to sustainability than the tendencies of such developments towards homogenisation. The mega city regions of Japan are analysed by Kaido and Kwon, and some comparisons made with European regions. From their findings, they argue that if city regions are to be 'world class' then they should balance quality of life, mainly accessibility to essential facilities, with economic and spatial growth, and sustainable urban forms. The rapid urban growth in China is considered by Jiaping Wu with a study of Shanghai's peri-urban area. Key issues of development zones and environmental impacts are considered, and it is suggested that lessons can be drawn from Shanghai by other fast-growing cities where peripheral growth rates are high. The final Asian example is the city of Taichung in Taiwan. Shih-wei Lo illustrates the attempts of the city administration to propel the city to 'world class' status by the commissioning of flagship buildings, a process akin to the regeneration of Bilbao and its place on the world's consciousness.

The second section concludes with two contrasting examples for North America – one with strong planning controls, and the other with *laissez-faire* governance. May So describes the way Vancouver has kept the concepts of quality and liveability at the heart of its policies, and how this policies such as EcoDensity are evolving into making Vancouver a sustainable city. By contrast Zhu Qian analyses Houston, a city with private sector driven *laissez-faire* planning, where controls are exercised, not through zoning, but often by private restrictive covenants. Notwithstanding, Houston is a world city, although unlikely to be a sustainable one.

The third section – *Aspects of Urban Fragmentation* – considers the propensity of polycentric forms to increase fragmentation, and illustrates through case studies the impacts of fragmentation on urban form. Kozak sets out the theoretical underpinning to the debate about urban fragmentation. He then illustrates its impacts in a case study of Buenos Aires, showing how fragmented urban forms, normally associated with peripheral development, have been imported into central city locations. It is not just the built environment that suffers from fragmentation, and Yang shows in Singapore, the natural environment can also be fragmented through development. He also shows the interaction between urban and natural forms and competing development through the extent of the materials used in the construction industry. The tensions over the use of limited space are also demonstrated in Cairo by Fahmi, who shows the complexity of competing demands from different stakeholders, and the impacts these have on urban form. A related point is shown in two studies of Bangkok, where the influence of the middle-class stakeholders can be seen in their living environments and use of transport. Karnchanaporn and Kasemsook examine the lifestyle aspirations that are sold through roadside billboards to the urban middle-class, and show the impact through the growth in numbers of gated communities. The choices for middle-class living is reflected in transport choices and opportunities in Bangkok, and Charoentrakulpeeti and Zimmermann show, despite the introduction of mass transportation, how policies and choices favour the middle-class and exclude the poor. The poor are the subject of the final two chapters in this section, and both graphically illustrate urban fragmentation through poverty. Pavel considers poverty and urban form in Paris and Bucharest, and shows that in both cities there are geographically fragmented areas of poverty and disadvantage, as well as gentrified areas that are the preserve of the wealthy. More extreme is the case of India portrayed by Neekhra, Onishi and Kidokoro. Whereas in many cities in the world slums are relatively small fragmented areas, in Mumbai 54% of the population live in slums. The authors question how the mega-cities of India can ever be considered 'world class' when such extremes of poverty and wealth exist side by side.

A common critique to the concept of sustainability argues that it has been used and stretched so extensively that it has become difficult to understand what ideas underlie this term. The same could be said, albeit to a lesser degree, for the concepts of polycentrism and urban fragmentation. Throughout this book these three concepts are systematically challenged, assessed and [re]defined. What is the significance of discussing these concepts? Are there any lessons to be learned from current urban practices? What are the commonalities of the global examples presented here? The conclusion – *The Form of Cities to Come?* – attempts to answer these questions, pull together the various themes and tackle the issue of whether there are any pathways or processes in the context of world, even 'world class', cities to achieving more sustainable urban forms.

References

Beaverstock, J., Smith, R. and Taylor, P. (1999) A roster of world cities. *Cities*, **16**(6): 445–458.

Brinkhoff, T. (2007) *City Population*, Retrieved in February 2007 from: http://www.citypopulation.de/World.html

Burgess, R. (2005) Technological Determinism and Urban Fragmentation: A Critical Analysis, In: *9th International Conference of the ALFA-IBIS Network on Urban Peripheries*, Pontificia Universidad Católica de Chile, Santiago de Chile, July 11th–13th 2005.

Fainstein, S. S., Gordon, I. and Harloe, M., eds (1992) *Divided cities: New York and London in the contemporary world*, Blackwell, Oxford.

Faludi, A., Waterhout, B. and Royal Town Planning Institute (2002) *The making of the European spatial development perspective: no masterplan*, Routledge, London.

Friedmann, J. (1986) The world city hypothesis. *Development and Change*, 17: 69–83.

Ghent Urban Studies Team (2002) *Post ex sub dis: urban fragmentations and constructions*, 010 Publishers, Rotterdam.

Graham, S. and Marvin, S. (2001) *Splintering urbanism: networked infrastructures, technological mobilities and the urban condition*, Routledge, London.

Hall, P. and Pain, K. (2006) *The Polycentric Metropolis: Learning from mega-city regions in Europe*, Earthscan, London.

Jenks, M., Burton, E. and Williams, K. (1996) *The Compact City: A sustainable urban form?*, E & FN Spon, London.

Jenks, M. and Burgess, R. (2000) *Compact Cities: Sustainable urban forms for developing countries*, Spon Press, London.

Marcuse, P. and van Kempen, R., eds (2002) *Of states and cities: the partitioning of urban space*, Oxford University Press, Oxford.

Mercer (2007) *Highlights from the 207 Quality of Living Survey*, Retrieved in February 2007 from: www.mercer.com

Sassen, S. (2001) *The Global City: New York, London, Tokyo*, Princeton University Press, Princeton [First published 1991].

Smith, D. (2003) *The State of the World Atlas*, Earthscan, London.

UNCTAD (2006) *World Investment Report 2006, United Nations*, Retrieved in February 2007 from:http://www.unctad.org/Templates/Page.asp?intItemID=3975&lang=1

United Nations (2006) *World Urbanization Prospects: The 2005 Revision, United Nations, DESA, Population Division*, Retrieved in February 2007 from: www.un.org/esa/population/unpop.htm

Section One

Theoretical Approaches in a Global Context

Ben Derudder and Frank Witlox

1

What is a 'World Class' City?
Comparing conceptual specifications of cities in the context of a global urban network

Introduction

Throughout the last two decades, a number of scholars have tried to specify the idea of 'world class' cities in the context of a global urban network. This chapter compares the most important contributions in this research field, i.e. the efforts undertaken by John Friedmann (world cities), Saskia Sassen (global cities), and Allen Scott (global city-regions). First, we argue that the variety of terms employed in this research field is not a trivial matter of semantics. Second, we show how empirical research nonetheless tends to condense these analytical differences into a series of fuzzy concepts. The major implication is that if research on 'world class' cities wants to have analytical purchase, it needs to draw on proper conceptual specifications rather than refer to vague if commonsensical descriptions.

The key question raised in this chapter stems from a recently held international conference in Bangkok,[1] which had the basic objective of clarifying and discussing some of the issues related to the rise of so-called 'world class' cities. The rationale for this 'world class' city theme was the observation that this term has gained substantial currency in urban research at large. There can be no doubt that this is indeed the case, but one obvious question immediately springs to mind: exactly what is a 'world class' city?

It can easily be argued that this is a somewhat trivial question in the light of such a large-scale conference: we do not necessarily need to agree on a precise definition to discuss some of the issues associated with it. Furthermore, it should not be very difficult to come up with a more or less knowledgeable description of this elusive term. Conceptually, for instance, the term 'world class' city clearly refers to the aspiration to join the league of 'major' cities in a globalized economy. From this vantage point, the term 'world class' city is simply a discursive imaginary that helps frame national and urban politics around a 'going-global' agenda. Empirically, the term invokes the presence of 'assets' that are deemed necessary to be taken 'seriously' in this global gold rush: being a 'world class' city at least entails the presence of well-connected international airports, major hotel chains, and a 'climate' that is somehow conducive to inviting and redirecting globalized capital. Furthermore, prestigious mega-projects (Moulaert *et al.*, 2003) or bidding for World Expos or the Olympics (Short, 2004) are clear-cut signs of a city's aspirations in this context.

It is, however, at the same time unacceptable to succumb to such a general and uncritical description. There are two reasons for this. First, without close scrutiny, the very idea of 'world class' cities may well imply a naive surrender to

Fig. (opposite page). Tokyo, a first rank Global City, attracting global brands and 'world class' design: the Prada Store by Herzog and de Meuron (2003), Aoyama district, Tokyo Source: Mike Jenks

Fig. 1. (left) Marketing campaign to put Hong Kong on the map as 'Asia's world city'.

Fig. 2. (right) Advertisement for producer service firm Deloitte on Schiphol Airport (Amsterdam, 14/4/2004).

neo-liberal 'development' agendas. Policies for pursuing 'world class' city-status are often socially regressive to say the least. Such themes reflect what Harvey (1989) depicted as a shift in urban politics from 'urban managerialism' to 'urban entrepreneurialism', with the latter form of urban politics being preoccupied with a concern over economic expansion and inter-urban rivalry. Sassen's (2001a) seminal work on 'global cities', in turn, shows how the rise of global urban connectivity goes hand in hand with increased social and economic inequality (see also Marcuse and Van Kempen, 2000). Smith (1996; 2002) has gone even further by arguing that contemporary forms of gentrification in major cities around the globe are deliberate urban strategies that are closely connected with neoliberal visions of the 'necessity' of inserting a city into circuits of global capital. Taken together, this implies that the very idea of 'world class' cities requires careful examination rather than merely accepting it as some sort of reasonable research mantra.

The second reason why the term 'world class' city requires further elaboration is that it represents by no means a scientific concept: an overview of some commonsensical characteristics and the commonplace observation that cities such as London, New York, Tokyo, and Paris are undoubtedly 'world class' cities makes little impression in conceptual terms. To paraphrase Markusen (1999): the idea of 'world class' cities may connote, but it does not denote; it is not really clear which processes and characteristics are captured (and which are not), and the ensuing lack of conceptual rigor may fuel the growth of a chaotic concept that does not inform us of anything at all. As a consequence, the whole idea of there being 'world class' cities is in and by itself an analytically meaningless starting point and more precise specification is required. The basic purpose of this chapter, therefore, is to provide a critical review of the most important accounts of globalized urbanization. This vantage point implies that we will largely ignore the research that focuses on the urban politics of 'going global'.

Conceptualizations of cities in a global economy

Throughout the last two decades, researchers have analyzed the emergence of a 'global urban network' centred on a number of key cities in the global economy (for an overview of recent advances, see Taylor *et al.*, 2006). Taken together, these studies are loosely united in their observation that 'world class' cities such as New York and London derive their importance from a privileged position in transnational networks of capital, information, and people. There is, in other words, a widespread consensus that under conditions of contemporary globalization an important city 'is no longer identifiable for its stable embeddedness in a given territorial milieu. It is instead a changing connective configuration with variable actors which can be thought of as 'nodes' of local and global networks' (Dematteis, 2000, p. 63). In this section we review the main ways in which cities have been conceptualized in the context of a global urban network, i.e. as world cities, as global cities, and as global city-regions. Based on this overview, we contrast some of the alleged characteristics of cities within each of these conceptualizations.

World cities

The world city concept can be traced back to two interrelated papers by Friedmann and Wolff (1982) and Friedmann (1986). Both texts framed the rise of a global urban network in the context of a major geographical transformation of the capitalist world-economy. This restructuring, most commonly referred to as the 'New International Division of Labor', was basically premised on the internationalization of production and the ensuing complexity in the organizational structure of multinational enterprises (MNEs). This increased economic-geographical complexity requires a limited number of control points in order to function, and world cities are deemed to be such points. The territorial basis of a world city is more than merely a CBD, since

> reference is to an economic definition. A city in these terms is a spatially integrated economic and social system at a given location or metropolitan region. For administrative purposes the region may be divided into smaller units which underlie, as a political or administrative space, the economic space of the region (Friedmann, 1986, p. 70).

Friedmann (1986) tries to give theoretical body to his 'framework for research' by subsuming it under the heading of Wallerstein's world-systems analysis, hence the title of Knox and Taylor's (1995) *World Cities in a World-System* and the hyphen in 'world-economy' (see also Saey, 1996; Taylor, 2000). As is well known, Wallerstein (1979) envisages capitalism as a system that involves a hierarchical and a spatial inequality of distribution based on the concentration of relatively monopolized and therefore high-profit production in a limited number of 'core' zones. The division of labour that characterizes this spatial inequality is materialized through a tripolar system consisting of core, semi-peripheral and peripheral zones. The prime purpose of world city research, now, is that it seeks to build an analytical framework that searches to deflect attention from the role of territorial states in the reproduction of this spatial inequality (Brenner, 1998, p. 4). Territorial states have, of course, been prime actors in the unfolding of this uneven development, but drawing on the work of Mann (1986) and Dodgshon (1998), it can be suggested that the world-economy is radially rather than territorially managed. This means that the economic and political power of core territories is in fact spatially structured along well-defined routes that link centres of control via available authorative and allocative resources. Hence, what is commonly labelled as 'core' in world-systems analysis does not necessarily consist of a series of 'strong' territorial states, but of a hierarchy of major and lesser centres (i.e. world cities) that thereupon diffuse their status and function over a wider area and at different scales (Dodgshon, 1998, p. 56).

In other words: despite 'being largely studied through its mosaic of states ... the modern world-system is defined by its networks' (Taylor, 2000, p. 20), and world cities are the key nodes in such networks of power and dominance. Apart from being the economic power-houses of the world-system, world cities are also locales from which other forms of command and control are exercised, e.g. geopolitical and/or ideological-symbolical control over specific (semi-)peripheral regions in the world-system. Miami's control position over Central America is a case in point (Grosfoguel, 1995). Friedmann (1986, p. 69) reminds us, however, that 'the economic variable is likely to be decisive for all attempts at explanation', whereby major importance attaches to corporate headquarters and international financial institutions and agencies. Although the presence of a business services sector and/or a well-developed infrastructure seems to be required, the latter are conceptually less important, since they are necessary but not sufficient conditions in the formation of a network of world cities.

Global cities

The global city concept can be traced back to the publication of Saskia Sassen's *The Global City* in 1991. Sassen looks afresh at the functional centrality of cities in the global economy, and does so by focusing on the attraction of producer service firms to major cities that offer knowledge-rich and technology-enabled environments. In the 1980s and 1990s, many such service firms followed their global clients to become important MNEs in their own right, albeit that service firms tend to be more susceptible to the attractions of agglomeration economies offered by city locations. These emerging producer service complexes are at the root of global city-formation, which implies a shift of attention to the advanced servicing of worldwide production. Hence, from a focus on formal command power in the world-system, the

> emphasis shifts to the practice of global control: the work of producing and reproducing the organization and management of a global production system and a global market-place for finance ... Power is essential in the organization of the world economy, but so is production: including the production of those inputs that constitute the capability for global control and the infrastructure of jobs involved in this production (Sassen, 1995, p. 63-64).

Through their transnational, city-centred spatial strategies, producer service firms have created worldwide office networks covering major cities in most or all world regions, and it is exactly the myriad connections between these service complexes that gives way, according to Sassen (2001a, p. xxi), to the 'formation of transnational urban systems'. This urban network, Sassen (1994, p. 4) argues, results in a new geography of centrality that may very well cut across existing north/south divides. Hence, rather than reproducing existing core/periphery patterns in the world economy, this network may break through these divides.

The focus on urban agglomeration economies has a major implication for the territorial demarcation of global cities. Rather than being structured in mutual dependence with a hinterland, the functional centrality of global cities becomes 'increasingly disconnected from their broader hinterlands or even their national

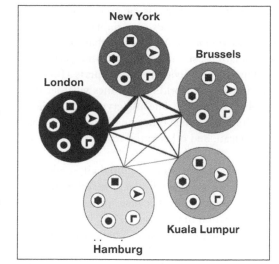

Key

r Arthur Andersen

■ Dresdner Bank

● Clifford Chance

◗ Booz Allen & Hamilton

► Lloyd's

economies' (Sassen, 2001a, p. xxi). To territorially demarcate global cities, Sassen (2001b, p. 80) therefore opts 'for an analytical strategy that emphasizes core dynamics rather than the unit of the city as a container – the latter being one that requires territorial boundary specification.' This does not necessarily imply that the functional centrality in global cities is a simple continuation of older centrality patterns as in New York City, since the territorial basis can consist of 'a metropolitan area in the form of a grid of nodes of intense business activity, as we see in Frankfurt and Zurich' (Sassen, 2001a, p. 123). It is nonetheless clear that the proper unit of analysis may very well be smaller than the 'metropolitan region'. Tokyo as a global city, for instance, is the 'Tokyo Metropolis' rather than the larger 'Tokyo Metropolitan Region' or the 'National Capital Region' (Sassen, 2001a, p. 371).

Global city-regions

Despite earlier contributions by Petrella (1995) and Veltz (1996), the global city-region concept is most commonly traced back to the work of Allen Scott (2001a; b; c), who has conceptualized global city-regions as the new key territorial units in a post-Fordist global economy. Following this lead, Scott maintains that the role of nation-states as chief territorial-organizational nexuses is increasingly being overtaken by an extensive archipelago of global city-regions. Perhaps ironically, a major driving force in this organizational transition can be found within the territorial scale which global city-regions are thought to replace. That is, the growth of global city-regions is crucially conditioned by a re-scaling of the national frameworks in which they are embedded. The latter re-scaling has been induced by new economic-geographic strategies of transnational-oriented capitalist firms, which are increasingly trying to circumvent and restructure the nationally organized Fordist–Keynesian régimes of accumulation, with their nationally organized forms of social, monetary and labour regulation (Brenner, 1999, p. 68).

In simple geographic terms, global city-regions consist of dense megalopolises that are bound up in intricate ways in intensifying far-flung transnational relationships.

They represent an outgrowth of large metropolitan areas or contiguous sets of metropolitan areas, together with surrounding hinterlands of variable extent which may themselves be sites of scattered urban settlements. Although global city-regions are invariably characterized by large populations, size is not their defining criterion. Rather, it is the extent to which the development of individual settlements is a function of the gravitational power of major regions of the world that defines whether they are contributing to the build-up of a global city-region. The archetypical example of a global city-region is the Pearl River Delta Area, which functionally connects Hong Kong, Shenzhen, Guangzhou, Zhuhai, Macau, and other small towns in the region to the global economy.

Scott *et al.* (2001, p. 17) stress that an identification of global city-regions in terms of population size is therefore not very adequate, but find it nonetheless feasible to refer to the world map of large metropolitan areas, because '[p]roduction and performance are ... raised by urban concentration in two ways. First, concentration secures overall efficiency of the economic system. Second, it intensifies creativity, learning, and innovation both by the increased flexibility of producers that makes it possible and by the enormous flows of ideas and knowledge that occur alongside the transactional links within localized production networks'. This hesitant crisscrossing between urban morphology and regionalized production networks make it unclear how a global city-region can be demarcated, but is clear that an initial threshold of morphological concentration must be surpassed to initiate a growth pole-like process.

Strictly speaking, there is no conceptual necessity to introduce a metropolitan region, since the requirement of a focal point is nowhere discussed, let alone demonstrated. Hence, although most global city-regions are constructed around a major city (in the functional and the morphological sense), there may well be global city-regions without a major city. Henton (2001, p. 398-399), for instance, addresses Silicon Valley as a 'global city-region', and paints a picture that is reminiscent of a region rather than a city: 'There were over twenty-seven local jurisdictions in the region and little cooperation. Silicon Valley has become so big and complex it had trouble dealing with regional challenge. No city by itself had the resources or authority to meet the big challenges.' This makes 'global city-regions' look very much like the 'regions' in Scott's (1998) earlier work, and it is therefore no surprise that some of the chapters in Scott's (2001c) edited book refer to regions rather than to cities or city-regions (e.g., Ohmae, 2001; Porter, 2001). The most important point for the present discussion is that a global city-region may stretch well beyond the confines of a metropolitan region.

The transnational network in which global city-regions are embedded has an 'archipelago' rather than a 'core/periphery' structure. Indeed, although 'not all large metropolitan areas are equally caught up in processes of globalization' (Scott, 2001b, p. 1), it can be noted that 'the process of urban and regional development we are describing here are not limited to the wealthiest countries'. As a consequence, Scott (2001a, p. 822) sees

> no reason – with due acknowledgment of the enormous difficulties posed by the vicious circles in which they are often caught – why at least some [metropolitan areas in the developing world] cannot benefit from the processes of urbanization and economic growth described above. These

processes suggest that some of the more urbanized regions in these countries will eventually accede as dynamic nodes to the expanding mosaic of global city-regions, just as city-regions like Seoul, Taipei, Hong Kong, Singapore, Mexico City, São Paulo, and others, have done, and are doing, before them.

Overview

Table 1 summarizes the gist of the three theoretical approaches discussed in this section. Although each concept has been refined and/or revised in other contributions, it seems fair to state that the table gives a balanced overview of the conceptual core of each term: (i) Friedmann's world cities are centres of dominance and power, (ii) Sassen's global cities are production centres for the inputs that constitute the capability for global control, and (iii) Scott's global city-regions are production nexuses in a global economy dominated by a post-Fordist accumulation régime. These different starting points give diverging perspectives on the main features of a city as a node in transnational networks: the city's prime function, the key agents in the urban network, the alleged structure of the network as a whole, and the territorial basis of the city-as-node.

One can argue back and forth on the profoundness of the differences summarized in Table 1, but it seems clear that there is an unambiguous need to distinguish between the three concepts if one wishes to grasp the rise of 'world class' cities. The territorial demarcation of the units of analysis, for instance, may differ significantly. While a global city is a single focal point that operates separately from its hinterland, a global city-region may consist of multiple cities and their hinterlands which may themselves be subject to urbanization processes. Put succinctly: the Pearl River Delta

Table 1. Main theoretical approaches in the study of a global urban network.

Source: Derudder and Witlox

	World cities	Global cities	Global city-regions
Key author	Friedmann	Sassen	Scott
Function	Power	Advanced servicing	Production
Key agents	Multinational corporations	Producer service firms	Firms embedded in post-Fordist production networks
Structure of the network	Reproduces (tripolar) spatial inequality in the capitalist world-system	New geography of centrality and marginality cutting across existing core/ periphery patterns	Archipelago structure replacing existing core/ periphery patterns
Territorial basis	Metropolitan region	Traditional CBD or a grid of intense business activity*	(Metropolitan) region**

* The spatial demarcation depends on the specific form of the territorialization of the core dynamics behind global city-formation. This implies that both the continuation of traditional CBDs (New York) as a new pattern centered on a grid of intense business activity (Zürich) is possible. However, the proper unit of analysis is clearly smaller than the 'metropolitan region' as a whole (see body of text for further elaboration).
**Although most global city-regions have one or more major cities at their core, there is no conceptual need for functional centrality (see body of text for further elaboration).

may very well be a global city-region, but it can most certainly not be a global city (and vice versa for Hong Kong). Another important difference lies in the anticipated structure of the overall network. While a network of world cities is expected to reproduce core/periphery-patterns across the world-economy, a network of global cities or global city-regions is expected to cut across such divides. In other words: it is not implausible that 'semi-peripheral' cities such as Mexico City, São Paulo, and Seoul are well-connected global service centres (i.e. global cities) without being major power centres in the world-economy (i.e. world cities). Hence, rankings of world cities and global cities may be expected to diverge rather than converge.

The creation of fuzzy concepts

In the previous section, we argued that some of the commonly employed terms in the study of globalized urbanization cover specific analytical constructs. However, although naming cities as nodes in a global urban network is thus clearly not a trivial play on words, there is a tendency in the literature to downplay the differences between key concepts. In this section, we demonstrate why it is problematic to condense the various terms into interchangeable notions that share a basic connotation of 'world class' cities. This is done through reviewing the blurry discourse in empirical studies of globalized urbanization: this empirical research has tried to assert its analytical importance by referring to various concepts, and we discuss to what degree these studies have employed the appropriate terminology. We begin by summarizing the assumptions behind the main empirical approaches, and then analyse the appropriate and employed terminologies for such studies.

Fig. 4. The GaWC inventory of 'world cities'.

Source: Beaverstock et al. *(1999)*

Main empirical approaches

Empirical research on global urban networks has relied on a wide variety of data sources to assess the importance of cities, but generally speaking the production

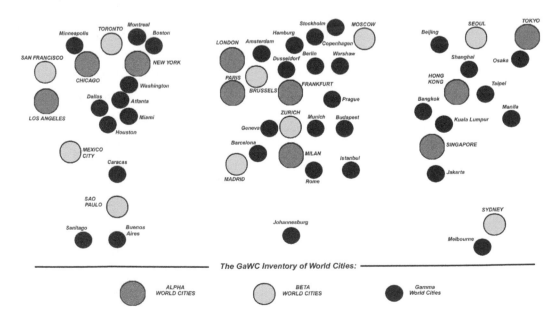

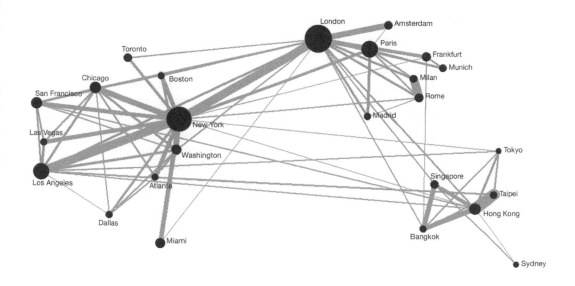

of these 'city rankings' has been premised upon two foundations; these may respectively be labelled (i) the corporate organization and (ii) the infrastructure approach (Derudder 2006).

The corporate organization approach starts from the observation that firms pursuing global strategies are the prime agents in the formation of urban networks. Two leading examples are the research pursued by the Globalization and World Cities group and network (GaWC, http://www.lboro.ac.uk/gawc) and a recent analysis by Alderson and Beckfield (2004; 2006). GaWC researchers have developed a methodology for studying transnational urban networks based on the assumption that advanced producer service firms 'interlock' cities through their intra-firm communications of information, knowledge, plans, directions and advice to create a network of global service centres (Taylor 2001). Building on this specification, information was gathered on the location strategies of 100 global service firms across 315 cities (Taylor *et al.*, 2002). Applying the formal social network methodology set out in Taylor (2001), this information was converted into a 315 x 315 matrix, which was then analyzed with standard network-analytical tools (Derudder and Taylor, 2005). Using a similar methodological approach, Alderson and Beckfield (2004) have analyzed links between 3,692 cities, based on the organizational geographies of 446 of the largest multinational firms and their subsidiaries for the year 2000. Despite some methodological differences, both studies base their city-centred spatial analysis on an assessment of the location strategies of firms with transnational fields of activity. In other words, it is suggested that a meaningful measurement of transnational inter-city relations can be derived from intra-firm connections between different parts of a firm's holdings: Alderson and Beckfield (2004, p. 813–814) consider this to be a 'key relation' in 'an MNE-generated city system', while Taylor (2004, p. 9) argues that it is 'firms through their office networks that have created the overall structure of the world city network.' The main difference between both approaches lies in the type of firms used throughout their analysis: GaWC researches focus on the location strategies of producer services firms, Alderson and Beckfield (2004) use information on the geography of multinational corporations irrespective of the exact nature of their activities (e.g. their table of the distribution of firms 'across industries' on page 821).

The gist of the infrastructure approach lies in the observation that advanced telecommunication and transportation infrastructures are unquestionably tied to key cities in the global economy. The most important cities also harbour the most important airports, while the extensive fibre backbone networks that support the internet have equally been deployed within and between major cities. This has created a vast planetary infrastructure network on which the global economy has come to depend almost as much as on physical transport networks (Rutherford *et al.*, 2004). These enabling (tele)communication and transportation networks are the fundament on which the connectivity of key cities is built, and it is therefore no surprise that the geography of these networks has been used to invoke a spatial imagery of a transnational urban network. Smith and Timberlake (2002, p. 139), for instance, have sought to describe the spatial patterning of a global urban network 'as indicated by their interrelations in the air passenger networks' (see also Derudder and Witlox, 2005), while Townsend (2001, p. 1700) has illustrated 'how global cities have fared in the rapid and massive deployment of Internet networks'. Both analyses claim that it is possible to devise an urban network based on the geography of infrastructure networks, whereby the main difference between both approaches lies in the type of infrastructure, i.e. physical transport versus telecommunications.

Appropriate and employed terminology

Table 2 summarizes the employed concepts and the appropriate terminologies within these empirical approaches. References to the employed concepts are, of course, primarily based on the directly used discourse, but some references are based on more implicit allusions. A case in point is Derudder *et al.* (2003), who present a global urban analysis of 234 cities based on the data gathered in Taylor *et al.* (2002). With the exception of a single mention of 'global cities' (Derudder *et al.*, 2003, p. 885), the authors employ the term 'world city', but it is quite clear that 'global cities' are deemed to be at least as relevant as a concept. For instance, Derudder *et al.* (2003, p. 877) ascertain that they follow 'Sassen in her treatise of cities as global service centers – locales where advanced producer services are concentrated for servicing their global corporate clients'.

The studies carried out within the corporate organization approach have a comprehensible conceptual background, and it is therefore fairly straightforward to pin down what their terms should be referred to. Studies drawing on the location strategies of producer service firms pertain to Sassen's global city-formation, while studies employing data on the organizational structure of large MNEs refer to Friedmann's world city-formation. In contrast to data on corporate organization, the importance of infrastructure data appeals more to a common sense approach than to a precise concept. For instance, a recent volume edited by Sassen (2002) features two consecutive chapters on infrastructure networks, in which Graham's (2002, p. 89) fibre network-analysis refers to 'global cities' and 'global city-regions' while Smith and Timberlake's (2002, p. 139) airline network analysis refers to 'world cities'. It remains, however, unclear how infrastructure data can distinguish between

Table 2.
Employed terminologies in empirical studies of a global urban network.
Source:
Derudder and Witlox

Empirical approach	Corporate organization		Infrastructure	
	Producer service firms	Multinational enterprises	Telecommunications	Physical transportation
Employed terminology	Beaverstock *et al.* (1999): world cities and global cities Taylor *et al.* (2002): world cities and global cities Derudder *et al.* (2003): world cities and global cities Derudder and Taylor (2005): world cities and global cities	Rodríguez-Pose and Zademach (2003): global cities Alderson and Beckfield (2004): world cities and global cities	Graham (2002): global cities and global city-regions Malecki (2002): world cities, global cities and 'large cities' Townsend (2001): global cities	Keeling (1995): world cities Smith and Timberlake (2001, 2002): world cities and global cities Derudder and Witlox (2005): world cities
	↓	↓	↓	↓
Appropriate terminology	Global cities	World Cities	(Global City-Regions)*	

* The term 'global city-regions' is placed in parentheses because further elaboration is required for showing that information on infrastructure networks represents a suitable operationalization of this concept.

these concepts. The employed terminologies in these infrastructure network-based studies are therefore not so much a starting point for a precise specification, but more of a vague attempt to suggest a credible conceptual backdrop.

If there is a meaningful analytical connection between theoretical concepts and infrastructure-based measurements, then the most relevant concept seems to be Scott's global city-regions. The main reason for this is that the connectivity of cities such as Frankfurt and Amsterdam cannot straightforwardly be traced back to a single, central location. Rather, the airports of these cities are in practice serving 'an emergent form in the global urban hierarchy' (Smith and Timberlake, 2001, p. 1671), which is a city-region that consists of one or more major cities together with their hinterlands. It is therefore misleading to conceive Frankfurt and Amsterdam as the 'real' units of analysis in their airline-based studies. Rather, it seems more fruitful to speak of the 'Rhine-Main conurbation' (ibid.) and the 'Randstad conurbation (which also includes Rotterdam and The Hague)' (ibid., p. 1672). The main point is that this suggests that the broader region is directly implicated in the connectivity within airline networks, since it is a function of both the city and its surrounding urban field. Hence the possible link to global city-regions, albeit that there is a clear necessity for further elaboration on the analytical connections between this concept and infrastructure data.

The confrontation of the employed and the appropriate discourse in recent empirical studies (Table 2) reveals that empirical studies only seldom refer to the appropriate terminologies. Taken together, this suggests that the various terms are treated as interchangeable notions that share a basic connotation of 'world class' cities.[2] This lack of analytical specification is, however, not a necessary feature of empirical research: influential scholars such as Olds and Yeung (2004, p. 515) and Brenner (1998, p. 29), for instance, equally start their theoretically-oriented articles with a footnote in which they state that the terms 'world city' and 'global city' will (and hence can) be used interchangeably. Although this does not necessarily compromise the relevance of the ensuing analyses, the broader point here is that such inapt discourse may progressively lead to the conclusion that the various 'approaches are perhaps not quite as distinct as they may seem' (Hall 2001, p. 61). The discussion in the previous section, however, has clearly shown that this is a problematic point of departure for this kind of research.

Discussion and conclusion

In this chapter, we have tried to show that Hall's (2001) suggestion – specific elaborations of 'world class' cities are not as distinct as they may seem – constitutes a questionable stance. It is, for instance, clear that Sassen's influential 'global city' concept should be treated as a specific analytical construct rather than as an attempt to refine existing conceptualizations of 'world class' cities. In the revised edition of *The Global City*, Sassen (2001a, p. xxi) maintains that 'when I first chose to use [the term] global city I did so knowingly – it was an attempt to make a difference'. This attempt to discriminate is further specified *vis-à-vis* two other important theorizations, i.e. Friedmann's 'world cities' and Scott's 'global city-regions'. Sassen (2001a, p. xxi) thereby stresses that although it may be the case that 'most of today's major global cities are also world cities' there may just as 'well be some global cities today that are not world cities in the full, rich sense of that term'. And, while global cities and global city-regions may 'as categories for analysis...share key propositions about economic globalization', it is clear that they 'overlap only partly in the features they each capture' (Sassen 2001a, p. 351). As a consequence, they have 'distinctive theoretical and empirical dimensions' (Sassen 2001b, p. 70).

The observation that cities such as London, New York, Tokyo, and Paris, invariably feature at the apex of the various rankings of world cities, global cities, and global city-regions does not imply that these concepts are interchangeable. Put simply: the answer to the question what exactly constitutes a 'world class' city depends on more specific conceptual underpinnings, and subsuming specific analytical constructions under the commonsensical flag of 'world class' cities is not necessarily a fruitful starting point for scientific enquiry.

Notes

1. The conference was the 7th International Symposium of the International Urban Planning and Environment Association (IUPEA), entitled 'World Class Cities: Environmental Impacts and Planning Opportunities?', held at the Chulabhorn Research Institute, Bangkok, 3–5 January 2007. It was hosted and organised by Kasetsart University under the auspices of IUPEA.
2. That said, we want to stress that this blurry discourse is not necessarily a sign of a sweeping conceptual conflation. For instance, despite favouring the term 'world city', GaWC researches unambiguously refer to Sassen's work on place and production in a global economy. Thus,

there can be little doubt that GaWC studies present an analysis of global cities, and passing judgment on the basis of an erroneous terminology would therefore be outright excessive.

References

Alderson, A.S. and Beckfield, J. (2004) Power and position in the world city system. *American Journal of Sociology*, **109**: 811–851.

Alderson, A.S. and Beckfield, J. (2006) Globalization and the World City System: Preliminary Results from a Longitudinal Dataset, in *Cities in Globalization: Theories, Policies, Practices* (eds P.J. Taylor, B. Derudder, P. Saey and F. Witlox), Routledge, London.

Beaverstock J.V., Smith, R.G. and Taylor, P.J. (1999) A roster of world cities. *Cities*, **16(6)**: 445–458.

Brenner, N. (1998) Global cities, glocal states: global city formation and state territorial restructuring in contemporary Europe. *Review of International Political Economy*, **5(1)**: 1–37.

Brenner, N. (1999) Beyond state-centrism? Space, territoriality and geographical scale in globalization studies. *Theory and Society*, **28**: 39–78.

Brenner, N. (2004) *New State Spaces: Urban Governance and the Rescaling of Statehood*, Oxford University Press, Oxford.

Brenner, N. and Theodore, N. (2002) Cities and the geographies of 'actually existing neoliberalism', in *Spaces of Neoliberalism: Urban Restructuring in North America and Western Europe* (eds N. Brenner and N. Theodore), Blackwell Publishers, Oxford.

Dematteis, G. (2000) Spatial images of European urbanization, in *Cities in Contemporary Europe* (eds A. Bagnasco and P. Le Galès), Cambridge University Press, Cambridge.

Derudder, B. (2006) On conceptual confusion in empirical analyses of a transnational urban network. *Urban Studies*, **43(11)**: 2027–2046.

Derudder, B., Taylor, P.J., Witlox, F. and Catalano, G. (2003) Hierarchical tendencies and regional patterns in the world city network: A global urban analysis of 234 cities. *Regional Studies*, **37(9)**: 875–886.

Derudder, B. and Taylor, P.J. (2005) The cliquishness of world cities. *Global Networks*, **5(1)**: 71–91.

Derudder, B. and Witlox, F. (2005) An appraisal of the use of airline data in assessing the world city network: A research note on data. *Urban Studies*, **42(13)**: 2371–2388.

Dodgshon, R.A. (1998) *Society in Time and Space: A Geographical Perspective on Change*, Cambridge University Press, Cambridge.

Friedmann, J. (1986) The world city hypothesis. *Development and Change*, **17**: 69–83.

Friedmann, J. and Wolff, G. (1982) World city formation: An agenda for research and action. *International Journal of Urban and Regional Research*, **3**: 309–344.

Graham, S. (2002) Communication grids: cities and infrastructure, in *Global Networks, Linked Cities* (ed. S. Sassen), Routledge, London.

Grosfoguel, R. (1995) Global logics in the Caribbean city system: the case of Miami, in *World Cities in a World-System* (eds P.L. Knox and P.J. Taylor), Cambridge University Press, Cambridge.

Harvey, D. (1989) From Managerialism to Entrepreneurialism: The Transformation in Urban Governance in Late Capitalism. *Geografiska Annaler B*, **71**: 3–17.

Hall, P. (2001) Global City-Regions in the Twenty-First Century, in *Global City-Regions: Trends, Theory, Policy* (ed. A. Scott), Oxford University Press, Oxford.

Henton, D. (2001) Lessons from Silicon Valley: governance in a global city-region, in *Global City-Regions: Trends, Theory, Policy* (ed. A. Scott), Oxford University Press, Oxford.

Keeling, D.J. (1995) Transportation and the world city paradigm, in *World Cities in a World-System* (eds P.L. Knox and P.J. Taylor), Cambridge University Press, Cambridge.

Knox, P.L. and Taylor, P.J. (1995) *World Cities in a World-System*, Cambridge University Press, Cambridge.

Malecki, E. (2002) The economic geography of the Internet's infrastructure. *Economic Geography*, **78**: 399–424.

Mann, M. (1986) *The Sources of Social Power* (Volume I: A history of Power from the Beginning to AD 1760), Cambridge University Press, Cambridge.

Marcuse, P. and van Kempen, R. (2000) *Globalizing Cities: A New Spatial Order?* Blackwell, Oxford.

Markusen, A. (1999) Fuzzy concepts, scanty evidence, policy distance: The case for rigour and policy relevance in critical regional studies. *Regional Studies*, **37**: 701–717.

Moulaert, F., Rodriguez, A. and Swyngedouw, E. (2003) *The Globalized City: Economic Restructuring and Social Polarization in European Cities*, Oxford University Press, Oxford.

Olds, K. and Yeung, H.W.C. (2004) Pathways to Global City Formation: A View from the Developmental City-State of Singapore. *Review of International Political Economy*, **11**(3): 489–521.

Ohmae, K. (2001) How to invite prosperity from the global economy into a region, in *Global City-Regions: Trends, Theory, Policy* (ed. A. Scott), Oxford University Press, Oxford.

Petrella, R. (1995) A global agora versus gated city-regions. *New Perspectives Quarterly*, Winter: 21–22.

Porter, M.E. (2001) Regions and the new economics of competition, in *Global City-Regions: Trends, Theory, Policy* (ed. A. Scott), Oxford University Press, Oxford.

Rodríguez-Pose, A. and Zademach, H.M. (2003) Rising metropoli: the geography of mergers and acquisitions in Germany. *Urban Studies*, **40**(10): 1895–1923.

Rutherford, J. Gillespie, A. and Richardson, R. (2004) The territoriality of pan-European telecommunications backbone networks. *Journal of Urban Technology*, **11**(3): 1–34.

Saey, P. (1996) Het wereldstedennetwerk: de nieuwe Hanze? *Vlaams Marxistisch Tijdschrift*, **30**(1): 120–123.

Sassen, S. (1991) *The Global City: New York, London, Tokyo*, Princeton University Press, Princeton.

Sassen, S. (1994) *Cities in a World Economy*, Pine Forge Press, Thousand Oaks.

Sassen, S. (1995) On concentration and centrality in the global city, in *World Cities in a World-System* (eds P.L. Knox and P.J. Taylor), Cambridge University Press, Cambridge.

Sassen, S. (2001a) *The Global City: New York, London, Tokyo* (2nd edition), Princeton University Press, Princeton.

Sassen, S. (2001b) Global cities and global city-regions: a comparison, in *Global City-Regions: Trends, Theory, Policy* (ed. A. Scott), Oxford University Press, Oxford.

Sassen, S. (2002) *Global Networks, Linked Cities*, Routledge, London.

Scott, A.J. (1998) *Regions and the World Economy: The Coming Shape of Global Production, Competition, and Political Order*, Oxford University Press, Oxford.

Scott, A.J. (2001a) Globalization and the rise of city-regions. *European Planning Studies*, **9**(7): 813–826.

Scott, A.J. (2001b) *Global City-Regions: Trends, Theory, Policy*, Oxford University Press, Oxford.

Scott, A.J. (2001c) *Introduction, in Global City-Regions: Trends, Theory, Policy* (ed. A. Scott), Oxford University Press, Oxford.

Scott, A.J., Agnew, J., Soja, E.W. and Storper, M. (2001) Global city-regions, in *Global City-Regions: Trends, Theory, Policy* (ed. A. Scott), Oxford University Press, Oxford.

Short, J.R. (2004) *Global Metropolitan: Globalizing Cities in a Capitalist World*, Routledge, London.

Smith, D.A. and Timberlake, M. (2001) World city networks and hierarchies, 1977–1997: An empirical analysis of global air travel links. *American Behavioral Scientist*, **44**(10): 1656–1678.

Smith, D.A. and Timberlake, M. (2002) Hierarchies of dominance among world cities: a network approach, in *Global Networks, Linked Cities* (ed. S. Sassen), Routledge, London.

Smith, N. (1996) *The New Urban Frontier: Gentrification and the Revanchist City*, Routledge, London.

Smith, N. (2002) New globalism, new urbanism: gentrification as global urban strategy. *Antipode*, **34**: 434–457.

Taylor, P.J. (2000) World cities and territorial states under conditions of contemporary globalization. *Political Geography*, **19**: 5–32.

Taylor, P.J. (2001) Specification of the world city network. *Geographical Analysis*, **33**(2): 181–194.

Taylor, P.J. (2004) *World City Network: A Global Urban Analysis*, Routledge, London.

Taylor, P.J., Catalano, G. and Walker, D.R.F. (2002) Measurement of the world city network. *Urban Studies*, **39**(13): 2367–2376.

Taylor, P.J., Derudder, B., Saey, P. and Witlox, F. (2006) *Cities in Globalization: Theories, Policies, Practices*, Routledge, London.

Townsend, A.M. (2001) Network cities and the global structure of the Internet. *American Behavioral Scientist*, **44**: 1697–1716.

Veltz, P. (1996) *Mondialisation, Villes et Territoires: L'Economie d'Archipel*, Presses Universitaires de France, Paris.

Wallerstein, I. (1979) *The Capitalist World-Economy*, Cambridge University Press, Cambridge.

2

Peter Marcuse

Globalization and the Forms of Cities

Introduction

'World class' cities are the winners among globalizing cities; they are the products of really existing globalization, globalization as it is practiced today.[1] But the processes of this globalization have contained a number of currents that can usefully be disaggregated, and whose separate influence can be seen both in the forms which cities as a whole take and the forms of the built structures within them.

The characteristics of really existing globalization, as they affect urban development, may be named briefly:[2]

1. A concentration of ownership and control in the hands of a decreasing number of overwhelmingly multi-national corporations.
2. A financialization of capital in which financial firms are increasingly the owners of, and control, the major large manufacturing and services firms, making key decisions for them whether or not they mange their day-to-day operations (Foster, 2007).
3. A shift in power relations between firms and their workers, with business profits and executive pay rising at a much faster rate than workers' wages (Cypher, 2007).[3]
4. A shift in power relations between firms and government, leading to the adoption of neo-liberal policies by most governments in developed countries (Harvey, 2005a; Brenner and Theodor, 2002).
5. Commitment of local governments to competition among cities for economically profitable businesses.
6. A rapid development of technology, particularly in communications and transportation and information processing, enabling much wider spans of control and networking among firms (Castells, 1998; Marcuse, 2002a).[4]
7. A rise in the extent of the services sector of the major economies, and in the most developed economies a decline in the presence of manufacturing; not to be confused with an absolute decline of manufacturing world-wide, but representing rather a relative decline of manufacturing in some cities resulting from their relocation rather than decline.
8. Most recently, a formal concern with security against the threat of terrorism (Graham, 2004).
9. Again recently, especially after the collapse of the Soviet Union, the

domination of the United States militarily, economically, financially, and culturally, throughout most of the world (Harvey, 2005b).

10. The set of social, economic, and political policies, and their attendant ideological grounding, generally subsumed under the name of neo-liberalism.

The impact of these forces on cities can be divided between the impact on cities as a whole, that is to say, on the process of urbanization, and the impact on specific aspects of the built environment within cities.

A word of caution: none of what is described here is historically without precedent, and in almost every case is not 'new', but represents rather the continuation and intensification of trends in urbanization and city development going on since at least the first half of the 19th century with the rise of industrialization, and most since the 18th century with the establishment of capitalism.

The forms of urbanization
Concentrated decentralization

The spatial pattern of business activities looks as if it is deconcentration, that is, a spreading out of activities over a broader landscape and a declining importance of central cities, but that is misleading. Concentration remains the pattern, but it is taking place both within and outside central cities. New York City is an example. The movement of major firms out of lower Manhattan is a pattern that has been increasingly evident since the 1970s, but has accelerated in recent years under the impact of globalization. The attack on the World Trade Center of September 11, 2001 ('9/11' hereafter) reinforced the trend (Marcuse, 2002b), but not just as to businesses that were directly affected by the World Trade Center destruction. Others that had earlier preferred lower Manhattan expanded or built up secondary or redundant centres elsewhere (ibid.), but primarily within the metropolitan region, and generally in locations where other similar businesses were clustered (ibid.).

That pattern of movement away from city centres may be called 'concentrated decentralization' (ibid), and it is generally recognized as having a long-standing history. Its older residential manifestation is suburban sprawl and in new towns (or older centres) built around manufacturing plants, often located there because of water power or extractive resources. Its newer spatial manifestation is in edge cities and the growth of business clusters outside central cities. For example, White Plains, Princeton Junction, Stamford serve as edge cities for New York City, but remain within metropolitan regions, if at varying distances from the centre. Its real estate manifestation is a sharp decline in property values in downtown areas in all but the largest and strongest cities. Its political manifestation is in policies of smart growth and in the economic revitalization of downtown areas through tax concessions, public loans and grants, limitations on sprawl, and the flexible application of land use controls (Bernstein, 2005).

These are mostly patterns of long duration whose present forms and extent are intimately connected with globalization, with the shift in economic activity from manufacturing to services, which changed the economic basis on which many downtowns had been built. They have to do with an increase in the centralization of control, enabled (but not caused) by technological advances that permitted an efficient increase in the span of control, and a process of globalization that permitted greater and greater accumulations of capital.[5] They have to do with the spatial patterns produced by that centralization of control, both within and among cities, so that a few downtowns grew rapidly while others declined abruptly. They have to do with the decentralization that communications and advanced transportation technologies made both possible and desirable. They have to do with real estate markets and labour markets that reflect these developments.

For businesses, and increasingly for residences, the pattern of decentralization is not one of shapeless sprawl. Agglomeration economies, the benefits of being near other activities in the same business sector, remain important and lead to the substantial clustering of business activities both inside and outside the central city of metropolitan areas. The availability of support services, such as accounting or law or financial firms, which are only profitable if there is sufficient market for them, the proximity of other like and related businesses for interchange of ideas reinforces concentration. On the residential side, the desirability of cultural, entertainment, environmental amenities, public facilities, access, that can only be provided where there is sufficient volume of demand. Thus, the pattern is appropriately called 'concentrated' decentralization, a pattern in which metropolitan areas grow as a whole, with growing clusters outside the centres, and decline of those centres themselves within all but the largest areas.

Regional patterns

Within this pattern of concentrated decentralization we find the spread of the so-called Edge Cities (Garreau, 1991) and edgeless cities and suburbanization and regionalization. There are differences in the forms that these phenomena take over time. To some extent, patterns of suburbanization have changed because of developments in transportation and communications technology, which has permitted longer commutes and enabled some work formerly done in central offices to be done in outlying parts of metropolitan regions and sometimes in residential communities. But the basic patterns of metropolitan concentration and suburbanization remain. There is evidence of tighter economic integration at the super-metropolitan, or regional, level, but the extent to which that has changed the pattern of urbanization remains unclear; as something that might be called 'megalopolis', it has been observed for some time (Gottmann, 1961).

Regionalization

Regionalization both of residences and workplaces is the general phenomenon of which edge cities are a major component. As Keil and Ronneberger (2000) describe it most specifically, globalization leads it to develop beyond the metropolitanization and sprawl that have long been known. It includes not only edge city development but also closer relationships between long-time independent cities to form economically integrated regions, as in the Frankfurt-Weisbaden corridor, the Ruhr, Silicon Alley, or the Randstad in the Netherlands. The 'insular configurations' around Frankfurt, Waley's (2000) description of the developments around Tokyo and Chakravorty's (2000) of 'new town' development around Calcutta reveal the regional nature of development 'at the edge'. This may well be seen as a further intensification of the earlier megalopolis formations.

Edge cities

Edge cities are a component of these new regions that may be seen as a totalized form of the suburban city that is significantly new. The definition of course cannot be that used by Garreau. He considers only those that have only recently been built, rather than all those that carry the function of edge cities, and owe their existence to simple consumer preferences, while in fact they are the results of much more complex processes in which consumer preferences initially play a very minor role. The reference here is to clusters of residences, businesses, commerce, and recreation, on an urban scale, removed from major central cities but related to them, whose independence in daily life from those central cities is in large part their reason for being. A regional view must be taken to understand properly the economic as well as social role of such edge cities. As opposed to metropolitan development, or mega-city growth, the point here is insulation, the down-playing of dependence, coupled with the development of activities that emulate and bring to the suburban location the same international business firms, the same professional consulting activities, the same cultural amenities, the same concerts and museums and theatres, the same religious institutions, that the central city has, if on a smaller scale.

Uneven development

Uneven development as a principle of geographic development has been long identified as a characteristic of capitalist economic organization (Smith, 1990). Its forms and extent change, but the logic driving it remains the same: the private market is the overwhelming determinant of urban development, and actors in the private market are concerned about the returns on investment, rather than the externalities their investments produce. Where those externalities are negative, it is left to the pubic sector to remedy them. Since capital flows to whatever region (and city and neighbourhood) is likely to produce the highest rate of return, and is free to move to new areas without regard for the costs of the abandoned structures left behind, uneven development measured in social terms is the inevitable result. In terms of this pattern as among cities, some cities will be winners and others losers.

Winner cities

'Global cities' are the winner cities in the increasing competition among cities. That competition takes place in a situation in which the advantages of spatial concentration of key economic activities is increasing dramatically. The literature making this point is by now enormous, and both the theoretical and the empirical basis for the trend are convincing, and by now well known, and will not be repeated here.

Loser cities

If there are winner cities, there must be losers. And indeed there are. In some cases, the shift of manufacturing to lower cost (generally in wages) locations, often overseas from the advanced economies, is a key factor; in others, it is simply falling behind in the competition to capture the growth in the financial sector and its related activities. The net result has been the phenomenon known as 'shrinking cities' in Europe,[6] with clear parallels in the United States. The absolute decline in population of cities such as Detroit, St. Louis or Newark in the United States, or Leipzig and Dresden in Germany, or Liverpool in England, has been extensively studied.

Loser cities, cities that are not world cities, are as much influenced by the ten processes we have identified with globalization as are winner cities, even if the extent of the globalized developments outlined here are quite different both in scale and proportion.

Mega-cities

The mega-cities of the Third World, the Mexico Cities, São Paulos, Nairobis, Mumbais of the world, are in their development a product of globalization also. They do not simply follow the development path of cities in the more technologically developed worlds; they are not simply at an earlier stage in a common path of development, but rather are the products of their own specific historical developments coupled with the strong influence of their positions within the world of globalization, with its threads of colonialization, uneven development, competition, division of labour, and exploitation. While no detailed examination of their differing patterns is possible here, the generalization might be hazarded that, while the patterns of globalized urbanization described in this section will be very different in such cities, most of the patterns of change in internal structure will also find their traces there.

Below the regional scale, the internal structure of cities has also been influenced strongly by the forces of globalization. This is true of structures: skyscrapers, citadels, barricaded places, as well as of the social strictures of neighbourhoods, with their varieties of ethnic enclaves and racial ghettoes and exclusive and gated communities, and of the treatment of various geographically defined features, such as waterfronts, central business districts, or brownfield sites. We now turn to a consideration of these internal developments in the form of cities.

The internal structure of cities

At the scale of the city itself, a wide number of changes accounted for by aspects of globalization can also be seen. These include the appearance of new typologies and large developments in central city and other central locations and their fringes; concentrations of ethnic areas and social housing; a 'global style' of architecture and design of public spaces.

The number of these changes leads directly to the further segregation of the city and the polarization of the location of activities within it; other changes are more geographically/topographically specific, and still others are the more recent products of the so-called 'war on terror' that has played a prominent role, at least in the United States, after 9/11. These are addressed under the three main headings: segregation and exclusion; soft locations; and, the impact of security.

Segregation and exclusion
Citadels

Citadels, in the form of hi-tech, generally hi-rise, mega-projects are becoming prevalent throughout the world, from London to Shanghai, Los Angeles to Kuala Lumpur, Detroit to São Paulo, Paris to Bandung. Their key features are a size and architectural form incorporating multiple uses in such fashion that the mega-project is separated, insulated, from the city in which it exists, and permits its users to remain sheltered in that project for the bulk of their daily activities and in many cases their night-time activities as well. Frankfurt, New York City, Sydney, Singapore, Brussels, contain classic examples, as does Tokyo, where citadels of government and of business share the skyline. Calcutta is on the way; Rio de Janeiro would like to be. The architectural style remains modern; it was dubbed the international style already in 1932, but has really earned that name now. Post-modern treatment of the edges (or more literally the tops) of such edifices do nothing to alter the modern technical rationality of their construction, nor the form that results from and embodies it. Fashions in styles may vary, but the representation of power, of wealth, of luxury, is inherent, as is the isolation, the separation, the distancing from the older urban surroundings. The grid of lower Manhattan may be carried into the street pattern of Battery Park City and visible to an interested observer in a helicopter, but the separation of the World Trade Center/Battery Park City complex from the rest of the city visually, and in terms of secure entrance, is obvious to all.

The use of the citadels is not however confined to those living in them; professionals, technicians, managers, administrators, need less well-off workers to carry out the functions assigned to them. And, layered in time, the janitors, parking garage attendants, security guards, needed for the effective operation of the citadel must be allowed in (Marcuse, 2002c). A few residents of the citadels, the city of the gentry, may walk to work if it has a foothold nearby; some may fly in by helicopter to land on the roof landing pads that characterize, for instance, almost every office building in São Paulo. But most will commute in by some form of limited transportation access, which is likely to be expensive but publicly subsidized, and likely to permit access even from a distance without treading on the ground of the rest of the more mundane city.

Skyscrapers and skyscraper clusters

The use (as opposed to the design) of skyscrapers for office purposes is of long standing. There are specific socio-economic conditions that promote such use, including the desire to capitalize on real estate values, to assemble within a single unit the many administrative and control functions of large business organizations. The prevalence of the skyscraper, both individually and as a cluster or citadel, in the mega-cities of developing countries and in the centre of international wealth such as Dubai and Abu Dhabi, not to speak of Shanghai, attest to its continued dominance as an image of wealth and power. Later in this chapter the analysis of the current uses of skyscrapers in central business districts is resumed.

Gentrified neighbourhoods.

Gentrified neighbourhoods are well known to the literature. Functionally, globalization has produced a class of professionals, managers, technicians, that may well be analogized to the gentry of earlier days in feudal systems. As they increase in numbers, so do they increase in importance, and in income, and the residential locations they choose become ever more clearly identified and separated. The gentrified city is often located in the inner parts of the older cities, or in adjacent neighbourhoods. The attractiveness of areas close to the inner cities, the places where 'urban' activity is centred, is particularly attractive to the gentry; and because these are often areas formerly occupied by the working class, the link between gentrification and displacement is close (Marcuse, 1985; 1986).

But the importance of gentrification has grown substantially with the processes of globalization, and one can see the separate impact of virtually all of its ten characteristics manifest in the process. Some, indeed, consider gentrification the umbrella process under which almost all the changes in the structure of cities in a period of globalization may be captured (Smith, 1990). Their relationship to citadels requires further detailed investigation. There is probably, in each urban area, a separation between those truly in control, and those who work for them, even at high levels within organizational structures. The separation is probably visible in the size and location of second homes, but very likely of first homes also. The top of the hierarchy is not likely to be involved in the process of gentrification, although, after a neighbourhood has been converted, units within it may serve them as *pieds-à-terre*. More likely, they will find more convenient accommodation whether within the citadels themselves or in the older fashionable upper-class neighbourhoods of the city.

Ghettoes

Ghettoes of course have long existed, but their formation/reinforcement under the impact of gentrification has produced new characteristics. They are today less exclusively dependent on race or ethnicity as the basis for their maintenance, although those factors continue to be a major sustaining force. Today the simple operation of the market, both the labour market and the real estate market, are sufficient to produce a process of ghettoization. And it is a ghetto of exclusion, rather than exploited service (Marcuse, 1997a; 1998). Yet the issue of race distinguishes United States ghettoes from those of most other countries, despite the common pressures of globalization.

Ghettoization as it is seen in the United States has taken a starker form than in most other countries because of the sharp impact of race. It takes a specific combination of 'racial' and economic circumstances to produce an American ghetto: specifically, a combination first, of a new form of urban poverty, long-lasting and deep and excluded from the expectation of conjunctural change, and second discrimination against a specific and identifiable (most readily by colour) group, discrimination with wide social prevalence and deep historical roots, but a group with strong formal claims to equality and full citizenship. Third, even given these two factors, firm governmental policies targeted to at least economic and political integration and the avoidance of spatial segregation can have a major effect on the resulting spatial patterns.

The first of these factors, the new form of urban poverty, exists not only in the United States, but has its parallel in almost every major city in the world (Mingione, 1996). In some cases, such as Calcutta or Rio de Janeiro, it is overlaid on a century or more of both abysmal poverty and social exclusion; in others, as in the Netherlands, Sydney, or Frankfurt, it is a new appearance, a matter of growing concern but yet nowhere near the dimensions of the first group. In New York City, it is well documented; in Tokyo, there is official denial of the existence of concentrated poverty (Waley, 2000). But the tendency to impoverishment and exclusion is detectable in all globalizing cities. The second characteristic, discrimination against an easily targeted group, is also increasingly visible in many places, and a source of concern in most. By and large that discrimination in countries other than the United States is against non-citizens, however, so that the claims that account for some of the tensions in 'race' relations in the United States are absent. Where there is full citizenship and concern about discriminatory treatment, as in West Germany against residents of the former East German state, there is no colour line to facilitate discrimination, and the spatial pattern tends to be more regionally based than intra-urban.

Thus the pattern of 'racial' exclusion and segregation as found in the United States is not at this time replicated elsewhere on any comparable scale. The forces that lead to economic polarization and the social relations attendant on immigration create tendencies similar to those in the United States; how far that will go remains a matter not yet determined.

Exclusionary enclaves[7]

Exclusionary enclaves are not new in the world, but their spread has been phenomenal in the last several decades. The walled communities of the rich, gated communities more closed from the rest of society than ever before, are now to be found, not only in the United States, but also in Johannesburg, Rio de Janeiro and many cities all over the world. Such luxury sites are still not very usual in European countries, but are becoming more and more important there also. We would expect to see a substantial expansion of these kinds of enclaves, housing many of those most directly benefiting from processes of globalization, business people, managers, leading artists and politicians, who have homes in many places of the world and are quite able to live isolated from their immediate surroundings. The residents of the luxury city live in walled communities, not spatially dependent on any particular geographical location in relation to the rest of the city. They rather create and control their own environment at the micro level.

Walling and gating by themselves are not sufficient to define a socio-spatial pattern, for the fact (or the symbol) of gating has spread to virtually all sectors of society; today one finds public housing projects in the United States, middle class suburbs, upper-middle-class enclaves, retirement communities, with walls of various sorts around them, or with the equivalent measures designed to provide physical protection against social dangers (Blakely and Snyder, 1995; Marcuse, 1997b). It is the extent of this development, with a specific focus on its appearance for communities of those made prosperous by the processes of economic change, to which demand attention here.

Exclusionary enclaves have been formed, not only by the gentry, but also around some areas of the rich, and are wished for, and often obtained, by residents of the suburban city, and even on occasion by those in the working class. Passing a certain stage in life, retirement communities may house (although separately) people with varying economic resources and earlier positions. Gentrified areas, almost by definition, cannot erect walls to define their boundaries, since they are encroachments on and reuses of areas previously occupied by poorer residents. Here the exclusion and control are accomplished by social, rather than physical, means (although individual buildings will have their own security systems, bars on windows, fences and gates at entry): the police presence will be enhanced, and private security guards will patrol.

Ethnic enclaves

Ethnic enclaves (to be distinguished both from exclusionary enclaves and from excluded ghettoes)[8] are perhaps the functional equivalent of the working class quarters of the traditional industrial cities of the nineteenth century. Whether we speak of Turkish areas in Frankfurt or the Netherlands, Pakistanis in Britain, Dominicans in New York City, immigrants in Tokyo, Vietnamese in Sydney, the pattern is the same (and only stringent government regulation prevents its reproduction in Singapore): new arrivals in the cities are used for lower-paid work, exploited more than their longer fellow-residents might tolerate, and residentially stay together for mutual support in difficult conditions. In time, such areas may lose their economic function because the pressures that prompted their residents to maintain them have abated, though residents of similar cultural or ethnic or religious background may stay near each other, as Logan (2000) shows for New York City. Yet the clustering such data shows may be seen as moving from the economic to cultural as its binding force. Singapore provides another example: as van Grunsven (2000) describes it, the re-clustering of Malays in social housing despite strenuous governmental efforts to produce integration shows the strength of cultural ties (as well, perhaps, of the lack of real economic and political integration).

Soft locations

The term 'soft' is associated by analogy to its use in zoning practice, where a 'soft' site is spoken of as one not developed to the limits its legal zoning permits, i.e. one viewed as ripe for change and new development. Such locations include:

- Central Business Districts
- Waterfronts
- Centrally located manufacturing locations
- Brownfield sites
- Concentrations of social housing
- Residential locations on the fringe of central business districts
- Historic buildings and sites
- Public spaces
- Suburbs, ethnic enclaves and excluded ghettoes

Beauregard and Haila (2000) look at three of these locations: waterfronts, brownfields and centrally located manufacturing areas (hollowed-out manufacturing zones), and suburbs (edge cities). They conclude that in each case there are changes, but that these follow earlier changes and are part of older patterns of change as well as new ones. That point can, of course be made, *pari passu*, of all of the soft locations mentioned above. They also conclude that spatial change in the existing built environment lags behind broader social and economic changes, and the question now becomes, as time goes in, will the changes grow and/or turn into something 'new'?

Central business districts

As mentioned above in Skyscrapers and skyscraper clusters, the development of central business district's clusters of skyscrapers is related to two aspects of globalization: the functions of skyscrapers, and their symbolism. There are specific socio-economic conditions that promote such development, including the desire to assemble within a single unit the many administrative and control functions of large business organizations, which makes skyscrapers a means to capitalize on real estate values. But there are also specific cultural features that make skyscraper clusters *per se* a desirable style of city planning. A good bit of this has to do with the dominance of U.S., and specifically New York, in the field of finance and international trade. Frankfurt, in developing its skyscrapers, was accused of imitating Manhattan, and the image of wealth and power that the skyscraper style represents clearly emanates from New York. Yet it fits badly into many European, and certainly German, city patterns. The reluctance of Berlin to allow very high buildings is in part a resistance to the U.S. influence that has much to do with very different cultural traditions.

Yet, given these differences, when skyscrapers are built in Germany, they seem to me to be modelled specifically on modern developments in the United States: American architects are in demand, and what is actually built does indeed continue to rival/compete with what is done in New York. Specific features of architectural design can no doubt be traced to this competition. And the symbolism of success in global competition for growth is surely a part of it.

Waterfronts

Two periods may be separated out in the evolution of the use of waterfronts in the last one hundred years. The first has to do with the role of shipping. In the most industrialized countries, waterfronts are no longer vital shipping or trans-shipping locations in many cities where it had once played a central role; shipping is rather concentrated in fewer locations, where its larger scale facilitates efficient modernization and inland transportation access is good. Thus New York City, originally located where it is because of its harbour, now finds most of its waterfront obsolete for shipping purposes. That is a phenomenon that has indeed been going on since the 1920s, and certainly since the 1950s.

Sometime thereafter, however, a second phase of change set in; in many parts of the third world where industrialization was taking off, modern port development changed the old nature of waterfront activity radically, in some cases replacing it with large-scale modern facilities, sometimes in areas distant from old port activities (London, Tokyo), in other cases reducing it in favour of major new port development outside the city (Calcutta, Rio de Janeiro). In the older cities in the industrialized world in which economic restructuring and globalization had reduced or eliminated industrial uses of the waterfront, major efforts developed, struggling against heavy past investment in the built environment, to transform the nature and uses of the waterfront. And closely related to the current process of globalization, and thus defining a new period in the evolution of waterfronts as locations, is the absorption of these previously neglected and under-utilized waterfront areas as adjuncts to the growing dominance of downtown service-oriented activities for the benefit of the new gentry. Waterfronts become amenities making downtowns more attractive. Whether the movement from the first phase to the second phase warrants the characterization of major change is a matter of definition.

Centrally located manufacturing locations

Again, there is a long-term process of change associated with 'new' developments. The movement of large-scale manufacturing from crowded central locations to the outskirts of cities has long been noted; green field sites are both physically and economically more advantageous for large-scale manufacturing, and the transportation and communication disadvantages of outlying sites have steadily been reduced or in fact reversed. The extension of this trend to small manufacturing, however, including to those involved in production directly related to other central city activities (e.g. printing, fashion dress production, repair facilities), is of more recent origin. These are activities conducted more efficiently in close proximity to the centre, unlike the earlier out-movers. But they also are being displaced, as the growth of pure service sector activities and their internationally-linked financial returns raises real estate prices to the point where socially incompatible uses are displaced, even if economically integral to the activities displacing them. The current dispute about the rezoning of the area occupied by downtown printing firms in the heart of the business district of New York City is an example.

Brownfield sites

Certainly the movement of manufacturing activities from less to more favourable locations (whether physically or economically judged) is not a new phenomenon. The difference between brownfield locations, generally not in the centres of cities, and those described above is that here the process is not of displacement, as in central locations, but rather simply of abandonment. Partially because of the difficulty of adapting a massive built form to other uses (although cases exist such as the transformation of warehouses to condominiums and artists' lofts) and partially because of enduring environmental pollution, such sites often find no re-use: hence become 'hollowed-out'. The scale of such hollowing-out has significantly accelerated in the period of globalization. It is also beginning to affect locations recently developed, where there is no physical problem of obsolescence, but simply shifts of international or national investment: thus automotive plants built in São Paulo on what were green field sites just thirty years ago now face abandonment; the pace of the process of industrial abandonment seems to be remarkably accelerated today.

Concentrations of social housing

Locations at which social developments are located have been subject to much study recently (Vale, 1993). Some of the issues are large-scale, as in the rehabilitation of the large developments built in the after-war years throughout Eastern Europe and in much of the West. Other concerns deal with inner-city high-rise developments, often deteriorating for political as well as physical and social reasons. The problems are not, of course, new, but the process of globalization and its accompanying economic changes and social impacts, in particular polarization of incomes and exclusion, put concentrated locations of social housing at the centre of issues of segregation and abandonment or gentrification. The locational aspects of these problems, and their relationship to the overall spatial structure of cities, are however as yet under-researched.

Residential locations on the fringe of central business districts

Gentrification is an aspect of the change in residential locations on the fringe of central business districts extensively considered, but such locations are subject to other forms of change besides gentrification: sometimes simply clearance to provide amenity benefits to the centre, sometimes changes of uses, from low-level to high-level services (warehousing to offices or residences), or from office to residence or the reverse, or for transportation infrastructure whose location and form have clear divisional effects. In general, the result is a shift in uses from residential to business, meaning generally a net reduction in the residential use of central areas.

Historic buildings and sites

The examples above suggest that an approach focusing on 'soft locations' deserves more thorough comparative study than they have hitherto received. The attitudes towards structures of historic meaning, for instance, have changed significantly in the last few decades, in ways not unrelated specifically to the pressures of globalization. As national boundaries are more readily and frequently crossed, the threat to local identity mounts; at the same time, the importance of identity, and specifically the linkage of identity to territory (in different ways both national and local territory) grows.

Public spaces

Public spaces have similarly undergone significant change, in form, usage, and control. The general movement is towards private control of what is done in public spaces. Sometimes that private control is exercised through pressure on public authorities, as in the 'cleaning up' of Times Square or 'amusement districts' in European cities; sometimes it is done directly privately, as in the use of private security guards by Business Improvement Districts or the managers of gated communities. Sometimes, ironically, the privatization of public space comes about through the offering of semi-public facilities in legally private spaces, e.g. malls or shopping centres. The net result is the same: the amount and openness of space for 'public' activities is eroded.

Suburbs, ethnic enclaves and excluded ghettoes

Suburbs, ethnic enclaves and excluded ghettoes were discussed above, where they were reviewed and defined by changes in the physical form of particular social components, that is, starting with their social composition and looking at their spatial reflection. But one might also start by holding location constant, and then examining what kinds of changes have taken place in those locations: how did locations that were ethnic enclaves or suburbs or ghettoes at the beginning change under the influence of more recent trends? A steady shift 'downward' of most older suburbs located close to the central city could be found, with frequently a conversion from suburb to ethnic enclave, and gentrification in a limited number with particular environmental amenity. One might find areas that were in the early years ethnic enclaves shift in the population they house (most likely from older to newer immigrants), with a visible impact on the exclusionary ghettoes near to which they are sometimes located. And one might find, as Badcock (2000) describes for Australia, but as is true of many cities, that areas of social housing that were middle class or even suburban in earlier days have become more and more ghettoes of the excluded. On the other hand, one might also find gentrification taking place even in standard suburbs.

The impact of security

9/11 has brought about significant changes in both the structure and use of spaces in major cities. What follows is drawn from experience in New York City (Marcuse, 2004). It should be true in diminishing strength as cities are less and less involved in the nexus of global relations, but that expected pattern is distorted by the way in which Homeland Security funds from the Federal government are distributed on a political, rather than a systematically assessed level-of-risk basis in the United States. The pattern also apparently holds with less strength in Europe and elsewhere, but no systematic comparative studies of the point have yet been undertaken. The changes have been three-fold: in the barricading of the external environment in cities, in the 'hardening' of architectural forms of buildings and built structures within those environments, and in the restrictions on the use both of the external and the built environment.

The barricading of the environment

The barricading of the environment is visible most prominently in the central business district. Its most egregious forms are the Jersey barriers, concrete dividers formerly used only to divide lanes on much-used highways. These heavy concrete barriers, some three feet high and 30 feet long, have been placed in front of both public and private buildings to prevent vehicles – presumably those carrying explosives – from coming close enough to a building to do damage. Some communities have applied sophisticated architectural design to making such barriers less obtrusive, e.g. by hollowing out their tops and making planters of them. Bollards (fortified sunken piers), that are relatively inconspicuous when lowered into the ground, can be electronically raised when danger is anticipated. At airports, traffic patterns are carefully controlled so that no motor vehicle can come too close to a sensitive building, and the location of garages increasingly takes into account the dangers of parking an explosive-laden vehicle in them.

The hardening of architecture and barricading of built structures and public areas

Security has had a profound influence on the shape of world cities. Its influence is visible in many ways: it has been recently most prominently symbolized by the police demands for changes in the design of the so-called Freedom Tower adjacent to the site of the former World Trade Center at Ground Zero. At the request of the New York Police Department's security experts, the base of this 1,776-foot-high symbol of freedom will be encased in concrete, with no windows or glass of any kind exposed to the outside. One design being considered is to encase this bunker-like base in panels of angled mirrors, to distract attention from the reality underneath.

These visible manifestations of the defence against terrorism, what Graham (2004) has called the 'splintering of the city', are clear not just in the 'Freedom Tower', but also in the proliferation of citadel constructions, and in more mundane and every-day structures such as the Jersey barriers, highway dividers placed in front of buildings or surrounding areas presumed vulnerable to attack. Similar, if not as extreme, modifications are being imposed by security concerns on entrances, parking garages, structural elements, elevators and stairways, and other presumed vulnerable components of new office buildings and high-rise privileged buildings in the metropolitan area. However, what is in some places done by physical structures is in other places done by social controls, as in the restrictions on demonstrations on public streets or in public places, and the restrictions on entry into both public and private buildings.

Restrictions on use

The restrictions on use imposed on public spaces and on access to private spaces have multiplied significantly since 9/11. The media routinely report on police restrictions on public assemblies in public spaces, whether they are protest demonstrations in front of City Hall or rallies against the war in Central Park. Permits are required for any marches, with more severe restrictions than before, and recently requirements for groups of bicycle riders to ride together through city streets have been required and upheld in the courts.

Conclusion

But in every case, both at the scale of patterns of urbanization and at the scale of internal structure, it is critical to remember that globalization is not a single homogenous external and inevitable process, historically new, but rather the outcome of clearly ascertainable historical processes of much longer duration than the last 50 years and largely traceable back to the 16th century and the rise of capitalism and the explosion of industrialization. Nor is globalization a process independent of the agency of actors benefiting from it and major social forces resisting it. To what extent globalization determines the future of world cities, and of all cities, is a matter within human control, and its ultimate direction is still subject to our influence. None of these measures are new, except perhaps in the extent and intensity, but in both those respects the progress of globalization has increased their impact significantly. The threat of terrorism has been used as not only to impose measures related rationally to such threat, but also to reinforce and legitimize trends already well under way before that, and discussed above in the context of the segregation of urban space, the construction of citadels and gated communities, and the retreat to the suburbs and edge cities. Security concerns based on the War on Terror have thus been used to fortify trends produced by globalization which in turn are accentuates of trends long present in the developed economies of the world.

Notes

1. I use the term 'really existing' to make it clear that what is being described is not some natural or automatic and inevitable process we witness today, involving the increasingly international character of business activities, culture, communications, and exchange, but simply the form that such internationalization is taking at the present time; other forms are imaginable.
2. I have discussed the definition of globalization in various publications, most extensively in Marcuse and van Kempen, 2000, pp. 1–21 and Marcuse, 2006.
3. See the extensive sources cited in Cypher, 2007.
4. Castells perhaps exaggerates the relative importance of this phenomenon; see my commentary in Marcuse, 2002a.
5. For a fuller discussion of the impact of globalization on city form see Marcuse and van Kempen, 2000, pp. 1–21.
6. See the Shrinking cities project (2002–2005) of the German Federal Cultural Foundation <http://www.kulturstiftung-bund.de/>, under the direction of Philip Oswalt (Berlin),

references and publications at http://www.shrinkingcities.com/publikationen.0.html?&L=1

7. For a detailed discussion, and formal definitions, of the categories of ghetto and enclave used here, see Marcuse, 2005.

8. See note 7 above.

References

Badcock, B. (2000) The Imprint of the Post-Fordist Transition on Australian Cities, in *Globalising Cities: A New Spatial Order?* (eds P. Marcuse and R. van Kempen), Blackwell, Oxford.

Beauregard, R and Haila, A (2000) The Unavoidable Continuities of the City, in *Globalising Cities: A New Spatial Order?* (eds P. Marcuse and R. van Kempen), Blackwell, Oxford.

Bernstein, F. A. (2005) In My Backyard, Please: The Infrastructure Beautiful Movement, *New York Times*, 27 February, Section 2, p. 37.

Blakely, E. J. and Snyder M. G. (1995) *Fortress America: Gated and Walled Communities in the United States*, Lincoln Institute of Land Policy Cambridge, Mass.

Brenner, N. and Theodor, N. (eds) (2002) *Spaces of Neoliberalism: Urban Restructuring in North America and Western Europe*, Blackwell Publishing, Oxford.

Castells, M. (1998) *The Information Age* (Volume I: The Rise of the Network Society. Volume II: The Power of Identity. Volume III: End of Millennium.) Blackwell, Oxford.

Chakravorty, S. (2000) From Colonial City to Globalising City? The Far-from-complete Spatial Transformation of Calcutta, in *Globalising Cities: A New Spatial Order?* (eds P. Marcuse and R. van Kempen), Blackwell, Oxford.

Cypher, J. (2007) Slicing Up at the Long Barbeque. *Dollars and Sense*, **269**: 30.

Foster, J. B. (2007) The Financialization of Capitalism. *Monthly Review*, 58(11): 13–33.

Garreau, J. (1991) *Edge City: Life on the New Frontier*, Doubleday, New York.

Gottmann, J. (1961) *Megalopolis: The Urbanized Northeastern Seaboard of the United States*, The Twentieth Century Fund, New York.

Graham, S. (ed.) (2004) *Cities, war, and terrorism: towards an urban geopolitics*, Blackwell Publishing, Malden, MA.

Harvey, D. (2005a) *A Brief History of Neoliberalism*, Oxford University Press, Oxford.

Harvey, D. (2005b) *The New Imperialism*, Oxford University Press, Oxford.

Keil, R. and Ronneberger, K. (2000) The Globalization of Frankfurt am Main: Core, Periphery and Social Conflict, in *Globalising Cities: A New Spatial Order?* (eds P. Marcuse and R. van Kempen), Blackwell, Oxford.

Logan, J. (2000) Still a Global City: The Racial and Ethnic Segmentation of New York, in *Globalising Cities: A New Spatial Order?* (eds P. Marcuse and R. van Kempen), Blackwell, Oxford.

Marcuse, P. (1985) Gentrification, Abandonment, and Displacement: Connections, Causes, and Policy Responses in New York City. *Journal of Urban and Contemporary Law*, St. Louis, Washington University, 28: 195–240.

Marcuse, P. (1986) Gentrification, Abandonment, and Displacement: Connections, Causes, and Policy Responses in New York City, in *Gentrification and the City* (eds N. Smith and P. Williams), Allen and Unwin, London.

Marcuse, P. (1997a) The Ghetto of Exclusion and the Fortified Enclave: New Patterns in the United States. *American Behavioral Scientist*, special issue 'The New Spatial Order of Cities', **41(3)**: 311–326.

Marcuse, P. (1997b) Walls of Fear and Walls of Support, in *Architecture of Fear* (ed. N. Ellin) Princeton University Press, Princeton, pp. 101–114.

Marcuse, P. (1998) The Ghetto of Exclusion and the Fortified Enclave: New Patterns in the United States, in *Towards Undivided Cities in Western Europe: New Challenges for Urban Policy*, Part 7: Comparative Analysis (eds H. Priemus, S. Musterd, and R. van Kempen), pp. 5–20.

Marcuse, P. (2002a) Depoliticizing Globalization: From Neo-Marxism to the Network Society of Manuel Castells, in *Understanding the City: Contemporary and Future Perspectives* (eds J. Eade and C. Mele), Blackwell, Oxford, pp. 131–158.

Marcuse, P. (2002b) Urban Form and Globalization after September 11: The View from New York. *International Journal of Urban and Regional Research*, 26(3): 591–596.

Marcuse, P. (2002c) The Layered City, in *The Urban Life World: Formation, Perception, Representation* (eds P. Madsen and R. Plunz), Routledge, New York and London.

Marcuse, P. (2004) The 'War on Terrorism' and Life in Cities after September 11, 2001, in

Cities, War and Terrorism: Towards an Urban Geopolitics (ed. S. Graham), Blackwell Publishing, Malden, MA.

Marcuse, P. (2005) Enclaves Yes, Ghettos No, in *Desegregating the City: Ghettos, Enclaves, and Inequality* (ed. David Varady), State University of New York Press, Albany, NY.

Marcuse, P. (2006) Space in the Globalizing City, in *The Global Cities Reader* (eds N. Brenner and R. Keil), Routledge, New York, pp. 26–269.

Marcuse, P. and van Kempen, R. (2000) Introduction, in *Globalising Cities: A New Spatial Order?* (eds P. Marcuse and R. van Kempen), Blackwell, Oxford, pp. 1–21.

Mingione, E. (ed.) (1996) *Urban Poverty and the Underclass*, Blackwell, Oxford.

Smith, N. (1990) *Uneven Development: Nature, Capital and the Production of Space,* 2nd edition, Basil Blackwell, Oxford [First published 1984].

Vale, L.J. (1993) Beyond the problem projects paradigm. *Housing Policy Debate,* **4**(2).

van Grunsven, L. (2000) Singapore: the Changing Residential Landscape in a Winner City, in *Globalising Cities: A New Spatial Order?* (eds P. Marcuse and R. van Kempen), Blackwell, Oxford.

Waley, P. (2000) Tokyo: Patterns of Familiarity and Patterns of difference, in *Globalising Cities: A New Spatial Order?* (eds P. Marcuse and R. van Kempen), Blackwell, Oxford.

3
Darko Radović

The World City Hypothesis Revisited:

Export and import of urbanity is a dangerous business

Introduction

The concepts of world cities and, in particular, of the implied 'world class' *quality*, are fundamentally about domination and power. This chapter explores some of the forces behind the 'world class' city label, and argues how cities, in their competition for the 'world class' aura, often end simply as tools in broader global power-games which have very little to do with actual urban quality. The key question is who defines the criteria of 'world class', and to what ends. It is clear that the criteria which are as commonly accepted today are reductivist in their nature and do not include imperatives of environmental and cultural sustainability.

When it comes to sustainability, for quite a long time academics and practitioners have been advised to avoid apocalyptic messages in their reports – the adage was that there is a need to be conciliatory and seek ways that engage, rather than alienate, those in power. Over the past twenty years there have been numerous efforts on that path (for one of my own recent contributions to appeasement see Low *et al.*, 2005). But the time has come to admit that such tactics do not work. All indicators show a grave and worsening environmental crisis. There is an urgent need for action.

This chapter questions globalized rules which dominate the world city and similar power-games; these have lead to the overall environmental and cultural crisis of today. It argues that it is high time to reconsider a profound paradigm shift and it puts forward proposals for 'elements of a new world city hypothesis' (which, for some, may go beyond what seems conceivable today, but for which the time has come).

Worlds, many and one

Michel de Certeau used to say that cities are the 'most immoderate of human texts' (de Certeau, 1984, p. 92), the most complex imprints of the humankind. They embody the totality of culture, the full spectrum of expressions which our communities are capable of generating. They indeed are spatial 'projection[s] of society on the ground', in which 'what is inscribed and projected is not only a far order, a social whole, a mode of production, a general code, it is also time, or rather, times, rhythms' (Lefebvre, 1996).

When he postulated that overarching definition, Henry Lefebvre, typically, could not let it stand alone, pretending to grasp the full complexity of *the urban*. True to his approach, he added 'another definition which perhaps does not destroy the first: the city *is the ensemble of differences* between cities', only to continue with another, that 'of plurality, coexistence and simultaneity in the urban of *patterns,* ways of living urban life' (ibid.). It is within that panoptic view of urbanity which gains special significance today, when globalization blurs boundaries between what 'city' and 'country' are and when truly global cities emerge, that I want to address the question of world cities and their effect on the agendas for a sustainable future.

There are many definitions of world city, what such a city is and what it should be about. The discourse usually begins with Friedmann's seminal 'World City Hypothesis' (1986). Many see that paper as a founding piece of an urban theory that not only recognised a new and emerging phenomenon, but is the one which inaugurated that phenomenon into a new field of study. An abundance of literature appeared in response to Friedmann's hypothesis, and reached its peak in the 1990s (Taylor, 2004). The obligatory supplements to the 'Hypothesis' are Sassen's *The Global City: New York, London, Tokyo* (1991) and *Cities in a World Economy* (2000). Some also add *The Rise of the Network Society* by Manuel Castells (1996). This chapter while pointing at those references draws upon much broader historical and ideological contexts.

Taylor (2004) gives a very thorough account of terminology associated with the world city debate. Also, Sassen succinctly explains her own selection of terms relevant to this chapter:

> The globalization of economic activity entails a new type of organizational structure ... a new type of conceptual architecture. Constructs such as the global city and the global city-region are, in my reading, important elements in this conceptual architecture. The activity of naming these elements is part of the conceptual framework ... When I first chose to use global city (1984) ... I did so knowingly – it was an attempt to name a difference: the specificity of the global as it gets structured in the contemporary period. I did not choose the obvious alternative, world city, because it had precisely the opposite attribute: it referred to a type of city which we have seen over the centuries ... In that regard it could be said that most of today's major global cities are also world cities, but that there may well be some global cities today that are not world cities in the full, rich sense of that term (Sassen, 2001, p. 79).

However, terminological nuances are not central to this chapter. There are some important distinctions between world city, global city, global city regions which will be left aside. As far as terminology is concerned, the focus here is on *the idea* of 'world class', and the hierarchical nature of relationships which are latently present in all of the above terms and which are generated by competition between cities for global dominance.

Of course, there is nothing new in the idea that cities compete. Historically and today, like lighting rods, they attract human activity, bring together disparate elements of society and, by assembling differences, they create energy-charged centralities. That quality establishes conditions for invention and creation of *the new.* By their very nodality and density, cities overcome spatial and temporal distances and generate situations that can transcend *the known.* As such, cities inevitably compete – both internally and with other cities. Important elements of urban competition are trade and commerce. Many urban types have indeed emerged from market-places, but the key 'markets' that cities are capable of creating were and remain those of human creativity. They offer places for exchange which go beyond simple monetary value of 'goods'.

In the Mediterranean region, ancient Greek cities, for example, were fierce competitors at many levels and at many scales – from those of their immediate contexts to the broad arenas which stretched far beyond their physical limits, to the edges of the then-known world. Their desire to compete was, literally, extending the limits of the Ancient world. The *polei*[1] fought for supremacy, for domination, for status. What to be 'the first' and 'the best' actually meant in those times was very place and time-specific; the distinctiveness and particularity of those small cities was always part of the equation. Difference itself established some of them as 'world class' within their own niches of excellence.

The *Pax Romana* saw a drastic cultural change, a consolidation of administrative and organised ways of generating, achieving and expressing domination and status of cities within a highly centralised world of the empire. Rome, the *Urbs*, was an unquestionable centre, a true and only world city of that era or, rather, of the 'known world'. At the same time, other capitals existed in their own worlds which were far enough away and unknown to the Romans. Those were the times of multiple worlds.

The mediaeval times led to a new wave of fragmentation in the Mediterranean region, with renewed struggles for differentiation, domination and status. The city states of Siena, Venice and Dubrovnik (to name only the first, and the truly exceptional two) treasured the same sense of uniqueness, pride, local patriotism and tendency to compete which characterised *polei* – forcefully stressing how their cities differ and where they excelled, in an effort to perpetuate and enhance plurality and difference as comparative advantages within the boundaries of their ambitions and abilities.

Similar tendencies (despite myriads of differences which result from unique local blends of environmental and indigenous cultural and societal expressions) are found in all parts of the world. Numerous archipelagos of rich and diverse urbanities existed in other parts of Europe, and in Asia, the Americas and Africa. A whole new field of inquiry, usually defined as Oriental Globalization, 'not merely critiques but overturns the conventional perspectives' of Eurocentrism, and 'implies a profound rethinking of world history' (Pieterse, 2006, p. 411). That body of work speaks about highly sophisticated expressions of a variety of cultures before and beyond the cradle of the now dominant, Western paradigms. Those were still the times which allowed the whole universe of many worlds.

What matters is that at the same time as the early developments in the Mediterranean, *urban excellence* existed in various other regions of the world. It meant different things, and the difference itself, in its many guises and shapes, has been acknowledged as quality. Those diverse regions possessed the synchronic urbanities, diverse city-cultures generated by their own local 'worlds', simultaneously in diverse mutual relationships, and variously aware (or blissfully unaware) of each other. Given the realisation of this complexity, the chapter now addresses the central questions: What are the criteria for definition of world cities (and/or of the implied world *culture*) today? And, what constitutes 'world class' today?

Darko Radović

Power, its flows and centre

In times of ever accelerating globalization, the complexity and richness, the very messiness (Low *et al.*, 2005) of what constituted 'the world' (or 'worlds') has been diminished, if not completely lost. The world, we now know, is a beautiful blue planet which, once embraced with a single glance, has suddenly shrunk. It became one and, thus, it became 'a thing', 'the other' that can be conquered and possessed. It became possible to conceive the power that would, for the first time, be truly global. Today, such power does exist. Like any power, it aims at totality. It can not be satisfied with fragments of what is known to exist. In its encounter with 'the other', power strives to dominate.

In a shrunken world, it is impossible to be unaware of differences. The newly acquired awareness of 'otherness' within our reach leaves an impression that the world is richer and more diverse than ever before. But, in reality, rampant globalization flattens differences. It seeks to comprehend in order to tame 'otherness', to swallow it and digest its juices (Derrida, in Hillis-Miller, 2001). The 'different' is exposed naked to the eye of those who have the means to see the whole world in a single glance. Within that new, single and simple world, definitions of what constitutes urban quality only seemingly remain the same. Cities are, arguably, still about competition, identity and pride. But the fundamental difference is that, by framing the world as a single 'panopticon', now a single meaning of those terms is dictated and demands simple sets of rules that would be understandable and acceptable primarily within the culture of the controlling power, and for the purpose of control.

At the end of the 20th century many argued that power (and control) became largely placeless, that there were no longer physical 'seats of power', but rather networks (Castells, 1996) and amorphous flows, which grow and contract as they seek, follow or create ever new crises and opportunities all around the world. In reality, though, the ruling power of today is not as de-centred as the theory may want it to look. Concentrated power does exist and it still commands most of the flows (Negri and Hardt, 2000).

In that pseudo-decentred environment, capitalism locates its activities opportunistically, to capitalise on the current potential of particular places and their natural and human resources. Once those resources get exhausted, the flows contract and vanish – only to appear elsewhere, where new opportunities arise. Definitions of business rarely include the moral and ethical criteria on which our cultures have been created. The new rules may draw upon some of the founding values only when they support the hunger for profit. Those are the situations when the 'placed' centres resurface as actual nodes of power.

The world we live in is very much based on new interpretations of some old power games. For global business, the most profitable of those games is the ugliest one – that of (carefully reframed) colonialism. The gaps between the rich and the poor, between the North and the South, between the East and the West are more pronounced than ever. An all-encompassing, totalising view of the world makes injustice visible. The little blue planet is seen to be geographically, culturally and racially divided between the winners and the losers. How the 'amorphous' capital allows it to be controlled and how it takes clear shapes and acquires distinctive, one-way traffic towards the centres, towards power, becomes glaringly obvious.

That partially answers the question: who defines global rules? – the rules in general and the specific ones discussed in this chapter. The definition of what constitutes world cities, global cities, what is 'world class' or, indeed, any other quality that matters to those in power, is strictly dominated by economic determinism – an offshoot of the ruling (non)ideology of free market and neo-liberalism.

Therefore, the world city phenomenon and the 'world class' syndrome of today are part of a complex, neo-colonial power-game which favours the 'West' (of a very articulate and narrow profile), the 'multinational capital' (with its distinctive accent), the 'US interests' – whatever we chose to call the dominant power of today.

World city, a hypothesis and construction of reality

Current definitions of world city are based on the above power relations and are the succinct spatial projections of the new social order. The world city is framed by the dominant ideology of the free-market, and world cities embody the multi-centred capitalism and its resulting inequities. Definitions are framed by the work of remarkable intellectuals: Friedmann, Sassen and their likes. Their tremendous analytic effort mainly identified *what was*, or *what seemed to be* (happening) in the 1980s and 1990s, but they never penetrated far *beyond* or *behind* the phenomena such as 'world class' and 'global class'.

Taylor (2004) stresses that the world city literature has attracted lots of scepticism [as exemplified by Short (1996), Beavestock *et al.* (1999) and others], and adds that 'the world city literature is vulnerable to criticism for its *dearth of evidence backing up its propositions*' [author's emphasis]. He quotes Short's term 'dirty little secret' which denotes what Taylor himself calls a 'data deficiency problem' and what Beavestock saw as 'the Achilles heel of the field':

> The problem largely concerns information on relations between cities. In fact, disquiet concerning *the paucity of evidence behind Friedmann's 'world city hypothesis'* [author's emphasis] surfaced in the initial published discussion of the paper and it might have been thought that this problem should have been sorted out by now. But no ... (Taylor, 2004, p. 32).

Taylor follows with a characteristic, succinct and thorough bibliographic survey. He concludes that there is a 'great paradox of this literature: there is a dearth of research on the connections between world cities and yet the latter cities' pre-eminence is based precisely upon those under-researched connections' (ibid., p. 33).

Without going into further details here, this hint at the possible myth-making is interesting in itself. Is there a scenario in which an 'invented' term (a hypothesis) created a new field of inquiry which, in turn, then framed the emergence of the 'real reality' of the world city? Friedmann himself identified the world city phenomenon as a *hypothesis*. His work projected that hypothesis as a research question into its *academic existence*. The hypothesis has found a very fertile soil and huge enthusiasm, both within academia and among the politically and economically powerful of the 1980s and 1990s. One possible reading of that success is that *the powers of the moment liked the hypothesis to such an extent that they made a chimera real*.

It is worth re-reading the words of Sassen quoted above: the constructs such as the global city and the global city-region appear to be elements in a new conceptual architecture of globalising economic activity. 'The *activity* of *naming* these elements is part of the conceptual framework' (Sassen, 2001, p. 9, original emphasis). History teaches us the equation: naming = being = owning (Rihtman-Augustin, 2000). The very act of research and naming hypotheticals may produce realities and, consequently, the desire for conquest and ownership of those new realities, of another 'new world'.

The consumption of 'world class' cities

From a broad, overall view of the world city phenomenon and 'world class' qualities, this chapter now takes a different perspective, asking how those cities will look. In order to get closer to an answer, four cities, which either aspire to become, or which already are considered to be 'world class', will be 'visited' – Barcelona, Shanghai, Bangkok and Tokyo. We will accompany the world class traveller, with his (or her) well-theorised mobility and networking skills, representing a new breed that both consumes the world cities and makes 'world class', at least as a myth, possible. For those global players, as Sassen argues, 'national attachment and identities are becoming weaker' (2001, p. 91). They are true postmodern (globalized) individuals who are 'forever acquiring new identities, creating new universes of realities, consuming whatever they think would satisfy their insatiable quest for meaning, identity and belonging: largely at the expense of the non-western cultures' (Sardar, 1998, p. 39). The world-class travellers are characteristically individualistic. Identifying their expectations provides insights into their ideology and values that underpin the very concept of 'world class'. A good introduction to what they want (and, perhaps, what they are) can be found in specialised travel guides. One of the latest, the *Wallpaper*[*2] series, offers 'a new concept in city guides, pocket sized, easy-to-use and discreet, so that you do not feel like a tourist'. The series is aimed at the particular market, at the self-conscious and self-proclaimed 'design-conscious traveller'. Such travellers have the means to travel, they want to experience the exotics (often within their business itineraries), without actually leaving the comfort of their own culture (which is, largely, a culture of globalized consumerism closely associated with 'world class' ideology). They are, thus, a new breed of armchair travellers, with a significant difference: they do not fantasise, they actually do travel. As they are not ready to part with their cushioned environments, their armchairs are always close by – in the first class seats and airport lounges, in the boardrooms, in the hotels which offer comforting sameness, with a pinch of cultural differences, with varying levels and kinds of local spice to chose from.

Barcelona

Since the death of the Spanish dictator Francisco Franco, and in particular since its very successful staging of the Olympic Games in 1992, Barcelona has been an undeniable success story. The city effortlessly climbed from one height to the next. From an oppressed provincial capital, pejoratively dubbed 'the Manchester of Spain', Barcelona became an important regional centre within the Spanish state, then regained its long lost status as an important node in the network of Mediterranean cities, becoming one of the key European capitals of culture, launching itself into the 'world class' orbit. That process was based on a strong desire to reconfirm local culture, local identity and pride. A charismatic Catalan leader, Pasqual Maragall was clear that with a strong Barcelona 'Spain is more respected, Catalonia freer,

Barcelona more of a city than ever' (Marshall, 2004, p. 81). In a single sentence he emphasised two aspects of a vision which proved to be critical for the forthcoming success – a strong emphasis on local identity and his firm belief in the necessity of the urban.

The spatial projection of that political vision was focused firmly on local issues and quality of life in Barcelona. 'Public spaces were said to be aimed, first of all, at solving the historical differences of the city, a city made too dense by the uncontrolled and speculative urban activity of the pre-democratic period' (Benach, 2004, p. 154). It was the power of leadership, channelled through institutions based on recognisably traditional models which generated new, undisputable quality. One of the keys to such success was an ability to energise and inspire the community, to help people formulate their initiatives and eventually, to make them proud of their own achievements. Barcelona, thus, started its journey towards the 'world class' in an old-fashioned Mediterranean way, by catering for its own citizens or, in more concrete terms – by *being* its citizens.

Success and global aspirations have brought global money to Barcelona (Marshall, 2004; Rowe, 2006), new riches with all the associated strings. The first sign that something has changed became apparent during the slow-down in the economy in the immediate aftermath of the Olympic boom. A Texan developer (the story goes) was around when needed, to provide funding for a major construction, at the intelligently chosen and previously purchased piece of land that borders *Avenida Diagonal*. The programme reflected the experience and values of the owner: an American style, suburban, shopping centre. After the economy of Barcelona picked up again, the global appetites which by now had become embedded in the business fabric of the Catalan capital, started to show up openly. That led to development of ever-newer and bigger capacities, which obviously aspire further towards global heights.

Barcelona of the early 21st century is as keen as ever to lead and to inspire (Radović, 2008 forthcoming), but the ambitions are now towards global horizons. The rules of urban development had to change accordingly. As Gugler explains, 'skylines distinguish world cities from other cities in poor countries. Office towers, five-star hotels, and luxury apartments stand out' (2004, p. 14). The skyline of Barcelona is changing in that direction rapidly. The city now aspires to attract the world-class traveller and his/her Wallpaper* Guide to Barcelona (2006) opens with an expected, reassuring sign of being in the right place. The first chapter, entitled 'Landmarks – the shape of the city skyline', says how Barcelona is 'growing at a dizzying rate'. It explains that the Western part of the city (the one where a recognisable shopping centre is located) is 'increasingly ... where business visitors find themselves', assuringly adding: 'not that the city heart has avoided change'. The brief introduction ends by stating how 'in the midst of all this upheaval, it can be comforting to cling to the familiar. *Casa Milà* was Gaudí's first purpose-built apartment block, and offers a gratifying selection of all his signature riffs, for those moments when the summer hawkers and handicams simply infect the *Sagrada Família*'.

In short, there is a New World emerging in Barcelona, but one can also consume some of the best known relics of its past. That can be done conveniently, without getting infected by contact with the 'summer hawkers'. The City Guide information flows on: after getting to know several facts about the *Santa Caterina* Market

(reconstruction of which, we learn, cost 13 million Euros), *Torre Agbar* (which is 142 metres tall and has 4,400 windows), *Montujïc* Communications Tower (whose designer Santiago Calatrava is 'the world's most crowd-pleasing architect after Frank Gehry') and *Casa Milà* (again) we move on to the twenty-four pages of 'Where to stay and which rooms to book'. That chapter briefly reviews several up-market hotels, only to proceed to eight pages of 'See the best of the city in just one day'. The traveller is advised to see 'an upmarket cantina … (breakfast served 08.00–11.30)', one cultural centre and one gallery and further doses of Gaudi – *Casa Batló* and *Sagrada Familia*. After a 'well earned siesta' you end up in a 'nightclub, lounge and cocktail bar that opens at 8 p.m.', which is conveniently located at the World Trade Centre. At ten you go to an 'eaterie (*sic*) favoured by the city's creative élite'.

Fig. 1. Images of 'world class' Barcelona Source: Darko Radović

The subheading entitled 'Urban Life' contains twenty-two pages of information about the trendiest of cafés, restaurants, bars and nightclubs of Barcelona. The 'Insider's Guide', on a single page, presents what makes one local fashion model (whose 'modelling career has taken her around the world') to 'always find time to return to her native Barcelona'. The 'Architour' adds a mix of Boffill, van der Rohe (his 'all marble, onyx and glass' pavilion), Foster, Herzog and de Meuron and EMBT. 'Shopping', 'Sports and spas' and 'Escapes – where to go if you want to leave town' are all there, too.

That is the 'world class' Barcelona, a city packaged for consumption by a particular traveller/reader.

Shanghai and Bangkok

The second visit is to two aspiring East and South East Asian cities – Shanghai and Bangkok. Shanghai is well known for its rapid urbanisation. The *lilongs* and other quarters of the former colonial concessions of Shanghai are giving way to new development, as if huge expanses of Pudong could not provide sufficient space for experimentation and appetite of the 'world class' capital.

The Wallpaper* Guide to that metropolis (2006) offers not only its standard, easy-to-use book structure, but also a selection of up-market hotels, cafés, bars and shopping destinations which are very similar to those in Barcelona. That should not be surprising. We are in yet another suburb of the global 'world class' city, we are busy and we do not have time for local surprises. For the 'world class' citizen, urban life of Shanghai is (as expected) all about cafés, restaurants, bars and nightclubs. While there may still be some parts of the 'messy' old town out there, the glamour

Fig. 2. Images of 'world class' Shanghai
Source: Darko Radović

on offer is themed 'Old Shanghai', 'China' and 'Asia'. Among places to eat is, we are assured, 'one of the best looking restaurants' which offers 'a merciful antidote to the acres of international beige and downlighting elsewhere', together with 'a kind of uncomplicated Asian fusion that is better than it has any right to be'. Our design conscious traveller will, therefore, not be exposed to local kitsch. The class which he gets for his money remains consistent with what Barcelona had to offer. The helpful insider in Shanghai is (we are led to believe) one of many local marketing executives. She is 'occasionally to be found perched on a stool' of one of the previously featured bars. An unmistakable sign of quality.

And, what about the Thai capital? At the very opening of the Wallpaper* City Guide to Bangkok (2006) we are warned that the city 'is a textbook example of

Fig. 3. Images of 'world class' Bangkok
Source: Davisi Boontharm

urban Asian sprawl'. 'It has abundant concrete, six-lane avenues in the wrong parts of the town and one-lane avenues in the right ones. There's barely a centre to speak of, and you can travel to areas full of teak house-lined canals and rice paddies an hour away and still be in the city'. But there is a ray of hope: 'If you break Bangkok down to how you want to shop, work and play, it can be manageable'. And then, the familiar Barcelona–Shanghai story continues. We learn, for instance, that this city is on the move. The indicators used to illustrate the point are, indeed, strong: 'Ten years ago', we learn, 'Bangkok had fewer than 100 7/11 stores and not a public transport or eight-lane highway in sight'. We also learn that 'just the thought of trying to see the city in 24 hours would send shudders through wide-eyed tourists and knowing locals alike'. But, due to the recent progress, now that seems to be becoming possible – '... things have begun to change in recent years'.

The complexity of Bangkok does not go unmentioned. 'The City of Angels is the city of the extremes and to explore properly, you should experience them all.' The Wallpaper* story rolls on. By the end of reading it one can only hope that the tropical climate will draw some sweat out of the world-class traveller, at least while venturing to see four pieces of architecture on offer: one with 'a touch of tropical Zen', the Elephant Tower, the Democracy Monument and/or the Bed Supperclub (which is, conveniently, like the Zen one, not only a sight to be seen but a restaurant and a nightclub).

The lifestyle of this consumer of the 'world class', as that of the rich foreigner in any colonial system of the past, assumes the poverty of the 'Other'. It needs the exotics, the savage, the oriental, to be experienced, consumed and looked down on.

The 'world class' consumer feeds off their sweat, while their Guides carefully edit the pain out of the actual experience and stay cushioned in their armchairs.

Tokyo

The final visit with the help of the Wallpaper* Guide is a city which usually tops the world city lists – Tokyo. It has all of the world city promises, and more: landmarks; hotels; 24/7 opening; urban life; an 'Architour'; shopping; sports; escapes – all that is listed in the trusted guide (Wallpaper*, 2006) again. 'There is no reason', we are told, 'why a single day in Tokyo can't serve multiple functions. Part of it, perhaps a large part, can be used for shopping, to find things that you simply won't find elsewhere …'. So, the emphasis is on local culture. If still, after all that, there is some spare time, you can escape Tokyo. A twelve-car sleeper train, with berths that are 'either a twin room or a suite, they come with a television, mini-closet, lavatory and the obligatory cotton *yukata* robe (a taste of local culture) and slippers. A fabulous Japanese or French meal is served in the dining car or in your cabin …'. And so on.

But, it should never be forgotten how Tokyo (and Japan) entered the process of globalization. That was through two dramatic 'encounters' with American arms – first in 1854, when the Black Ships of Commodore Matthew Perry entered the Tokyo Bay and requested an immediate opening of the country (as a market) to trade; the second time was after the carpet-bombings and victorious entry of the American troops in 1945. Those two events forced Japanese 'acceptance' of the West. The mid 19th and mid 20th century were the moments which critically marked later development and economic success of the country. Those historic snapshots again invite (a biased?) re-reading of de Certeau: the ordinary culture of the 'Other' was ordered and normalized through the act of opening to commercial domination. In this operation, the culture of the 'Other' was remaindered (paraphrased from Highmore, 2006, p. 16). The moments which combined repression (erasure) with power (inscription) are exactly what Tokyo was subjected to post 1945. Those were the times of forceful exorcism of many things Japanese and introduction of many things Western (American) which included (to name one of the more bizarre and culturally brutal ones) a school diet of bread, instead of traditional rice. That was the period of deliberate humiliation, remembered in the local catchphrase *shirakata ga nai*, or 'nothing can be done about it'.

Fig. 4. Images of 'world class' Tokyo
Source: Darko Radović

In a very different form, something similar is happening to Shanghai, Bangkok, Barcelona and many other aspiring cities of today. That is the continuum of erasure and inscription – the erasure of local and indigenous, and the inscription of 'global'

and foreign. That is the process in which 'world class' cities, through an importation of and intense exposure to the already vulgar (e.g. McDonalds, Starbucks, etc.), become sad places of deculturation and vulgarisation of the local.

Questioning 'world class' (city)

Much of the 'world class' city research bears strong sociolect of the globally dominant culture that sees cities primarily, or even exclusively, as domains of economy. It affirms dominant globalizing trends, and that is why it is so strongly supported by the exponents of global power. It conveniently reduces not only cities, but the whole of existence, to the rule of economy; citizens and nationals become mere consumers. That world is too narrow, and too poor to accept.

The reductionist views often come from the well-intended, but too narrow expert knowledge. For example, Sassen (2001, p. 85) asks 'have the new technologies and organizational forms altered the spatial correlates of centrality?' only to proceed with a statement that 'today there is no longer a simple straightforward relation between centrality and such geographic entities as the downtown, or the central business district'. There is, however, a counter-question: should not such statements be more culture-specific (and less self-confident), especially in a discussion of supposedly global relevance? Some cultures were never based on centrality; some cities, while themselves being nodes, were never centrically organised. Many historic metropoles of Asia, for instance, were conceived as decidedly poly-nodal. Such cities and their cultures first, certainly and logically, never followed the models of European origin. They were the legitimate centres, the capitals in and of that old world of many cultures, many centres and manifold qualities. Even the terminology which Sassen uses is indicative of a single, very particular cultural model. That language bears a recognisable idiolect, the one which shaped 'the CBDs', 'donut development' and similar business-defined expressions. While talking global, Sassen seems to think about specific regions and cultures first, then she generalises and projects that sociolect and culture-specific knowledge on all regions and cultures. When delivered with authority such inscriptions conquer the 'Other' and annihilate *the possibility* of sustainable difference. By generalising culture-specific views and by suggesting how to measure 'success', this approach defines the destiny of the 'Other'. Sassen, for instance, explains that 'we are seeing a sharp growth in the number of centres that are part of the network as countries deregulate their economies. São Paulo and Bombay for instance ...' (ibid. p. 87). That is, most likely, meant to sound encouraging. One way or the other – the message is clear. Those who play by the rules (embodied in the IMF, the World Bank and similar keepers of the 'order', and proponents of the 'economics without history and anthropology' (Pieterse, 2006, p. 413)), can get into the Club, very close to (although never really to) the very top.

Such academic 'verdicts' easily translate into reality for millions of real people who inhabit other cities and cities of the 'Other'. They are based on imported criteria. Those criteria and the whole process of their conception and projection are highly questionable – unless one accepts that the global economy is the *only force* which will shape the urbanities of future. Such determinism is not acceptable.

A short survey of criteria which form the commonly accepted and frequently used definitions of the world city[3] include: presence of international financial institutions, law firms, headquarters of (especially multinational) corporations, stock exchanges, advanced communications infrastructure on which modern transnational corporations rely, active influence and participation in international events and world affairs, presence of powerful and influential media with an international reach. Beaverstock *et al.* (1999) 'scored cities as global centres in terms of the significant presence of major firms providing services in accountancy, advertising, banking/finance, and law' (Gugler, 2004, p.7). Others mapped the presence of various command functions such as stock-market strength or international flights (Gugler, 2004). More exclusive categories include presence of the US dollar billionaires.

There is a clear pattern and an obvious value-system embodied in that list, a decidedly 'first world' perspective. Much of that literature also projects an air of inevitability around globalization and its various manifestations. The perceived inevitability suggests that, even if we dislike, question and fear the totalising globalizing processes, we can not do much about it. That is a reflection of the broad aura of 'free-market economy which was, over the prolonged period of time, promoted into an ideology without alternative, a force that announces the end of history' (Fukuyama, 1992). Similar is the supporting discourse on global democracy, on one-style-suits-all human rights, increasing dependence on the first-world technologies and many other practices of 'a false ideological universality, which masks and legitimises a concrete politics of Western imperialism, military interventions and neo-colonialism' (Zizek, 2005, p. 128).

It is clear that

> when we speak about a global economy, we have in mind one single structure, underlying economic exchange in any place on the planet ... The same can be said of the technological sphere: it is marked by the unity of technologies – computer, satellites, electric or nuclear energy. But, would that make sense, understanding the cultural theme in the same way? Could we speak of 'one' global culture, or 'one' global identity in the same manner we consider the economic and technological levels? Surely not ... (Ortiz, 2006, p. 402).

The world city debate can, and it should be reframed and seen through the lens which identifies processes that are much broader than the economy and capital. In order to question that dominant order it is worth asking questions such as: How would the list of 'world class' city criteria look if, for instance, the basis for their ranking and for measuring their success was a triple, rather than a single bottom line (Elkington, 1998)? Where would New York, London, Paris, Tokyo end up if, besides (not instead!) of their economy, ecological and social qualities of their environments were included? How would the calculation of their ecological footprint affect the equation?

That is where environmental agenda comes strongly into the discussions of 'world class'. The ecological footprint of London, being 'the total area of productive land and water required on a continuous basis to produce the resources consumed and to assimilate the wastes produced by that population, wherever on Earth the land is located' (Hall, 2005, p. 155), is 125 times larger than its surface area. Tokyo demands between 1.2 and 3.6 times the area of the whole of Japan to sustain its current pattern of existence.[4] Those huge footprints cause damage in far-away places, in other, 'less fortunate' parts of the world.

Is that 'world class'?! Is that what other cities should aspire to achieve? Does anybody care about the carrying capacity of the little blue planet?

The paradigm shift – parallel worlds of class and difference

What to do? We have seen how a hypothesis can become reality. The world city discourse needs a new agenda, new theory and new definitions which would be relevant now and in the foreseeable future. There is a need for complexity, for sensitivity capable to comprehend that in urbanism, we deal with an elusive, immoderate, even messy whole. The process of globalization in the late 20th and early 21st centuries does not show such sensibility. It is necessary to establish critical qualifications of difference. One of those would be that difference between *globalization* and *mondialisation*. While the term globalization 'may well be applied to the economic and technological spheres, 'mondialisation' adapts itself better to the cultural universe. The 'mundi' (world) category is then articulated to both dimensions' (Ortiz, 2006, p. 401). There is a need for true mondialism, 'unity but not an imagined uniformity. Unity instead must be expressed through the multiplicities of diversity' (Eisenstein, 2004, p. 55). Such

> ... mundialised culture promotes a cultural pattern without imposing the uniformity of all; it disseminates a pattern bound to the development of world modernity itself. Its width certainly involves other cultural manifestations, but it is important to emphasise that it is specific, founding a new way of 'being-in-the-world' and establishing new values and legitimisations. And that is the reason why *there is not and there will not be a single global culture, identical in all places. The globalized world implies a plurality of world-views* ... And this means that *globalization/mondialisation is one and diverse at the same time* (Ortiz, 2006, p. 403, author's emphasis).

In that process, environmental sustainability should have a decisive role. Environmental sustainability should always be coupled with cultural sustainability, forming a dialectical pair which frames the thinking about, and the realities of, the urban. The self-destructive, unsustainable patterns of development (such as those exemplified by London and Tokyo) are possible and even hailed as 'world class', because of globalization's bias towards the metropolitan centre.

The time has come for a paradigm shift. We need a *new 'world class' city hypothesis*, one that radically questions domination of econocentric assessments of human and urban development, and which includes the palette of diverse and sensitive mechanisms (some of which may deny the very possibility of measurement (Low *et al.*, 2005)). The new 'world class' city hypothesis needs:

- to redefine urbanity, so that it embraces environmental responsibility (Radović, 1994; 2005). If good manners, which are central to the original meaning of the term urbanity (Ramage, 1973), are to extend beyond interpersonal relations towards the whole of the city – then we can speak about decidedly *urbane environments*. If the same attitude is extended further, towards the whole of our environment – we can speak about a truly *sustainable environment;*[5]
- to reinforce Lefebvre's powerful requests for the *droit à la ville*, the renewed right to the city, and for *le droit à la différence*, the right to difference – the rights to be empowered and to be different (against forces of homogenisation, fragmentation, and hierarchically organized power (Soja, 1999), and which are experienced as globalization;

- to reclaim *the rights to (each particular) city and to (each particular) urbanity*, as fully-developed local cultures. Each city needs to have its own rights, the right of and the right to (its own, particular) urbanity. We need to reclaim an ability to live in multiple worlds and for that we need multiple practices that generate, and multiply the rules which define excellence – *parallel worlds of class and excellence.*

Notes

1. Ancient Greek term for cities or city-states (polis in singular)
2. www.phaidon.com
3. See, for example, Wikipedia.com
4. http://www.gdrc.org/uem/observatory/jp-tokyo.html
5. Developing this concept further, the author has introduced the concept of *eco-urbanity*. For details see http://www.ecourbanity.org and http://csur.t.u-tokyo.ac.j/ecourbanity

References

Beaverstock, J.V., Smith, R.G. and Taylor, P.J. (1999) A roster of world cities. *Cities* **16**(6): 445–58.

Benach, N. (2004) Public Spaces in Barcelona 1980-2000, in *Transforming Barcelona* (ed. Marshall, T.), Routledge, London, pp. 151–160.

Castells, M. (1996) *The rise of the network society*, Blackwell Publishers, Malden, MA.

De Certeau, M. (1984) *The Practice of Everyday Life*, University of California Press, Berkeley.

Elkington, J. (1998) *Cannibals with Forks: The Triple Bottom Line in 21st Century Businesses,* New Society Publishers, Gabriola Island, BC, Canada.

Eisenstein, Z. (2004) *Against Empire*, Zed Books, London, New York.

Friedmann, J. (1986) The World City Hypothesis. *Development and Change* 17: 69–83.

Fukuyama, F. (1992) *End of History*, Free Press, New York.

Gugler, S. ed. (2004) *World Cities Beyond the West*, Cambridge University Press, Cambridge.

Hall, T. (2005) *Urban Geography*, Routledge, London.

Highmore, B. (2006) *Michel de Certeau, Analysing Culture*, Continuum International Publishing Group, London, New York.

Hillis-Miller, J. (2001) *Others*, Princeton University Press, Princeton and Oxford.

Lefebvre, H. (1996) *Writings on Cities*, Blackwell, Cambridge, MA.

Low, J., Gleeson, B., Green, R. and Radović, D (2005), *The Green City: Sustainable Homes, Sustainable Suburbs*, University of New South Wales Press, Sydney, Australia.

Marshall, T. ed. (2004) *Transforming Barcelona*, Routledge, London.

Negri, A. and Hardt, M. (2000) *Empire*, Harvard University Press, New York.

Ortiz, R. (2006) Mundialization/Globalization. *Theory, Culture & Society*, **23**(2–3): 401–403.

Pieterse, J. (2006) Oriental Globalization. *Theory, Culture & Society*, 23(2-3): 411–413.

Radović, D. (1994) Urbanity and Ecologically Sustainable Development, in *Architectural Science: Its influence on the Built Environment*, Deakin University, Geelong, pp. 165–169.

Radović, D. (2005) Urbanity and Sustainability: CH2 in the urban space of Melbourne, in *Sbo5 Conference Proceedings*, SB05, Tokyo.

Radović, D. (2008 forthcoming) The Barcelona Intersection – where environmental and cultural meet, in *Eco Edge* (ed. E. Charlesworth), Architectural Press, Oxford.

Ramage, E.S. (1973) *Urbanitas, Ancient Sophistication and Refinement*, University of Oklahoma, Norman.

Rihtman-Augustin, (2000) *Ulice moga grada*, Biblioteka XX vek, Beograd.

Rowe, P. (2006) *Building Barcelona – a second Renaixença*, Barcelona regional, Actar, Barcelona.

Sardar, Z. (1998) *Postmodernism and the Other*, Pluto Press, London, Chicago.

Sassen S. (1991) *The Global City: New York, London, Tokyo*, Princeton University Press, Princeton.

Sassen, S. (1997) Cities in Global Economy. *International Journal of Urban Sciences* **1**(1): 11–31.

Sassen, S. (2000) *Cities in a World Economy, Sociology for a New Century*, Pine Forge Press, Thosuand Oaks, CA, London, New Delhi.
Sassen, S. (2001) Global Cities and Global City-Regions: A Comparison, in *Global City-Regions* (Scott, A. ed.), Oxford University Press, Oxford.
Short, J.R. (1996) *The Urban Order: An Introduction to Urban Geography*, Blackwell, Cambridge, MA.
Soja, E. (1999) *Postmodern Geographies*, Verso, London, New York.
Taylor (2004) *World City Network – a global urban analysis*, Routledge, London, New York.
Wallpaper* (2006) *Bangkok*, Phaidon Press Ltd, London.
Wallpaper* (2006) *Barcelona*, Phaidon Press Ltd, London.
Wallpaper* (2006) *Shanghai*, Phaidon Press Ltd, London.
Wallpaper* (2006)*Tokyo*, Phaidon Press Ltd, London.
Zizek, S. (2005) Against human rights. *New Left Review*, 34: 115–131.

Sustainability and the 'World Class' City:

What is being sustained and for whom?

Introduction

'Sustainability' is a key promotional strategy for many 'world class' cities – because, after all, who could imagine a 'world class' city that was not aspiring to be 'sustainable'? Sustainability with its emphasis on the social, economic and environmental dimensions of development is presented as an antidote to many of the less desirable impacts of globalization on various localities (both urban and rural) across the world. For instance, 'world class' cities are often associated with highly polarized labour markets. This polarization is reflected not only in terms of income distribution, and increasing income insecurity for the less well off, but is also played out spatially, where low income people are either forced into unaffordable housing at the centre or by necessity to the periphery along with the industries not associated with the global economy (Fainstein, 2006, p. 116). These processes are clearly evident in Australia ' … where opportunity and vitality has become more localized in a smaller part of the country', particularly in the major capital cities, and where ' … reduced or more limited opportunities have meant that many people and places have suffered' (O'Connor *et al.*, 2001, p. 60). Sustainability in rhetoric at least provides frameworks that suggest that these, and other, inequitable impacts can be addressed or 'solved' while ensuring economic growth without neglecting environmental care. The centrality of 'environment' in sustainability discourse is also identified as holding out the possibility of addressing ' … the un-ecological conduct and anti-environmental practice of living in global cities' (Luke, 2006, p. 281).

While the concept of world city has been used in a variety of ways and is itself contested, the three most important criteria for such status as outlined by Simon (2006) arguably could be:

- The existence of a sophisticated financial and service complex serving a global clientele of international agencies, transnational corporations (TNCs), governments and national corporations, and NGOs;[1]
- The development of a hub of international networks of capital and information and communication flows embracing TNCs, IGOs,[2] and NGOs; and
- A quality of life conducive to attracting and retaining skilled international migrants i.e. professionals, managers, bureaucrats, and diplomats. In this sense, quality of life embraces not only physical and aesthetic aspects of the environment but also broader considerations

such as perceived economic and political stability, cosmopolitanism, and cultural life (Simon, 2006, p. 206).

Sustainability potentially adds two dimensions to 'world class' city criteria – social equity and environmental responsibility. This chapter explores these two dimensions of sustainability: equity and environment through the interrogation of the question – sustainability of what and for whom? It asks: has sustainability simply been captured in world class or global cities rhetoric as a marketing tactic to 'sell' the image of cities with little claim to either equity or environmental care (as O'Riordan and Church have suggested (2001, p. 12)) or does it hold out the possibility of thinking about, and acting on the social and environmental impacts of globalization on various localities across the globe? Or put another way does sustainable city discourse simply play neatly into the 'boosterism' that has become increasingly associated with 'world class' city status (Marcuse, 2006, p. 366) or does it offer an alternative vision that could lead to more equitable and environmentally sustainable cities?

The chapter begins with a brief overview of the way in which the concept 'sustainability' is currently understood both more broadly and within the context of sustainable city discourse. This is followed by a case study of how the sustainable city is currently spoken and written about in Australia, drawing in particular on the recent House of Representatives Standing Committee on Environment and Heritage's inquiry into sustainable cities 2025. The case study explores the way in which the key sustainability challenges facing Australian cities are framed and also who and what are included and excluded in the discourse.

Sustainability

The most commonly used definition of sustainability and sustainable development is Brundtland's where it is defined as 'development that meets the needs of the present without compromising the ability of future generations to meet their own needs' (WCED, 1987, p. 8). Central to this definition is the need to maintain the resource base while ensuring both economic growth and inter and intra generational equity.

In the literature sustainability and sustainable development are often used interchangeably to denote a 'new' way of thinking about the relationship between environment, society and economy; ' ... to define the ills and articulate the answers of society' (Myerson and Rydin, 2004, p. 98). As a concept, sustainability therefore promises to do many things. It compels 'us' to consider the social, economic and environmental dimensions of 'a problem' or 'problems' and incorporate all of these dimensions into planning and 'problem solving'. Sustainability is therefore purportedly integrative but also visionary, requiring long term thinking because ultimately it is about future making. As a result, sustainability has been described as a process of change towards an unknown future requiring reflexive thinking. As O'Riordan expresses it:

> Sustainability can never be achieved, for it defines itself by its own pathways. Each stage is a reflection of its predecessor and a prognosis for the next (O'Riordan, 2001, p. xix).

What the work of O'Riordan, Myerson and Rydin suggests is that far from having a fixed meaning, sustainability remains relatively undefined and is continuously evolving. The process of definition, redefinition and evolution occurs in and through discourse where meanings are debated, contested and also where consensus is reached. Understood as discourse, sustainability becomes a site of struggle over words, meanings and knowledge(s).

Storylines

The approach to discourse analysis adopted in this chapter draws on the work of Hajer (1995) who argued that sustainable development should be analysed as a storyline that has created ' ... the first global discourse–coalition in environmental politics' (Hajer 1995, p. 14). This coalition shares a way of talking about environmental matters 'by virtue of its rather vague story-lines' (ibid., p. 14).

Story lines hold fragmented or contradictory discourses together through suggesting a common understanding. They are narratives on social reality that allow the construction of 'a problem' as well as discursive closure (ibid., p. 62). Analysis based on storylines therefore has the potential to explain how, given the complexity, the contradictions and the contestation there is also a degree of consensus in the literature around what constitutes a 'sustainable city'. So while there are a whole range of actors involved in sustainable city discourse, often with competing interests and agendas there are common threads, common understandings that makes the discourse make sense – 'it sounds right' (ibid.). And because it 'sounds right' it holds together as a storyline making it quite compelling but also at the same time making it difficult to challenge or disrupt.

Discourse analysis based on storylines requires looking in detail at the specific practices through which common understandings are produced and transformed. It ' ... investigates the boundaries between the clean and the dirty, the moral and the efficient, or how a particular framing of the discussion makes certain elements appear fixed or appropriate while other elements appear problematic' (ibid., p. 54). In order to do this one needs to show whether definitions 'homogenize' a problem or make a problem understandable, or whether definitions 'heterogenize' or open up established discursive strategies.

This chapter attempts to disrupt the dominant storyline because as Sandilands has argued

> ... it is through a process of shaking and disrupting hegemonic discourses and practices of so-called 'urban sustainability' that spaces may be created for the promotion of alternative urban socioecological relations (Sandilands 1996, p. 125).

The Sustainable City

Whitehead (2003) has noted that despite considerable debate over the extent and severity of the socio-ecological problems facing urban areas ' ... there does appear to be a considerable degree of consensus over how the international political community should address the complex hybrid of social, economic and ecological problems which face urban areas' (Whitehead, 2003, p. 1184). He notes the focus in contemporary research on the practical implementation of sustainable development as a policy goal, and the lack of analysis of the sustainable city as an object of political contestation and struggle (ibid., p. 1184). Focusing on implementation of sustainable urban development has tended to reduce it ' ... to a technical matter of institutional restructuring, traffic management, architectural design and the development of green technologies' (ibid., p. 1187) around which there is a great deal of international consensus suggesting that ' ... either the process of defining sustainable development has been completed or that it can no longer proceed at the conceptual level but must be achieved through specifying it, preferably quantitatively' (Myerson and Rydin, 2004, p. 101).

During the 1980s and 1990s this global consensus focused in particular on compact urban forms as offering the most sustainable future (Williams *et al.*, 2000, p. 7) in contrast to dispersed settlement patterns which were generally considered unsustainable. The well rehearsed benefits of compact cities included reducing the need to travel, increased use of public transport and active transport modes (with associated health benefits) preserving biodiversity and farmland on the urban rural fringe and greater efficiencies in energy and water use (see for instance Hillman, 1996, p. 36). In contrast the dispersed or sprawling city was considered to lead to car dependence, pollution, inactivity, loss of agricultural land and biodiversity on the urban rural fringe, inefficient use of energy and water, and inequality. In the Australian context this dichotomy found expression in the 1996 State of the Environment Report – which characterized the two positions as the oppositional – 'suburbanisers' versus 'reurbanisers'.

Suburbanisers are defined as ' ... those who favour continuing low density suburban development' and 'are less inclined to see serious or intractable problems' in suburban and outer suburban areas (State of the Environment Advisory Committee, 1996, p. 3.24). Reurbanisers in contrast identify serious and intractible problems in outer suburban areas, including long term poverty, discontent, vulnerability and deprivation. Reurbanisers also point to the adverse impact of car use in these areas on the social, built and natural environments (ibid., pp. 3.25-3.26). The report was clearly in support of reurbanisation as the most appropriate approach for moving towards more ecologically sustainable cities (ibid., 1996). Built into this acceptance is the common assumption that social issues can be addressed simply by changing urban form.

Even though research throughout the 1990s suggested that the relationships between urban form and other sustainability benefits were unsubstantiated or dependent on a range of other variables more significant than urban form (Breheny, 1996, p.13; Williams *et al.*, 2000, p. 7), because of the apparent urgency the compact city was 'hastily' implemented across the globe (Breheny, 1996, p. 13).

The dichotomy between compact versus dispersal (or sprawl) persists today in most discussions about sustainable cities. The compact city model has since been adapted to place emphasis on multi modal cities, polycentric cities or urban villages as a way of delivering more services to people living outside of urban inner areas, while reducing infrastructure costs and land take. While these can take a number of forms from urban infill, to multi-nodal, or corridor growth models all have the characteristics of mixed use higher density development around transport infrastructure and services.

The distinction between compact or high density cities and dispersed or sprawling cities is often presented as contrasting future visions or storylines (Hajer, 1995) – an unsustainable or dystopic one, on the one hand, and the alternative a sustainable (almost) utopian one (Prugh *et al.*, 2000, p. 41). 'At stake is nothing less than the future of Western Lifestyles' (Breheny, 1996 p. 13). In sustainable city discourse these two visions are based around a series of dualisms or binary oppositions (Hattingh Smith, 2005) – beginning with compact versus dispersed but also including sustainable/unsustainable, efficiency/waste, compact/sprawl, lively/

boring, viability/decline, responsible/irresponsible. These binary oppositions are so powerful in the discourse that often ' … by uttering a specific element one effectively reinvokes the storyline as a whole' (Hajer, 1995, p. 63).

Sustainable Cities 2025: an overview

This chapter considers the debate about 'world class' and sustainable cities by analyzing and identifying the storylines in a key Australian report. The House of Representatives Standing Committee on Environment and Heritage's Inquiry into Sustainable Cities 2025 began in August 2003 and the final report was tabled in parliament on 12th September, 2005. 196 submissions were received and 15 public hearings with representatives from organizations, industry, researchers and individuals were held. The purpose of the inquiry was not to set specific actions but to provide a 'national map' of the issues and approaches (House of Representatives Standing Committee on Environment and Heritage, 2003, p. 2) as a way of informing future policy. Consequently the discussion paper was written as a scoping document and a way of canvassing input.

The discussion paper begins with the statement 'Cities of the Future must be sustainable cities'. It then goes on to define the sustainable city of the future as one which will:

> … integrate the built and natural environment. The sustainable city will assist in retaining the biodiversity of Australia, have a developed infrastructure that gives efficient and equitable access to services and utilities, preserve the essentials of the 'Australian lifestyle' and contribute to the economic wealth of the nation (ibid., p. 4).

These are all, of course, laudable aims clearly linking the need for cities to be economically competitive while retaining biodiversity and ensuring equitable access to services. So this 'future vision' is one that 'we' should all share, and what is needed is planning, a clearly articulated strategy and a holistic national approach (ibid., p. 4).

The discussion paper is structured around a series of questions based on seven 'visionary objectives':

1. Preserve bushland,[3] significant heritage and urban green zones;
2. Ensure equitable access to and efficient use of energy, including renewable energy sources;
3. Establish an integrated sustainable water and stormwater management system addressing capture, consumption, treatment and re-use opportunities;
4. Manage and minimise domestic and industrial waste;
5. Develop sustainable transport networks, nodal complementarity and logistics;
6. Incorporate eco-efficiency principles into new buildings and housing; and
7. Provide urban plans that accommodate lifestyle and business opportunities (ibid., p. 4).

The emphasis is therefore on efficient use of resources and the minimization of waste to accommodate lifestyle and business opportunities. It is notable that 'equitable access' is defined here within the context of energy efficiency.

Because the focus of the inquiry is on discussion and input 'from a wide range of professions, community groups, local and state governments, researchers, businesses, industry associations and individuals' (ibid., p. 2) one would expect a wide range of divergent views about what constitutes a sustainable city. However, there is instead a great deal of commonality and agreement. One of the reasons for this is that the boundaries around how these discussions should proceed are already established. For instance, the terms of reference for the committee leave little question about what a sustainable city would be like spatially:

The terms of reference for the Committee were to inquire into and report on the following:

- The environmental and social impacts of sprawling urban development;
- The major determinants of urban settlement patterns and desirable patterns of development for the growth of Australian cities;
- A 'blueprint' for ecologically sustainable patterns of settlement, with particular reference to eco-efficiency and equity in provision of services and infrastructure;
- Measures to reduce the environmental, social and economic costs of continuing urban expansion; and
- Mechanisms for the Commonwealth to bring about urban development reform and promote ecologically sustainable patterns of settlement (ibid., p. 3).

All of these terms of reference refer to the need to contain urban expansion or 'sprawl' as a way of ensuring social equity and eco-efficiency and so most of the discussion throughout the inquiry focused on how to achieve this desired outcome. Social equity is here conflated with eco-efficiency and the term 'sprawl' operates metaphorically to denote the current unsustainability of Australian cities.

Of the 196 submissions to the inquiry[4] the majority were from private individuals, followed by non government organisations (principally from transport, environmental groups, or those with a concern with built heritage), and local government. Other submissions came from consultancy firms, federal government departments, state governments, academics, industry, industry associations, professional associations and business associations. Absent were any submissions from the social welfare sector with the exception of the federal Department of Health and Ageing, and from two Sydney-based health services. Despite the number of private submissions they were under-represented during the public hearing phase. Over the 15 days and 52 sessions of public hearings in five of the seven Australian capital cities evidence was heard from only four private individuals.[5] Noticeably absent were NGOs like 'Save our Suburbs' and the National Trust who spoke in opposition to urban consolidation policies in their submissions.

Nevertheless, the Inquiry process brought together a range of diverse interests and discourses and the final report, tabled in Parliament in August 2005, attempted to distil all of this down into a common national vision or approach for Australian cities to the year 2025. The first recommendation was that the Australian Government establish an Australian Sustainability Charter ' … that sets key national targets across a number of areas, including water, transport, energy, building design and planning' (House of Representatives Standing Committee on Environment and Heritage,

2005, p. xvii). So from the outset the focus is almost entirely on resource use and targets to reduce consumption of those resources as evidenced in the structure of the final report, in the specified targets and in the report's opening sentence:

> Australian cities are facing a number of critical issues. Water shortages, congested transport, and demands placed on energy and urban development must be addressed (ibid, p.1).

Not one of the 32 recommendations in the report directly targets issues of social equity.[6]

Sustainability of what?

Throughout the Inquiry the 'environment' and environmental problems are most often understood as a problem of consumption – using too many scarce resources inefficiently. This required the framing of environment as resource base, paving the way for market-based solutions (Escobar, 1996). The focus on resource consumption was reflected initially in the structure of the discussion paper with its emphasis on settlement patterns (or sprawl), green zones, energy, water, waste and transport It is also reflected in the structure of the final report which is similarly organised with the addition of governance and research and feedback.

So dominant was the discourse around reducing consumption through efficiency gains, even those groups who spoke out against the construction of the sustainable city as compact and efficient still used the language of efficiency because to not do so would have meant using the opposite – inefficiency. As an example the Automobile Association of Australia (House of Representatives Standing Committee on Environment and Heritage, 2003–2004, submission 121), in arguing for increased funding for roads suggested that this would help address increasing congestion levels, reduce travel times and improve safety. The Community Group 'Save Our Suburbs NSW' (ibid, submission 23) in arguing against high density developments also used the language of resource efficiency. Their submission suggested that the impacts of higher density housing included increased atmospheric pollution, increased hard surfaces leading to storm water pollution, increased noise, increased energy use because of dryers, lifts, air conditioners and lighted public areas and the destruction of viable residential housing.

The main focus throughout the Inquiry was, however, on changing consumer behaviour through education and awareness. Representatives from the building industry for instance suggested that ' … most home owners are not sophisticated enough to fully understand the implications of the benefit of higher initial capital costs' (House of Representatives Standing Committee on Environment and Heritage, 2004–2005, 1st April p. EH11). But, as noted by an energy provider it is also poor consumer choice – '[t]raditionally, when people come into our gas shops they are much more interested in what the colour is and whether it fits' (ibid., 8th June, p. EH53).

'Education' was the key for many witnesses and once educated and aware 'people' are expected to behave responsibly. This means that 'everyone' must be involved. As one local government witness expressed it, ' … it seems we have built a mandate for regulation … [b]ut from the individual's point of view it is: "I am not going to be the well-intentioned loner. I want one in, all in"' (ibid., 6th April, p. EH34). So no one is excluded from the process because ultimately it is about 'our' collective and individual well being. This meant the construction of an 'environmental citizenship' involving the 'environmental mobilisation of the entire population' and the 'normalisation of every single individual' (Darier 1996, p. 596).

In the case of the Inquiry this process of normalisation and control extends beyond people's consumption patterns right into their homes and how they care for their bodies. As one local government councilor put it – ' ... When it starts to get into people's homes and affects their day-to-day living, you feel that you are really having an impact' (House of Representatives Standing Committee on Environment and Heritage, 2004–2005, 6th April, p. EH37).

As in the broader sustainability discourse another dominant theme in the submissions and during evidence was on the need to reduce car dependence which was blamed for a range of urban ills including suburban sprawl and associated with this the loss of productive agricultural land and biodiversity on the urban rural fringe, obesity, air pollution, a lack of interaction and sense of 'community', isolation, marginalisation and loneliness. And in line with the dominant storyline the solution, in most of the submissions and in evidence is compact urban form which is seen as the solution not only to automobile dependence but for a whole litany of urban ills. The cited benefits of moving towards compact urban form are greater efficiency in resource use, less congestion and a more 'liveable' and attractive city. Liveability is associated with vibrancy, liveliness, interaction and a 'sense of community'; qualities which are essential, as several submissions noted, to boost the competitive advantage of cities.

What is interesting is that in the final report the Committee refused to be drawn into the compact versus sprawl debate, arguing instead that 'there is no simple solution to the expansion or consolidation of our cities. Neither concept is in itself the answer to a more sustainable city, nor intrinsically an indicator of an unsustainable city' (House of Representatives Standing Committee on Environment and Heritage, 2005, p. 48). What remains unquestioned however is that the solution is a spatial one; higher density housing in mixed use neighbourhoods close to public transport in both inner and outer areas as a way of reducing the need to travel and hence enhancing the liveability of cities.

Fig. 1.
Agriculture: One of the dominant themes in submissions to the Inquiry was loss of productive agricultural land on the urban rural fringe.
Source: Judy Rogers

*Fig. 2.
Development:
Loss of
biodiversity on
the urban rural
fringe as a result
of development
was another key
theme discussed
throughout the
inquiry.
Source: Judy
Rogers*

So in responding to the question – *what is to be sustained*? clearly the dominant focus was on the resources available to the city's inhabitants, along with the image of the Australian cities as vibrant, lively, and attractive places to live, work, play and invest. To achieve this requires a fully informed, educated and 'responsible' public who all fully embrace the sustainability imperative and a compact urban form.

Throughout the Inquiry there was very little acknowledgement of the equity impacts of some of these proposals, even though as Marcuse points out the costs of moving towards environmentally sustainability will not be borne by everyone equally and, in fact, measures are likely to aggravate income inequities (Marcuse, 1998, p. 108). One submission from an environmental NGO in fact argued that ' ... housing affordability cannot be a driving force in major inner cities. Policies need to be implemented for the socially disadvantaged but this is a separate issue' (House of Representatives Standing Committee on Environment and Heritage, 2003–2004, submission 81, p. 2). Only one submission argued for some level of income redistribution. So who is the sustainable city for?

Sustainability for Whom?

In considering the question sustainability for whom it is perhaps useful to first consider how (and if) social equity is currently understood in sustainability discourse, because as Dobson points out the connection between the two is ' ... more often based on ... wishful thinking than on clear-sighted analysis ... ' (Dobson, 2003, p. 83). He points out that there is no 'hard empirical evidence' to support the conclusion that they are compatible and that, in fact, their objectives differ in quite fundamental ways. The proposition is assumed to be true; it has become part of the dominant storyline. For Dobson theoretically informed, empirical research must focus on the meaning(s) of social justice and sustainability embedded in sustainability discourse and to examine their compatibility over the full range of meanings.

So how is social equity understood throughout the Inquiry? To begin with the discussion paper is surprisingly quiet on the topic; a silence noted in a number of submissions, particularly from a number of local governments who pointed out the focus on environmental sustainability without due consideration of social and economic sustainability. This perceived emphasis on environment over other aspects of sustainability was seen as a failure to acknowledge the need for integration of the social and economic aspects of sustainability with the environmental, and unless the former factors were considered then there would ' ... continue to be a low uptake of real environmental sustainability outcomes, even if people desire it' (House of Representatives Standing Committee on Environment and Heritage, 2003–2004, submission 161 p. 1). The National Museum of Australia in noting the omission of social sustainability in the discussion paper argued that:

> Strong, variegated communities are necessary to build the capacities that can generate more sustainable futures, especially in cities that face complex problems relating to their economic, social and environmental bases. All too often, social sustainability falls off the agenda or is considered subordinate to environmental and economic issues, despite the fact that it underwrites individual and collective creativity and innovation that can deal with the challenges posed by increasing pressures from global, regional and local forces (ibid., submission 43 p. 3).

The focus in this extract and in most considerations of social sustainability was on 'community' capacity building, or building human capital to encourage 'creativity and innovation' rather than issues of distributional equity. Influenced by submissions and evidence from the health sector the final report linked 'community identity with ' ... more active lifestyles and social interaction' (House of Representatives Standing Committee 2005, p. 52) again with a focus on building community capacity. To achieve this capacity there is an emphasis on the 'local' because as the final report noted:

> As cities grow, it was suggested that people will identify more with the local area than with the larger city, and so local connections and community interactions are critical in establishing a sense of well being and identity (ibid., 2005, p. 24).

So it is a local, spatially bounded 'community' that is privileged in the discourse. The concept 'community' has powerful metaphorical importance in the construction of the sustainable city and it is used in several ways. Firstly, it denotes the responsibility 'we' all have, because sustainability is the responsibility of all Australians – industry, business, government and community. This language of inclusion, so dominant in sustainability discourse means that the community is taken as homogenous – 'we are all in this together' and it could be argued that the language of inclusion effectively writes out the possibility of thinking about differences between people not only in the way they live, where they travel, where they work but also in terms of their access to resources, socio economic circumstances, gender or culture, except under the very loose concept of 'lifestyle'. Secondly 'community' is represented in sustainable city discourse as the antidote (or opposite) to the disadvantage, loneliness, isolation and dreariness of life in the sprawling suburbs or on the urban rural fringe which some of the submissions noted can lead to social conflict. For example, the Environment Business Australia suggested: 'Cities need to be designed to minimize friction between different cultural and socio-economic groups' (House of Representatives Standing Committee on Environment and Heritage, 2003–2004, submission 92, p. 9) and for them denser cities provide a greater sense of community. However, who this community is, is never entirely specified. And finally, community is represented as an ideal; in terms of 'the promise of community'. As the storyline goes, living at higher densities will provide more opportunities for face-to-face contact, more interaction resulting in a 'sense of community', or a sense of belonging and identity. While many of the submissions promoted this as a goal, others were quite specific. The Ecodemocrats[7] for instance in their submission argued for urban centres with ' ... outdoor lounge-rooms as convivial public space ... that allow people to enjoy each other's company, bump into friends, form acquaintances' (ibid., submission 88, p. 4).

Here, once again 'the community' is constructed as a homogenous whole. This somewhat romantic vision of communitarianism with its privileging face-to-face relationships has been criticized by many commentators. Young for instance has argued that while the ideal of community is an understandable dream of social wholeness, symmetry and identity, it has serious political consequences because it ' ... denies and represses social difference' (Young, 1990, p. 227). These sentiments are echoed by David Harvey who in his critique of new urbanism argued:

> Community has ever been one of the key sites of social control and surveillance, bordering on overt social repression. Well-founded communities often exclude, define themselves against others, erect all sorts of keep-out signs ... (Harvey, 1997, p. 3).

For Harvey appeals to 'community' exclude and divide because of a refusal to confront the political economy of power (ibid., p. 3). Romanticized ideas about homogenous, closely knit communities effectively 'hide' or hinder the possibility of thinking about current and emerging inequities in cities within the framework of sustainability.

Fig. 3. Public Transport: The final report suggested that swift, reliable and affordable public transport would go some way towards addressing social equity concerns.

Source: Judy Rogers

Equity concerns did not, however, fall entirely off the agenda. The most commonly found expression in concerns about affordable housing and the link between affordable housing and development on the urban rural fringe. These concerns were however, quickly dispensed with in the final report which suggested that a large part of the answer lay in swift, reliable and affordable public transport. Broader concerns about social polarization which are emerging in Australia as a result of globalization were discussed in only two of the submissions but remained unacknowledged in the final report.

So in answering the question for whom – the answer is clearly all of 'us' – 'us' being a homogenous, inclusive environmentally responsible citizenry who live, work and play locally to avoid the need to travel and who consume resources efficiently. But this construction of the sustainable citizen in sustainable city discourse effectively writes out the possibility of writing and speaking about difference, about social equity and about distributional issues in a way that could in fact lead to a more equitable and sustainable city for all.

Conclusion

As this brief case study demonstrates two opposing storylines dominate sustainable city discourse in Australia and these storylines may mirror the construction of the sustainable city globally. Within these storylines there is a powerful binary at work based on the choice between a sustainable future on the one hand, and an unsustainable one on the other. The sustainable city storyline begins with the 'crisis' of consumption and the need to recognize biophysical limits, consequently it is up to each and every Australian to change their wasteful and irresponsible behaviour. This means that 'we' need to change the way we live, what we consume and the way we travel to live more 'sustainable' lifestyles. The outcome, or the promise, will be a lively, vibrant, healthy, efficient future; effectively the opposite of the way 'we' now live. So compelling is this storyline and so dominant the conclusions appear as self evident.

It could be argued that measures to improve the 'sustainability' of cities may well deliver more attractive, liveable cities for a global professional class and for more affluent local people and consequently, sustainable city discourse in Australia at least clearly plays into the 'boosterism' that has increasingly become associated with 'world class' city status (Marcuse, 2006, p. 366). What is at stake is the image of Australian cities as attractive, 'vibrant' places to live, work and invest. But what is lost as a result? The case study demonstrates that social equity concerns lay outside of the dominant storyline and so they could not, and were not, considered.

This would suggest that discourse around the sustainable city cannot serve as an anecdote to many of the less desirable impacts of globalization on localities particularly in terms of social equity unless discussions confront head on issues of inequality and power. While this may well disrupt the dominant storyline it holds out the possibility of delivering more equitable and environmentally sustainable cities for all.

Notes

1. Non-governmental Organisation
2. Intergovernmental Organisation
3. Bushland is a distinctly Australian term referring to land that is uncultivated.

4. Of these all but one are publicly available at http://www.aph.gov.au/house/committee/environ/cities/subs.htm
5. Other groups represented included local government, consultants, academics, Commonwealth, State and Territory Government departments, professional and business representative bodies, industry, particularly from the water energy and housing sectors, and environment and transport NGOs. The public consultation phase also included one roundtable of health professionals and two roundtables with local government.
6. Of the 32 recommendations three relate to governance, one to planning and settlement patterns, seven to transport, five to water, four to energy and three to research and feedback. The first three recommendations relating to Governance or leadership are considered by the chairperson to be the most important (House of Representatives Standing Committee on Environment and Heritage 2005, p. ix) requiring a 'new overarching framework for sustainability governance', so that the principles of sustainability can be put on the national agenda.
7. The Ecodemocrats are a network of environmentalists within the Australian democrats, a minor political party in Australia.

References

Breheny, M. (1996) Centrists, Decentrists and Compromisers: Views on the Future of Urban Form, in *The Compact City: A Sustainable Urban Form?* (eds Jenks, M., Burton, E. and Williams, K.), E & FN Spon, Oxford.

Darier, E. (1996) Environmental governmentality: the case of Canada's Green Plan. *Environmental Politics*, 5(4):585–606.

Dobson, A. (2003) Social Justice and Environmental Sustainability: Ne'er the twain shall meet?, in *Just Sustainabilities: Development in an unequal World* (eds Agyeman, J. and Bullard, R.D), MIT Press, Cambridge.

Escobar, A. (1996) Constructing Nature: Elements for a Poststructural Political Ecology, in *Liberation Ecologies: Environment, Development, Social Movements* (eds Peet, D and Watts, M), Routledge USA and Canada.

Fainstein, S. (2006) Extract from Inequality, in Global City Regions in *The Global Cities Reader* (eds Brenner, N. and Keil, R.), Routledge, Oxon, USA and Canada.

Hajer, M.A. (1995) *The Politics of Environmental Discourse: Ecological Modernisation and the Policy Process*, Claremont Press, Oxford.

Harvey, D. (1997) The New Urbanism and the Communitarian Trap. *Harvard Design Magazine*, Winter/Spring: 1–3.

Hattingh Smith, A. (2005) Constructions of sustainability and spatial planning; the case of Dalton Flatts, Durham, Planning Inquiry. *The Town Planning Review*, 76(3), Proquest Social Science Journals.

Hillman, M. (1996) In Favour of the Compact City, in *The Compact City: A Sustainable Urban Form?* (eds Jenks, M., Burton, E. and Williams, K), E & FN Spon, Oxford.

House of Representatives Standing Committee on Environment and Heritage (2003) *Discussion Paper Sustainable Cities 2025*, Commonwealth of Australia, Canberra.

House of Representatives Standing Committee on Environment and Heritage (2003–2004) *Inquiry into Sustainable Cities: Submissions*. Retrieved from: http://www.aph.gov.au/house/committee/environ/cities/subs.htm

House of Representatives Standing Committee on Environment and Heritage (2004–2005) *Inquiry into Sustainable Cities*: Transcripts. Retrieved from: http://www.aph.gov.au/house/committee/environ/cities/hearings.htm

House of Representatives Standing Committee on Environment and Heritage (2005) *Sustainable Cities*, Commonwealth of Australia, Canberra.

Luke , T. (2006) 'Global Cities' vs. 'global cities'; Rethinking Contemporary Urbanism as Public Ecology, in *The Global Cities Reader* (eds Brenner, N. and Keil, R.), Routledge, Oxon, USA and Canada.

Marcuse, P. (1998) Sustainability is not enough. *Environment and Urbanisation*, 10(2):103–111.

Marcuse, P. (2006) Space in the Globalising City, in *The Global Cities Reader* (eds Brenner, N. and Keil, R.), Routledge, Oxon, USA and Canada.

Myerson, G. and Rydin, Y. (2004) *The Language of Environment: A new rhetoric*, Routledge, London and New York [First published 1996].

O'Connor, K., Stimson, R. and Daly, M. (2001) *Australia's Changing Economic Geography: A Society Dividing*, Oxford University Press, Australia.

O'Riordan, T. and Church, C. (2001) *Synthesis and Context in Globalism, Localism and*

Identity: Fresh Perspectives on the Transition to Sustainability (ed. O'Riordan, T.), Earthscan, U.K. and USA.

O'Riordan, T. ed. (2001) *Globalism, Localism and Identity: Fresh Perspectives on the Transition to Sustainability,* Earthscan, U.K. and USA.

Prugh, T., Constanza, R. and Daly, H. (2000) *The Local Politics of Global Sustainability,* Island Press, Washington D.C.

Sandilands, C. (1996) The shaky ground of Urban Sustainability: A Comment on Ecopolitics and uncertainty, in *Local Places in the Age of the Global City* (eds Keil, R. Wekerle, G.R. and Bell, D. V. J.), Black Rose Books, Montreal.

Simon, D. (2006) *The World City Hypothesis: Reflections from the Periphery in The Global Cities Reader* (eds Brenner, N. and Keil, R.), Routledge, Oxon, USA and Canada.

State of the Environment Advisory Committee (1996) *Australia: State of the Environment 1996,* CSIRO Publishing, Collingwood, Victoria.

Whitehead, M. (2003) Re-analysising the Sustainable City: Nature, Urbanisation and Regulation of Socio-environmental Relations in the UK. *Urban Studies,* 40(7): 1183–1206.

Williams, K., Burton, E. and Jenks, M. eds (2000) *Achieving Sustainable Urban Form,* E & FN Spon, London and New York.

World Commission on Environment and Development (WCED) (1987) *Our Common Future,* Oxford University Press, Oxford.

Young, I.M. (1990) *Justice and the Politics of Difference,* Princeton University Press, Princeton.

5

Mike Jenks and Daniel Kozak

Polycentrism and 'Defragmentation':
Towards a more sustainable urban form?

Introduction: urban form and sustainability

With the rapid increase in urbanisation worldwide, particularly in Asia, and the growth in the number of very large cities and metropolitan regions, the need to achieve more sustainable urban forms is becoming critical. The sheer scale of the changes from a rural to a predominantly urban world population leads to some questions about the adequacy of our existing knowledge and the paradigms of urban sustainability. There has, of course, been a considerable amount of research to determine what defines and characterises the form of the sustainable city, and which forms may most affect sustainability. This is a complex issue.

There can be little doubt that concepts such as the 'compact city' have permeated thinking about urban sustainability in the Western world for over 15 years. In 1990 the EU published an important Green Paper strongly advocating the principles of sustainable urban form (CEC, 1990). Since then a considerable body of knowledge has built up in the West, with much research and the publication of numerous influential books (e.g. Breheny, 1992; Jenks *et al.*, 1996; Williams *et al.*, 2000; Jenks and Dempsey, 2005). In the UK ideas about sustainability have been incorporated into policy (e.g. DETR, 2000), and into numerous government publications giving planning guidance (e.g. ODPM, 2004a and b), into publications from the Commission for Architecture and the Built Environment (CABE). And although terms such as the 'compact city' may not be universal, similar concepts aiming to describe and achieve urban sustainability can be found in, for example, the USA, with New Urbanism and Smart Growth initiatives. All tend to advocate urban forms that are high density and mixed use, and which are contained in order to reduce travel distance and dependence on private transport, as well as being socially diverse and economically viable.

The arguments about sustainable urban forms appear on the surface to be convincing. The claims are that compact urban forms are spatially sustainable, environmentally sound, efficient for transport, socially beneficial and economically viable. Spatially, the argument goes, urban areas are kept within clear boundaries and the spread of low density suburbs is contained. This form has the benefit of preserving valuable agricultural land, and ensuring that land in existing urban areas is intensified and used more efficiently. Such compact forms mean that the home, work and leisure activities are likely to be in close proximity, which should reduce the need to travel, particularly by car, and that more sustainable modes of transport such as walking and cycling will be encouraged. Transport and environmental benefits can then result. Reduced dependence on car travel reduces harmful greenhouse gas emissions and pollution. The higher densities found in compact forms mean that public transport becomes more economically viable, and so too does less polluting power generation as local combined heat and power schemes also become more viable. Similarly, more people concentrated in an urban area help sustain local businesses, facilities and services, and the shorter travel distances mean these functions are more easily accessed by all who live there. Yet, however widespread and persuasive these arguments may be, compact city concepts are largely untested in practice.

Nevertheless, at least in the UK, Europe and much of the Western world, the idea of a more compact urban form provides a comforting model that is rapidly being implemented. However, there remain questions about the concept's validity. For every positive claim, negative impacts are also registered. Higher densities may lead to perceptions of overcrowding, more traffic, and may not be the favoured choice for residents wishing to purchase homes – population trends still tend to favour rural or suburban locations. Also, the relationships between social and economic factors and urban form tend to be indirect: many other factors are more important in achieving social and economic sustainability.

It is surprising, therefore, that the term 'compact city' has resonated beyond the Western world, and has been taken up by countries where the urban context is very different (e.g. Jenks and Burgess, 2000). Across the globe there are some 60 metropolitan regions with populations of more than 5 million inhabitants. Of these regions, 46% have populations in excess of 10 million, the largest being the Tokyo Metropolitan Region with more than 36 million people. Most of these large regions, some 62%, are to be found in Asia where the growth rate is fast (World Gazetteer, 2005). These vast regions, with very large cities at their core, suggest that the idea of a compact city could be seen as a contradiction in terms. So why is there so much interest in the idea, especially in developing countries? Are there some aspects of compact form that might help contain growth and lead to more sustainable urban forms?

Evidence is beginning to emerge that some policies and forms that could be suggested as appropriate for world cities and their regions, and for cities experiencing rapid growth through in-migration of rural populations. At the regional level, networks of cities and polycentrism have claims to interconnections that, although characteristic of globalization, may have some potential for sustainability. At the city level polycentric development may have, or at least is claimed to have, the potential to achieve more sustainable forms, provided all the 'nodes' are linked with efficient public transport (e.g. see Gilbert, 1996; Lambregts and Zonneveld, 2003; Urban Task Force, 1999). At this level, and within urban areas, promising urban forms may include 'transit oriented development' (e.g. Boarnet and Crane, 2001; Calthorpe, 1993; Cervero, 1998; Low and Gleeson, 2003). Linking public transport and development may take a number of different forms, including the intensification (or densification) of development around transport interchanges – transit development zones (TDZ) (e.g. see TOD/TDZ website, 2004). TDZs may include higher density residential development with commercial facilities and other services. Major transport routes may also provide the opportunity for denser development along the routes, and this has been successfully undertaken in

Curitiba in Brazil (Acioly, 2000; Curitiba websites) and in Bogota, Colombia, with its *Transmilenio* bus rapid transit system.

So is polycentrism a form that can help world cities become more sustainable, or is it another hopeful planning ideal that has credence in theory but problems in practice? The rest of this chapter considers the concepts associated with polycentrism and some of the implications of this emerging urban form. Aspects of two world cities, Buenos Aires and Bangkok, are then examined in more detail to uncover the nature of their polycentric forms and whether these represent the fragmentation of the cities, or contribute towards sustainability. Finally some ideas are proposed about 'defragmenting' the city to achieve more sustainable urban forms in world, mega- or very large-cities.

Cities, regions and polycentrism

The concept or phenomenon of polycentrism has become increasingly important within the current urban debate. It can refer to different scales of the built environment whether at the world, regional or city level, and has been characterised by a varied terminology – e.g. polycentric regions, polycentric urban systems, multimodal urban systems (these being either centres, sub-centres or nodes), and networks, urban networks or polycentric networks (being a system of linkages between such sub-centres and nodes). Polycentrism can also be characterised as a phenomenon that has evolved over time, or one that is new, resulting from globalization and urban growth, or as an objective planners wish to achieve (e.g. the European Spatial Development Perspective (ESDP) encouraging polycentrism as a central policy objective (Faludi *et al.*, 2002, p. 13)). It is worth giving a little consideration to some of the debates surrounding these terms and issues.

At the large, regional scale it has been argued that 'the polycentric mega-city region' is a 'new phenomenon [that] is emerging in the most highly urbanized parts of the world' (Hall and Pain, 2006, p. 3). Furthermore, '[i]t is no exaggeration to say that this is the emerging urban form at the start of the 21st century' (ibid.). *Polycentrism* has been defined as 'a new form' which is

> a series of anything between 10 and 50 cities and towns, physically separate but functionally networked, clustered around one or more larger central cities, and drawing enormous economic strength from a new functional division of labour (ibid., p. 3).

This definition mainly applies to polycentrism at a regional scale, in this case to different regions in Europe. But as Hall and Pain acknowledge, 'the entire concept of polycentricity proves highly scale-dependant' (ibid., p. 4). Indeed, polycentrism is not only applied to urban systems at a continental-regional scale but also at the urban and metropolitan scales, as well as at national-regional and global scales. It is similar with the use of the term *network*, an urban metaphor that is closely associated with the concept of polycentrism. It can be used either to describe the increasingly interconnectivity of some cities in the world, 'the global city network' (e.g. Knox and Taylor, 1995; Sassen, [1991] 2001), or applied to a certain territorial and operational organisation, 'the network city' (e.g. Dupuy, [1991] 1998; Graham and Marvin, 2001).

Both the terms polycentrism and urban networks are inseparable from discussions about globalization. Hall and Pain (2006, p. 4) argue that underlying the 'spatial processes' associated with polycentrism are 'two basic and parallel shifts': 'the globalization of the world economy' and its 'informationalization'. They explain the latter (after Castells (1989; [1996] 2000)) as 'the shift in advanced economies away from manufacturing and goods-handling and towards service production, particularly into advanced services that handle information' (Hall and Pain, 2006, p. 4). 'Globalization and informationalization together', they say, 'result in the increasing importance of cities at the very top of the hierarchy, the so-called *world cities* or *global cities*' (ibid., p. 6, original emphasis). Indeed, world cities are commonly understood as nodes of a global, polycentric network.

It is apparent that the same logic is used to conceptualise different scales. Geographical distance is disregarded in favour of other criteria of closeness such as information exchange, commercial links and transport connections. The implications of understanding a city as polycentric because it is either part of a polycentric system or, because it constitutes a polycentric system in itself, are certainly different. The use of the same term applied at both scales may therefore obscure their meaning. What follows is concerned with the polycentric system in itself, and in particular considers the relationship between polycentrism, fragmentation and sustainability at this scale of urban form.

Polycentrism: urban fragmentation and sprawl, or sustainable urban structure?

There are some inherent tensions at the regional and city scale; between polycentric forms that have evolved and developed over time with few constraints, and those that have evolved but with conscious policies to attempt to integrate them. Two general examples are considered here from the literature; Los Angeles, characterising 'urbanised' suburban sprawl and fragmentation, and the Randstad in the Netherlands, representing attempts to integrate the polycentric structures. In theory at least, of the two extremes, the latter should exhibit a higher potential to achieve a more sustainable urban form.

Despite Hall and Pain's affirmation that polycentrism is a 'new form' they note that 'Jean Gottmann originally identified it as long ago as 1961 in his pioneering study of Megalopolis' (2006, p. 3). The urbanised area which Gottmann (1961) named *Megalopolis*, being the vast region from Boston to Washington D.C., was the result of a gradual coalescence of extended metropolitan areas previously regarded as independent. This concept was a qualification of the idea of their being the sum of different conurbations, as Patrick Geddes ([1915] 1968) defined them at the beginning of the twentieth century. Looking back at the form of Geddes's conurbation maps it is possible to find different types of centres or nodes. The extended stain, the amorphous sprawl, associated with the term conurbation could also be interpreted as the basis of a polycentric system. Conurbation and sprawl suggest suburban homogeneity, but this homogeneity has never been complete. The seeming oxymoron 'suburban centre' refers to a predominantly commercial node (e.g. shopping centre, cinema, petrol station) surrounded by residential suburbs and/ or an increase of density in a suburban area. In contrast, the idea of polycentrism refers to a hierarchical territorial extension; a 'network' of interconnected nodes. But the boundaries of these nodes are not always clear cut and they often extend and blend with one another. It is thus pertinent to ask: what are the differences between polycentric development and sprawl? Is polycentric development just a form of organised sprawl?

The concept of urban sprawl has recently been reassessed (e.g. Betsky, 2002; Soja, 2002): for example Soja argues that 'sprawl today is no longer what it used to be':

> the very nature and meaning of both urban sprawl and sustainable development have changed so dramatically over the past thirty years that even the best conceptualisations and practices associated with them must be significantly rethought and revised (Soja, 2002, p. 77).

Soja takes the case of Los Angeles to develop his argument: the 'urban region of Los Angeles, long considered the sprawling antithesis to sustainable urban development [is] today the densest large metropolis in the [US]' (ibid.). Soja's point is that since the 1950s many urban regions in the US, such as that of New York but primarily that of Los Angeles, have gone through a process of densification, which he calls the 'wonderfully in-built paradox' of the 'urbanization of suburbia' (ibid., p. 79), to the extent that many places previously considered suburbs now 'deserve to be called cities in themselves' (ibid.). Thus, he describes the 'postmetropolis of Los Angeles' as a 'highly dispersed and polycentric global city-region' (ibid., p. 83). What was once depicted as 'sixty suburbs in search of a city', Soja says, should now be described 'as 200 cities that are redefining the industrial capitalist metropolis' (ibid.).

It is interesting to compare the examples that Soja presents as nodes of the 'polycentric global city-region' of Los Angeles and contrast them with the European cases presented by Hall and Pain. In the former, the paradigmatic case is Irvine, the 'core city of the Orange City County Technopolis':

> a compact collection of new housing developments, numerous industrial parks and office complexes, vast shopping malls, the adjacent John Wayne International Airport, and, in the south, the University of California-Irvine ... Clustered around Irvine like a bunch of grapes are at least a dozen other cities of between 100–400,000, forming a composite posturban metropolis, or as some call it, a county-city of around 2.5 million ... There is no clearly identifiable downtown to Irvine or to the larger city cluster, and mapping population densities would show no obvious density gradients ... (Soja, 2002, p. 84).

Along with Irvine, Soja analyses Mission Viejo and Moreno Valley. The first is a 'themed-parked patchwork of housing developments' (ibid., p. 85) that includes 'themed' gated communities that range from a 'recreation of Cervantes' Spain' with 'buildings sharing an appropriately simulated design and colour', to 'Florence Joyner Olympiad Village', 'a place designed to attract parents who are determined to make their children become Olympic quality athletes' (ibid.). By contrast, Moreno Valley 'has few theme-parked or gated communities, but consists mainly of relatively affordable family housing constructed by various developers' (ibid., p. 86). While in Irvine, 'there are probably more jobs concentrated ... than there are households with workers that commute outside municipal boundaries' (ibid., p. 84), in Moreno Valley 'nearly one-fifth of the workers ... travel at least two and a half hours to their jobs, causing extraordinary social disruption and pathology, including much higher than average rates of divorce, suicide, spouse and child abuse' (ibid., p. 86). 'Although it looks like a thriving new suburb … beneath the surface are psychological, social, and economic problems as serious and as difficult to respond to as those of the worst inner-city slums' (ibid.).

The picture that Soja presents under the label of 'polycentric global city-region', themed gated communities, shopping malls, business parks and outer-city slums, mainly connected – but also separated – by highways, differs from the polycentric ideal that arises from the European example.

The Randstad in the Netherlands is often used as a paradigmatic example of European polycentric development. It embraces the cities of Amsterdam, The Hague, Rotterdam and Utrecht (Fig. 1), and extends into a wider region known as Deltametropolis to include the towns of Amersfoort and 'the new city of Almere in the reclaimed polders east of Amsterdam' (Hall and Pain, 2006, p. 12). It is a well-connected polycentric form, with a high speed rail network making the major cities accessible within one hour's travel (Okabe, 2005). Although some argue that the Randstad is not a single metropolis (e.g. de Boer, 1996), many planners and theorists discuss and promote it as one single polycentric metropolis. That is indeed the position of the Dutch state, put forward by the *Deltametropolis Association*.[1] It is a position further reinforced by government spatial planning policy. This is based upon a number of polycentric urban networks, including the Randstad, and requires: integrated programmes related to citizens' preferences and urban character; development around transport nodes; a 'highly varied concentration combination of uses'; and the avoidance of competition between nodes and promotion of 'complementarity and coherence with other centres' (VROM, 2001).

It is therefore not surprising that the Randstad has been used as a model. Jacobs (2005), for example, used it to develop a theory of Multinodal Metropolises; a term which he explains is coterminous to 'multicentred, polycentric or polynuclear'

Fig. 1. (below–left)
The Randstad
Source: Daniel Kozak

Fig. 2. (below–right)
'Two cities within each other's influence growing into a multinodal system'
Source: Adapted from Jacobs (2005)

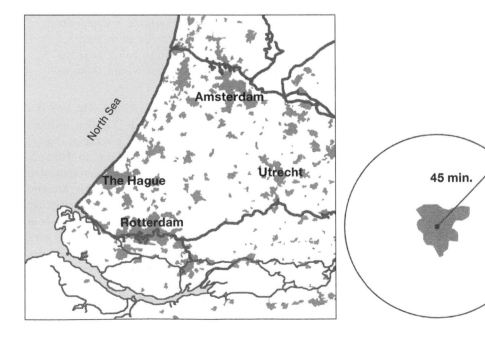

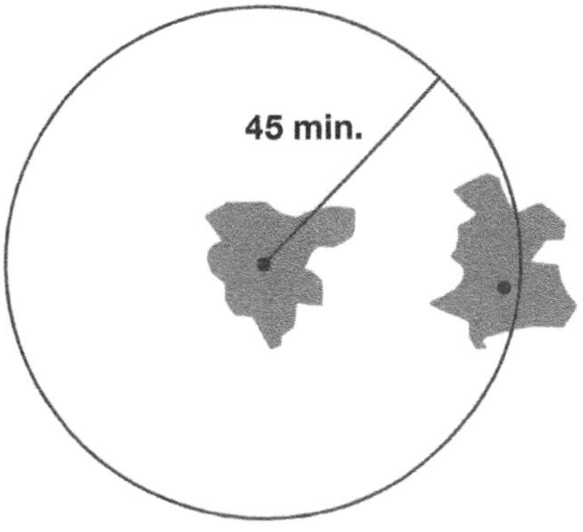

Fig. 3.
'Transformation
from a single
urban system
to a multinodal
system'
Source: Adapted
from Jacobs
(2005)

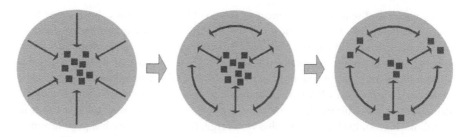

(ibid., p. 370). He identifies three cases 'for the development of a multinodal urban system'. The first case is 'the transformation from a single [nodal] urban system to a multinodal system'; the second is the construction *ex novo* of a city or town based on 'the multinodal system'; and the third is 'the development of a multinodal urban system' as a result of the growing together, functionally, of 'two or more previously separate urban systems' (ibid., pp. 371-372) (Figs 2 and 3).

It could be argued that Los Angeles and the Randstad share some morphological and structural commonalities that explain why both are referred to as polycentric metropolises despite their great differences. It can be also argued that they represent two completely different types of polycentrism and that thus this concept cannot be necessarily categorised as an urban type. It is reasonably clear from the literature that Los Angeles, despite the densification of its sprawling suburbs, is a model of urban fragmentation. The Randstad, by contrast, attempts to integrate its nodes together in a rational and planned way.

Polycentrism, fragmentation, sustainability?

With these ideas in mind an analysis and assessment of the concept of polycentrism and its relationship with sustainability and fragmentation can proceed. Two cases have been chosen – Buenos Aires and Bangkok. These are not cities that spring to mind as 'models' that have received attention from academics and professionals, such as Los Angeles and the Randstad above. Rather they represent, perhaps, a messier reality from parts of the world where urbanisation has been, and continues

Fig. 4.
Average
percentage
annual rate
of population
change – Buenos
Aires and
Bangkok
Source: United
Nations (2006)

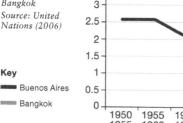

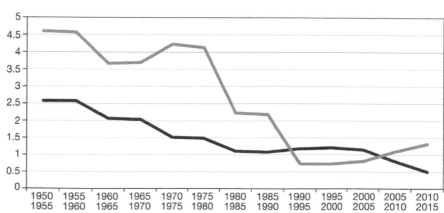

to be, rapid, and which contain most of the world's largest cities. Both cities have been classified by the Globalisation and World Cities network as gamma world cities (Beaverstock *et.al.*, 1999), and both have grown from monocentric cities into forms that exhibit a certain polycentric spatial structure. In the case of Buenos Aires this is in an incipient trend (Abba and Laborda, 2006) while in Bangkok the polycentric structure is more consolidated. Both cities have also experienced rapid growth, although the growth rate has now slowed down, more so in Buenos Aires that in Bangkok (Fig. 4). Both cities are also very large, Buenos Aires with a population of around 12 million and Bangkok of some 6.5 million (United Nations, 2006), but with the addition of an estimated 4 million unregistered daytime citizens making the Bangkok total nearer to 10 million (Sundaravej, 2001; BMA, 2007).

Buenos Aires

The Metropolitan Area of Buenos Aires (MABA) covers 3,833 km^2 and includes two different jurisdictions: the Autonomous City of Buenos Aires (ACBA), federal district of Argentina, with an area of 203km^2, and its conurbation governed by the Province of Buenos Aires, which stretches over 3,630km^2. The population of the former is 2,776,138 and its average density is 13,680 persons per km^2. The population of the latter is 9,270,661 and its average density 2,554 persons per km^2 (INDEC, 2001). Thus, the MABA with a total population of 12,046,799 and an overall density of 3,143 persons per km^2 is the third largest urban agglomeration in Latin America after Mexico City and São Paulo.[2]

Buenos Aires was founded in 1580, and after three centuries of continuous but slow growth it was still a 'large village'; both in population and physical structure (Scobie, 1974, p. 11). In 1887, the year in which Buenos Aires became the capital of Argentina, the first Municipal Census registered a population of 437,875. Five decades later, in 1936 the Municipal Census registered 2,4154,43 inhabitants (Romero and Romero, [1983] 2000), a population increase of 552 per cent (Fig. 5).

Despite disagreements about the forces that shaped this extraordinary metropolitan explosion and why Buenos Aires took its particular form,[3] most authors agree that the period 1880-1940 was the first metropolitan cycle in Buenos Aires which set up the urban structure that still underlies the ACBA and the main axes of its conurbation. Population growth during this period came from successive waves of immigration from Europe, which practically ceased in 1930 (Torres, 1993, p. 3). In 1940–1960, the new population was mostly the result of internal migration from Argentine provinces, and from neighbouring countries (ibid.). This new population that predominantly settled in the periphery was attracted by the new industries that emerged in the outskirts of Buenos Aires. The 1960 census showed for the first time that the population in the surroundings (3,772,000) exceeded that in Buenos Aires City (3,263,000) (Sargent, 1974, p. 146). The total population of the metropolis in 1960 was 7,035,000 – 34.5% of the population of Argentina (ibid.). The growth rate of the metropolitan region slowed down during 1960–1980, and with a more modest growth rate reached a population of 12,046,799 in 2001, increasing to 12,550,000 by 2005 (INDEC, 2001; United Nations, 2006) (Table 1).

Throughout these periods the primacy of the historic centre of the city remained unchallenged. Unlike most Latin-American cities, this is still today the main financial, commercial and administrative centre of the metropolis and country.

Table 1.
Buenos Aires's population 1955–2005
Source: *United Nations (2006)*

Date	Population '000s
1955	5,799
1965	7,317
1975	8,745
1985	9,959
1995	11,154
2005	12,550

Fig. 5. Urban growth in Buenos Aires from 1887 to 1936

Source: Based on Gorelik (2004) and Sargent (1974)

Key

--- Municipal limits until 1887

=== Limits of the Federal Capital (now ACBA) traced in 1888

▮ Built area, *ca.* 1887

▮ Built area, *ca.* 1936

The radial-concentric and strongly monocentric structure delineated in the first metropolitan cycle was reinforced during the second period, in which centres and sub-centres were densified along the railway system. Until 1950 Buenos Aires was still a compact city, but since then the metropolis has grown to become increasingly dispersed (Ainstein, 2001). The pattern of urbanisation in the second half of the twentieth century, which completely lacked the public guidance and control that characterised the first metropolitan cycle, departed from the railway axes. Supported by an extended network of bus and coach services, the metropolis expanded to the interstices between the previously consolidated urban corridors. The expansion of the 1960s–1990s did not create new centralities. The newly urbanised territory usually lacked adequate access to infrastructure and remained highly dependent on the previous existing centres, leaving the system of centralities virtually unchanged (Fig. 6).

The trend since the 1990s in the MABA has moved from the traditional monocentric form with a 'tree structure', towards a more polycentric form with a 'network structure' (Abba and Laborda, 2006). Encouraged by a privatised and expanded structure of highways, new centralities have emerged and grown at an extraordinarily high rate: between 1994 and 2004 although the traditional centres grew by 30%, the new centralities grew by 148% (ibid.). These new metropolitan centralities generally group together a number of gated communities, business parks, shopping malls, hypermarkets, recreational centres such as multiplex cinemas and sometimes even more sophisticated services such as concert halls and universities. Given that they are predominantly accessible by private car, they are mostly exclusively inhabited and used by the upper and upper-middle classes (Fig. 7).

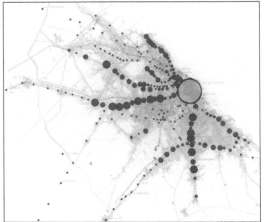
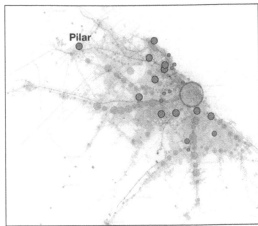

Pilar: a node in a polycentric system?

The evolution of the urban form of Buenos Aires towards a polycentric one, and the rise of numerous new centralities raises a key question. Are these centralities nodes that are integrated into a polycentric system, or are they just fragmented sub-centres set within what appears to be a polycentric form? One such new centrality, called Pilar, is considered below.

Pilar, situated 50km from the city centre, is the most paradigmatic case of these emerging new centralities. Between 1980 and 1991 Pilar's population increased by 46%, and between 1991 and 2001 by 61.4%, when it reached 233,508 inhabitants (INDEC, 2001). That is by far the highest growth rate in the metropolis. Much of this increase is related to the appearance of new permanent gated communities;[4] a phenomenon that started in the 1980s in the north of the MABA and reached a peak between 1996 and 1997 (Arizaga, 2005, p. 25). By 1999 the sum of the total area in the metropolis covered by gated communities was 300km² (Ciccolella, 1999, p. 10), and this had grown to 350km² by 2007; that is 1.7 times the total area of the city (Clarin Newspaper, 2007). There are now 461 gated communities in the MABA (ibid., 2006), one third of which are located in Pilar (Thuillier, 2005, p. 13) (Fig. 8).

Although approximately only 6% of the population of Pilar actually live in gated communities (ibid.) their appearance and multiplication brought in a significantly higher number of inhabitants that either work in those gated communities or in the new service sector in Pilar. Some live in Pilar's historical centre or in social housing neighbourhoods, and many in squatter settlements nearby.

The differences between those who live inside the gated communities and those who do not are manifold. Fernández Wagner and Varela (2003, p. 71) note that the 'privatisation of urban infrastructure' is reaching unprecedented levels. For example, three gated communities in Pilar are in the process of constructing their own private wastewater treatment plant, while most of the very low income population who live surrounding these gated communities do not have their lavatories connected to any type of wastewater network (ibid.).

Fig. 6. (above-left) Traditional centres in the MABA
Source: Adapted from Welch Guerra (2005)

Key

Dominant historic centre

Traditional centres according to railway ticket sales

Fig. 7. (above-right) New centralities in the MABA
Source: Adapted from Welch Guerra (2005)

Key

New centralities

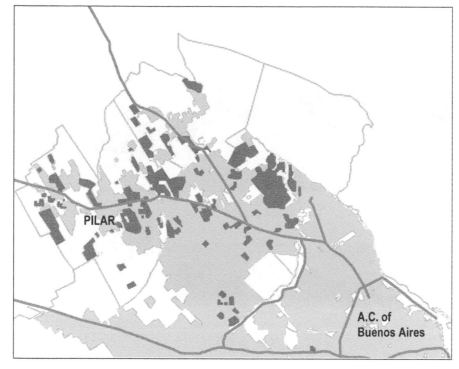

Key

Open grid built area

Gated communities

Highways

Pilar is also paradigmatic because its governance seems to adapt to a current political-economic model. According to Fernández Wagner and Varela (2003, p. 63) the Planning Code of the Municipality of Pilar is continually adjusted in order to fit the development of new gated communities. The size and autonomy of these gated communities range from small groups of houses with few common spaces to 'mega-developments' that include a large variety of services. An example of the latter is the so-called 'green city' of Pilar del Este (http://www.pilardeleste.com.ar), which contains in 550 hectares various 'gated neighbourhoods' that are in turn all enclosed by a second encompassing fence. According to its brochure, Pilar del Este – which is still under development – will include an 'important civic and shopping centre', 'high-level primary and secondary schools', 'recreational centres' and 'sport clubs'.

The model which the development of Pilar follows is closer to the type of polycentrism in the USA such as Orange County, than to the European model exemplified by the Randstad. It is based on insularities connected by highways accessible only to those who can afford them (Fig. 9). There is a vast array of recent studies that depict the Pilar as one of the clearest examples of urban fragmentation in Buenos Aires (e.g. Ciccolella, 1999; Prévôt Schapira, 2002; Szajnberg, 2001; Thuillier, 2005; Torres, 2001).

Is Pilar a node of a polycentric system? There is a significant concentration of jobs, with business and industrial parks distributed along the highway linking Pilar to the ACBA, and there is a vast array of services that range from schools, universities, shopping malls and cinemas to the recent inauguration of a branch of the *Argentine Mozarteum* (La Nación, 1999). It is therefore not possible to reduce Pilar to a simple suburb. It has considerable autonomy. But it is not a transport node.

Perhaps its greatest flaw is that it is not linked to an efficient mass transport system. Its lack of integration is to a large extent related to the fact that many places in Pilar are only accessible by private transport. So Pilar is a new centrality that is an urban fragment, disconnected to the other urban centres. Although geographically placed within the seemingly polycentric form of Buenos Aires, is not part of an integrated system. Pilar is not atypical, and is an example of the trend of the other new centralities towards a structure of urban fragmentation.

Thus, it is pertinent to examine and contrast the case of Buenos Aires with Bangkok, a metropolis which, in the recent past, has incorporated a mass transport system that links some of its new centralities.

Fig. 9. Scenes from the 'Acceso Norte' highway: Business and industrial parks, amusement parks, shopping centres, gated communities and squatter settlements

Source: Daniel Kozak

Bangkok

Whereas Buenos Aires has only recently developed a tendency towards a more polycentric structure, Bangkok is now largely considered a polycentric metropolis. While the new centralities in Buenos Aires are fragmented rather than integrated with the rest of the metropolis, the difference between Buenos Aires and the new centralities in Bangkok is due to the relatively recent introduction of a mass rapid transit system that links some of Bangkok's nodes together. This section considers whether this might form an integrative and therefore more sustainable form.

Bangkok Metropolitan area comprises 1,569 km² with a population density of around 3,607 persons per km² (Alpha Research, 2005; BMA, 2007). The city was founded in 1782 and grew around its traditional centre, expanding to cover 18km² with a population of some 600,000. This monocentric city form was structured around water-borne transport with the river forming the main artery, and a network of canals. The major expansion of population occurred after the Second World War,

*Table 2.
Bangkok's
population 1955
– 2005
Source: United
Nations (2006)*

Date	Population '000s
1955	1,712
1965	2,584
1975	3,842
1985	5,279
1995	6,106
2005	6,593

supported by rail and new roads radiating from the traditional centre out into the surrounding countryside (Table 2), leading to many new centralities and a more polycentric urban form (Fig. 10, source Sakdicumduang, 2004).

Over the past 30 years, new infrastructure has been built with new road systems, expressways, a mass rapid transit system and a new airport, reinforcing existing and adding some new centralities to the metropolitan area. In 1990 Bangkok's first mass rapid transit scheme was approved by the Minister of Transportation and Communication. The idea was a 60km system to link the main railway terminus in central Bangkok with the airport and the site of the 1998 Asian Games at Rangsit. It was called the Bangkok Elevated Transport System (BETS) otherwise known as the Hopewell project. The project failed to materialize other than in a long line of concrete pillars, and the contract was cancelled in 1998 (Khang, 1998). Despite the Hopewell failure, in the early 1990s two elevated rail concessions were let to construct Bangkok's BTS SkyTrain, a network of 23km which was opened for service in December 1999 (Moor and Rees, 2000; BMA, accessed 2007). An underground metro (MRT) began construction in November 1996, and the 20km line was opened in July 2004 (BMCL, accessed 2007). The existing railway network and the mass rapid transit linked together a number of sub-centres in Bangkok, in particular where the lines intersected. Rail and MRT intersected at the main railway station of Hua Lamphong and at Bang Sue, the SkyTrain and MRT intersected at Asoke/Sukhumvit and Sala Daeng/Si Lom (two of the key night-market and entertainment areas in Bangkok), and the two SkyTrain lines intersect at Siam Square (the main shopping

*Fig. 10.
Polycentric
Bangkok
Source: after
Sakdicumduang
(2004)*

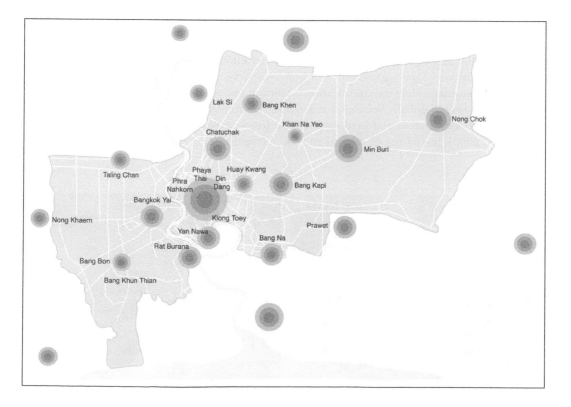

area). In theory at least this should provide a measure of integration to a number of centralities in Bangkok, making transport easy and bringing people together. So does the introduction of a mass rapid transit system help turn a polycentric but fragmented structure into one that is polycentric and integrated?

In the limited system introduced in Bangkok to date, the short answer is no. There are several reasons why practice does not bear out the theory. The first is capacity and cost. Public transportation in Bangkok carries around 2.5 million passengers a day, the majority of whom (1.8 million) travel by bus. There are 3,578 transport authority buses, 3,415 private (joint service) buses and 3,195 mini-buses and small buses (BMTA, 2006). The SkyTrain, after a shaky start, by 2005 carried some 500,000 per day (at its full capacity) and the MRT carries around 200,000 per day although its capacity is 400,000. But if the rapid transit has the capacity to carry only less than a third of the daily passengers, it also excludes the majority of the working population in terms of cost. The non air-conditioned buses charge between 6–8 baht[5] for an unlimited distance, rising to 12–22 baht for air-conditioned buses which charge by distance. By contrast the SkyTrain charges by distance and its fares range from 15–40 baht, as does the MRT with fares from 14–36 baht for single journeys (Fig. 11).

The inherent stratification by capacity and cost, which excludes the majority, represents a fragmentation of transportation. This fragmentation is reinforced by the design and location of its stations (nodes), especially that of the elevated SkyTrain. The tracks follow the existing street pattern below, and they serve a large array of Bangkok's international and tourist centres including seven cinema complexes, six convention centres, four museums and/or monuments, 30 major shopping complexes and 49 international hotels. Because of SkyTrain's elevation, these international venues are separated from the streets of Bangkok below, and many of them are linked directly at high level by 'sky bridges'. This form of connection is being developed rapidly, and there are now 23 'sky bridges', and even a kilometre long 'sky walk' linking two stations – Siam Square and Chit Lom – to serve a large concentration of shopping complexes. Here the stratification is cultural and social, separating the international world of globally branded stores and entertainments at high level from the streets with local traders and hawkers and Thai temples and shrines at ground level (Jenks, 2003). It is a form of cultural and social fragmentation (Fig. 12).

*Fig. 11. Buses for the poor, mass rapid transport for the better-off: a fragmentation of transport
Source: Mike Jenks*

Fig. 12. (above–upper) Tradition on the streets, a globalized world above: cultural and social fragmentation

Fig. 13. (above–lower) A project with global ambitions in a poor and undeveloped environment: potential economic fragmentation

Sources: Suraswadi (2004); Mike Jenks

Where the transportation modes intersect, either new development is planned or is taking place. Major new development is being constructed at, for example, Siam Square (Central World), and plans for development have been drawn up for Bang Sue where the metro meets the mainline railway. The development of these existing or new centralities further reinforces the tendency towards fragmentation, separating poorer citizens from developments that are designed to make money and serve the rich and middle classes. The plans for Bang Sue illustrate the point. The new metro station is adjacent to a suburban rail station, and the main land uses around the station are dominated by the large industrial complex of the Siam Cement Works. The proposed development shows the transformation of the area with a transit interchange, convention centre, offices, flats and a park, clearly targeted at a wealthy clientele (Suraswadi, 2004; Fig. 13). This, like many of the other nodes of the mass rapid transit system, represents a form of economic stratification and reinforces urban fragmentation.

Bangkok, like Buenos Aires, has a spatial structure that has polycentric centralities that are not well integrated together. The introduction of mass rapid transport systems in Bangkok should have helped to link the new centralities together, but this has not happened. Instead the system has reinforced fragmentation in terms of transport, culture and economics. For those who can afford it the system integrates a wealthy stratified 'global' world, leaving behind (and below) the reality of congested but vibrant Bangkok. Of course it may change. In 2004 the Commission for the Management of Land and Traffic (CMLT) approved a Mass Rapid Transit Master Plan to extend the existing network of 47.3km by a further 243.7km to serve Bangkok and its vicinities, opening up the system to the more populous areas of the city. This would physically link more of the polycentric nodes, and also provide the potential for access by a wider social stratum of the population (Figs. 14a and

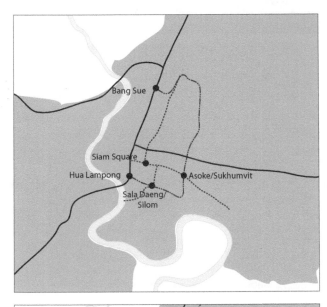

Fig. 14a.
Bangkok:
existing Mass
Transit Routes
Source: after
http://2bangkok.
com/2bangkok/
MassTransit/
map/map.shtml

Fig. 14b.
Bangkok:
planned
expansion Mass
Transit Routes
Source: after
http://2bangkok.
com/2bangkok/
MassTransit/
map/map.shtml

Key

·—··· MRT-Metro

······· BTS-SkyTrain

—— SRT-Railway

b). If the critical mass of people using the mass rapid transit system makes it more economic and the disparity between the system's fares and those of buses becomes closer, then Bangkok might get closer to the theory and move towards a more sustainable, less fragmented, polycentric form.

Conclusion: towards a 'defragmented city'

In world cities, mega-cities or very large urban regions or agglomerations, there appears to be an inexorable process of growth and development that transforms them from monocentric to polycentric forms. Achieving urban sustainability in the face of such size and growth is a major problem. For those cities that retain a predominantly monocentric form, compact city concepts may stretch the theory beyond its capacity to achieve what is claimed to be a sustainable urban form. Where polycentric forms predominate, then in theory at least, if the polycentric nodes take the form of high density small 'compact cities', and if these nodes are inter-connected by efficient and environmentally clean public transport, then polycentrism holds out some hope. However, the process may just accelerate and reinforce unsustainable and undesirable urban fragmentation. While the two cities chosen as examples in this chapter may not be scientifically representative, they do illustrate some of the trends and problems. So what do they say about the relationship between polycentrism and sustainability?

Both examples show a trend towards a polycentric urban form. Buenos Aires retains a strong monocentric centre but its aggressive growth over the decades has led to many new centralities and polycentric forms at its peripheries. Bangkok's traditional centre has reduced in importance and many new centres have grown both within the city and at its peripheries. Buenos Aires illustrates that there is no equitable integration in a mega-city without an efficient mass transport system, while Bangkok shows that if the transport system is not accessible to a large proportion of the population it increases inequalities and fragmentation.

Neither city can make many claims to sustainability. Both are car-dependent and rely either fully or mostly on heavily polluting buses for their public transport. In both cases social and economic inequities abound. The polycentric forms largely represent a process of fragmentation with, in the case of Bangkok, largely ineffective integration of a mass rapid transport scheme.

While the examples here give little comfort to the ideal of sustainability resulting from their urban forms, they provide an insight into what may be needed to achieve it. The form of a city may help to promote or discourage different types of social (and transport) integration but it cannot assure it. Urban form cannot compensate for most types of socio-economic inequalities, and so there is a clear limit to the concept of urban form as a sustainability provider. Even at the scale of the mega-city, both polycentric and monocentric urban forms can have tendencies towards either integration (and thus likely to be more sustainable), or fragmentation and unsustainability (Fig. 15).

As cities expand, particularly through rapid growth, the emerging new centralities will form the nodes of a more polycentric urban form. Such nodes, which could be as large as towns with their own autonomy, may be considered sustainable in their own right. A polycentric node may follow compact city concepts, and its transport and energy may be efficiently managed and encourage walking or cycling, and so it may at least have stronger claims to sustainability than if its

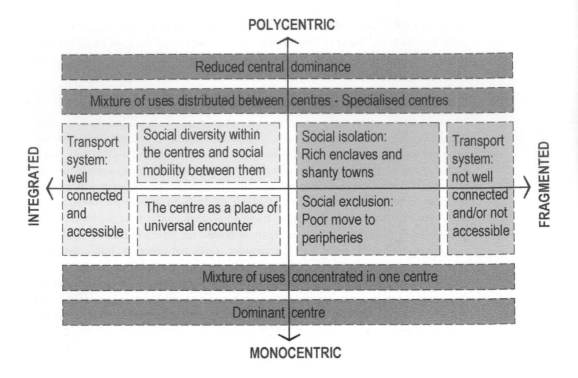

*Fig. 15.
Urban form:
polycentric,
monocentric,
integrated and
fragmented
Source: Jenks
and Kozak*

growth were unrestrained. However, if it is immersed in a network which is mainly connected by highways or where the social contrasts are blatant, it can hardly be considered as a sustainable urban form. Whatever the 'sustainability' of the node might be in itself, the whole system has to be considered. Thus the geographic rise of polycentrism within very large cities does not necessarily of itself create a form that is more sustainable than the monocentric form. Often, polycentric forms are more likely to intensify fragmentation than to reverse it. The use of transportation to link polycentric nodes should, in theory, help make the form more sustainable. In practice, at least as this limited study shows, it could serve to further reinforce fragmentation, and all the negative consequences that might have.

If the tendency is towards urban fragmentation, what can be done to create a more integrated spatial structure? Furthermore, if most contemporary metropolises are currently fragmented, as the large extent of literature dedicated to this subject seems to suggest, how can this be reversed? Is it possible to 'defragment'[6] existing fragmented cities?

The examples show that fragmentation occurs both between the new centralities (nodes), and within them. This has both social and economic consequences of exclusion and inequalities. It is unlikely that any manipulation of urban form will achieve much in solving these particular problems – and indeed polycentric forms may exacerbate them. Nevertheless the imperative is to ensure that there is social diversity within the nodes and social mobility between them: perhaps the Dutch policies of 'complementarity and coherence' point to a potential solution (VROM, 2001). But some elements of urban form may assist this process. By their nature, the nodes in a polycentric system are likely to be formed of higher density concentrations. This gives the opportunity to increase the density of development and to distribute densities in a way that allows the feasibility of an economic mass transport system. The nodes, as evidenced in Buenos Aires, can be sufficiently large to support a whole range of functions and uses, and do already provide jobs. This change too can be built on to intensify development to ensure there is a sufficient mixture of uses to allow access to jobs without the need to commute long distances to find employment. With nodes that are economically strong and have the critical mass to support good public transport, these nodes can be connected within the polycentric form, making it more of an integrated system. These mass transport systems must be accessible, efficient and connect all the nodes of the system, and also be affordable to the majority. But most important, all the population, rich and poor, need to be included in the system and no urban areas bypassed or ignored.

However, these suggestions are unexceptional, being an extension in scale of ideas currently embodied in some policies and enlightened practice related to compact city concepts. There is a long way to go if very large cities are to achieve more sustainable forms, or at least, less unsustainable urban forms. Yet, with the rapid growth in urbanisation, especially in Asia and Africa, the problems are pressing if global warming and climate change are to be ameliorated before it is too late. An interesting and diverse bibliography has been built over the last few years which describes, explains and conceptualises urban fragmentation. However, there is no equivalent body of research which studies what policies should be implemented to prevent and reverse these trends. In other words, a research agenda that focuses on 'urban defragmentation' is needed.

This chapter suggests that such an agenda should concentrate less on whether the form is polycentric or not, and more on the way that urban form can support social and economic equity, as well as physically encourage integration. Figure 15 suggests it is possible for both monocentric and polycentric forms to achieve a more sustainable spatial form. But this is a complex issue, involving many factors, and it is clear there is not just one route to achieving sustainable urban form. There are many routes, and these are mediated by the social, economic and cultural context. These routes may be found in a process of 'defragmentation' – a mixture of physical intervention and policy initiatives that may vary from place to, from culture to culture. Above all, the phenomenal urban growth of the early 21st century needs to be matched with an even faster growth in knowledge that is evidence-based, practical, and possible to implement.

Notes

1. www.deltametropool.nl
2. These figures correspond to the National Census of Population, Households and Housing 2001 performed by the National Institute of Statistics and Censuses (INDEC, 2001).
3. For example Scobie (1974, p. 70) considers three determinant factors which 'pulled and tugged at the shape of Buenos Aires during these years': the location of the new port, the design of the railway system, and the location of the national capital. In a similar way, for Sargent (1974, p. 45) the 'principal determinants of the rapid planting and slow settlement of the Buenos Aires metropolitan fringe ... were the expansion and improvement in service railroads, urban land speculation, and ... growth of local industry'. However, scholars more recently have challenged these views by arguing that 'it was not the mere will of land speculation and technological modernization' (Gorelik 2004, p. 27) which shaped the form of Buenos Aires between 1887 and 1936, but the determination of a reformist state. The early public definition of detailed urbanisation plans for the vast territory surrounding the traditional city, which took five decades to accomplish, upholds this argument.

4. Previous to the permanent gated communities there existed some weekend country clubs; the older ones from the 1940s. In turn, some of its members became permanent residents (mostly since the end of the 1980s), but the novelty of the new gated communities is that they were planned from the beginning for permanent residency (Arizaga, 2005, p. 25).

5. 1 US Dollar = 31.4 Baht; 1 Euro = 42.8 Baht; 1 Pound Sterling = 63.2 Baht on 8 July 2007 (www.xe.com/ucc/convert.cgi)

6. The term 'defragment' is used as a loose analogy with the process of defragmentation of a computer. This process involves, inter alia, physical reorganisation, contiguity, and the avoidance of fragmentation through compaction (http://en.wikipedia.org/wiki/Main_Page).

References

Abba, A. P. and Laborda, M. (2006) *Centralidades urbanas*. Retrieved in December 2006 from: www.atlasdebuenosaires.gov.ar

Acioly, C. (2000) Can Urban Management Deliver the Sustainable City?, in *Compact Cities: Sustainable Urban Forms for Developing Countries* (eds Jenks, M. and Burgess, R.), Spon Press, London.

Ainstein, L. (2001) Buenos Aires: nuevos criterios para la reestructuración de sectores urbanos. *Estudios del Hábitat*, 2(6): 51–64.

Alpha Research (2005) *Thailand in Figures*, Alpha Research Co. Ltd., Bangkok, Thailand.

Arizaga, C. (2005) *El mito de la comunidad en la ciudad mundializada. Estilos de vida y nuevas clases medias en urbanizaciones cerradas*, El Cielo por Asalto, Buenos Aires.

Betsky, A. (2002) Making ourselves at home in sprawl, in *Post ex sub dis: urban fragmentations and constructions* (eds Ghent Urban Studies Team), 010 Publishers, Rotterdam, pp. 89–99.

BMCL (Bangkok Metro Public Company Limited) (2007) Retrieved in December 2006 from: www.bangkokmetro.co.th

BMCL (Bangkok Metro Public Company Limited) (2005) *M.R.T. Route Map*, BMCL, Bangkok.

Beaverstock, J., Smith, R. and Taylor, P. (1999) A Roster of World Cities. *Cities*, 16(6): 445–58.

BMA (Bangkok Metropolitan Authority) Retrieved in March 2007 from: www.bma.go.th

BMTA (Bangkok Mass Transit Authority) (2006) *Bus operations in Fiscal years 1992–2006* Retrieved in March 2007 from: www.bmta.co.th

Boarnet, M., and Crane, R. (2001) *Travel by Design: The influence of urban form on travel*, Oxford University Press, New York.

Breheny, M., ed. (1992) *Sustainable Development and Urban Form*, Pion, London.

Calthorpe, P. (1993) *The Next American Metropolis: Ecology, community and the American dream*, Princeton Architectural Press, New York.

Castells, M. (1989) *The informational city: information technology, economic restructuring, and the urban-regional process*, Basil Blackwell, Oxford.

Castells, M. (2000) *The Rise of the Network Society*, Blackwell, Oxford [First published 1996].

Cervero, R. (1998) *The Transit Metropolis: A global inquiry*, Island Press, Washington D.C.

Ciccolella, P. (1999) Globalización y dualización en la Región Metropolitana de Buenos Aires: Grandes inversiones y reestructuración socioterritorial en los años noventa. *EURE*, 25(76): 5–27.

Clarin Newspaper (2006) *Suplemento Countries*, 18 March, pp. 8–10.

Clarin Newspaper (2007) *La población de los barrios cerrados, en ascenso continuo*. Retrieved in April 2007 from: http://www.clarin.com

Commission of the European Communities (CEC) (1990) *Green Paper on the Urban Environment*, European Commission, Brussels.

Curitiba websites: http://solstice.crest.org/sustainable/curitiba; http://www.solutions-site.org/artman/publish/article_62.shtml

de Boer, N. A. (1996) *De Randstad bestaat niet*, Nai, Rotterdam.

Department of the Environment, Transport and the Regions (DETR) (2000) *The Urban White Paper: Our Towns and Cities*, DETR, London.

Dupuy, G. (1998) *El Urbanismo de las Redes. Teorías y Métodos*, Oikos-Tau, Madrid [First published 1991].

Faludi, A., Waterhout, B. and Instute Royal Town Planning (2002) *The making of the European spatial development perspective: no masterplan*, Routledge, London.

Fernández Wagner, R. and Varela, O. (2003) Mercantilización de los servicios habitacionales y privatización de la ciudad: Un cambio histórico en los patrones de expansión residencial de Buenos Aires a partir de los noventa, in *La cuestión urbana en los noventa en la Región Metropolitana de Buenos Aires* (eds Catenazzi, A. and Lombardo, J. D.), Instituto del Conurbano, Universidad Nacional de General Sarmiento, Buenos Aires, pp. 43–74.

Geddes, P. (1968) *Cities in Evolution: An Introduction to the Town Planning Movement and to the Study of Civics*, Ernest Benn Limited, London [First published 1915].

Gilbert, A. ed. (1996) *The Mega-City in Latin America*, United Nations University, Tokyo.

Gorelik, A. (2004) *La grilla y el parque. Espacio público y cultura urbana en Buenos Aires, 1887–1936*, Universidad Nacional de Quilmas Editorial, Buenos Aires.

Gottmann, J. (1961) *Megalopolis: the urbanized northeastern seaboard of the United States*, Twentieth Century Fund, New York.

Graham, S. and Marvin, S. (2001) *Splintering urbanism: networked infrastructures, technological mobilities and the urban condition*, Routledge, London.

Hall, P. G. and Pain, K. (2006) *The polycentric metropolis: learning from mega-city regions in Europe*, Earthscan, London.

INDEC (2001) *Censo Nacional de Población, Hogares y Viviendas del año 2001*. Retrieved in May 2005 from: http://www.indec.mecon.ar

Jacobs, M. (2005) Building a multinodal metropolis: a short guide, in (eds Hulsbergen, E. D., Klaasen, I. T. and Kriens, I.), *Shifting sense in spatial planning: looking back to the future*, Techne Press, Delft , pp. 369–384.

Jenks, M., Burton, E and Williams, K. eds (1996) *The Compact City: A Sustainable Urban Form?*, E & FN Spon, London.

Jenks, M. and Burgess, R. eds (2000) *Compact Cities: Sustainable Urban Forms for Developing Countries*, Spon Press, London.

Jenks, M. (2003) Above and below the line: Globalisation and urban form in Bangkok. *The Annals of Regional Science*, 37(3):547–557.

Jenks, M. and Dempsey, N. eds (2005) *Future Forms and Design for Sustainable Cities*, Architectural Press, Oxford.

Khang, D. (1998) Hopewell's Bangkok Elevated Transport System (BETS), *School of Management*, AIT, Bangkok.

Knox, P. L. & Taylor, P. J., eds (1995) *World cities in a world-system*, Cambridge University Press, Cambridge.

Lambregts, B and Zonneveld, W. (2003) *Polynuclear Urban Regions and the Transnational Dimension of Spatial Planning*, Delft University Press, Delft.

La Nación (1999) *Los Percusionistas de Estrasburgo en la Filial Pilar del Mozarteum*. Retrieved in May 2007 from: www.lanacion.com.ar

Low, N. and Gleeson, B. eds (2003) *Making Urban Transport Sustainable*, Palgrave Macmillan, Basingstoke.

Moor, M. and Rees, C. (2000) Bangkok Mass Transit Development Zones, in (eds Jenks, M. and Burgess, R.) *Compact Cities: Sustainable Urban Forms for Developing Countries*, Spon Press, London.

Okabe, A. (2005) Towards the Spatial Sustainability of City-regions: A comparative study of Tokyo and the Randstad, in (eds M. Jenks and N. Dempsey) *Future Forms and Design for Sustainable Cities*, Architectural Press, Oxford, pp. 55–71.

ODPM (2004a) *Planning Policy Guidance 3: Housing*, Office of the Deputy Prime Minister, London.

ODPM (2004b) *Planning Policy Guidance 13: Transport*, Office of the Deputy Prime Minister, London.

Prévôt Schapira, M. F. (2002) Buenos Aires en los años '90: metropolización y desigualdades, *EURE* 28(85): 31–50.

Romero, J. L. and Romero, L. A. (2000) *Buenos Aires: Historia de Cuatro Siglos (Tomo 1)*, Altamira, Buenos Aires [First published 1983].

Sakdicumduang, T. (2004) *Rediscovering Sustainable Urban Form, Creative Development Co. Ltd*, Bangkok. At Compact Cities International Seminar, Thammasat University Bangkok, 31 March 2004.

Sargent, C. (1974) *The Spatial Evolution of Greater Buenos Aires, Argentina, 1870–1930*, Center of Latin American Studies, Arizona State University, Tempe, Arizona.

Sassen, S. (2001) *The global city: New York, London, Tokyo*, Princeton University Press, Princeton [First published 1991].

Scobie, J. R. (1974) *Buenos Aires: plaza to suburb, 1870–1910*, Oxford University Press, New York.

Soja, E. (2002) Sprawl is no longer what it used to be, in *Post ex sub dis: urban fragmentations and constructions* (eds Ghent Urban Studies Team), 010 Publishers, Rotterdam, pp. 76–88.

Sundaravej, S. (2001) Keynote speech at the international conference *Achieving Sustainable Cities in the SE Asian Region*, Bangkok, Thailand.

Suraswadi, K. (2004) *The Compact City in a Regional Context*, Office of Transport and Traffic Policy and Planning, Bangkok. At Compact Cities International Seminar, Thammasat University Bangkok, 31 March 2004.

Szajnberg, D. (2001) *Guettos de ricos en Buenos Aires*. Retrieved in May 2004 from: www.mundourbano.unq.edu.ar

TOD/TDZ website: http//www.smartgrowth.umd.edu/InternationalConference/Conference Papers/

Thuillier, G. (2005) El impacto socio-espacial de las urbanizaciones cerradas: el caso de la Región Metropolitana de Buenos Aires. *EURE*, **31(939)**: 5–20.

Torres, H. (1993) *El Mapa Social de Buenos Aires (1940–1990)*, FADU-UBA, Buenos Aires.

Torres, H. (2001) Cambios socioterritoriales en Buenos Aires durante la década de 1990. Sanitago: *EURE*, **27(80)**: 33–56.

United Nations (2006) *World Urbanization Prospects: The 2005 Revision*, United Nations, DESA, Population Division. Retrieved from: www.un.org/esa/population/unpop.htm

Urban Task Force (1999) *Towards and Urban Renaissance*, E & FN Spon, London.

VROM (Ministry of Housing, Spatial Planning and the Environment) (2001) *Summary: Making space, sharing space, Fifth National Policy Document on Spatial Planning 2000/2020*, National Spatial Planning Agency, VROM, The Hague, Netherlands.

Welch Guerra, M., ed. (2005) *Buenos Aires a la deriva*, Biblos, Buenos Aires.

Williams, K., Burton, E. and Jenks, M. eds (2000) *Achieving Sustainable Urban Form*, E & FN Spon, London.

World Gazetteer (2005) http://www.world-gazetteer.com

Section Two

Polycentric Regions and Cities:
Perspectives from
Europe, Asia and North America

6

Olli Maijala and Rauno Sairinen

Promoting Sustainable Urban Form:

Implementing urban consolidation policies around the Helsinki Metropolitan Region

Introduction

Urban sprawl is recognised as a major environmental, economic and social challenge in many growing European cities, likewise in the Helsinki Metropolitan Region. Urban sprawl and anti-sprawl policies are in many ways linked to the questions of urban sustainability, increasing fragmentation and polycentric development. In the world city debate the practical experiences of these linkages are needed.

This chapter assesses the ways in which development is steered towards consolidating the urban structure in the margins of the Helsinki Metropolitan Area. The research focuses on two satellite towns, Hyvinkää and Lohja, whose development is very much connected to the development of the Helsinki region. Hyvinkää has both rail and motorway connections to Helsinki, while Lohja has only a motorway. The chapter focuses specifically on the policies and procedures of these two cities.

The aim is to ascertain why some municipalities seem to have been successful in developing an integral, consolidated urban structure while others have not. In the case of the Helsinki Metropolitan Area, an assessment is made as to whether it is possible that satellite cities might promote sustainable regional growth (anti-sprawl). Of particular interest are the measures that are taken locally by municipalities to promote urban consolidation in planning, the guiding instruments that are used, and the problems that might get in the way of a consolidation policy. Collectively such implementation measures represent an operative policy for consolidation, and the term 'urban consolidation policy' refers to an urban anti-sprawl policy by which both urban structure and the quality of the living environment are improved.

Fig. (opposite page). Den Haag – a polycentric node of government offices and sustainable modes of transport, and a main rail station close by, providing efficient public transport interconnecting with the Randstad metropolitan region

Source: Roberto Rocco and Devisari Tunas

The concept of urban consolidation policy in Finland

In the context of the international debate on sustainable urban form, official urban planning recommendations in Finland have for a long time promoted a densification-oriented policy. By the late 1990s, densification policy was replaced by the concept of *urban consolidation*, which was considered, officially at least, to give a greater emphasis to aspects of quality related to urban compaction. Despite the emphasis that the concept receives in official goal setting, it has not been explained; no official definition, or any kind of white paper, exists about either the concept or the policy.

The most important legal land-use guidance document that does include the concept is the National Land-Use Guidelines of Finland, one of the main objectives of which is 'a more consolidated urban structure and the quality of the living environment'. The guidelines state that

> the well-functioning and the economy of living environments is to be promoted by means of utilising and consolidating existing urban structures. In the consolidation of urban structures, the quality of the living environment should be improved. Urban structures are to be developed so as to make services and workplaces available to different groups of the population ... (Ministry of the Environment, 2002).

This suggests that there might exist a large variety of different interpretations of the concept and the policy at the local level, and also between different actors in the same local policy network.

The term 'urban consolidation policy' is used here to refer to an urban development policy where both urban structure and the quality of the living environment are improved. Thus, the target of an urban consolidation policy is to:

- improve the eco-efficiency of the urban structure (e.g. through densification);
- diminish negative environmental impacts and;
- improve the quality of services.

The quality of services can, in this context, imply a 'good living environment', 'good accessibility of services' or 'good public transport'. So the concept of consolidation concerns all three aspects of sustainability: ecological, economic and social/cultural.

The content of a consolidation policy can be illustrated by placing its various dimensions in a grid, where the X axis is the quality of services, and the Y axis the mitigation of negative ecological impacts. Urban consolidation would, in the best-case scenario, be located in the upper right-hand cell of the grid. The other typical concepts and situations in consolidation policy discourse can also be placed in the grid (Fig. 1).

The principal operational concept of urban consolidation is accessibility. It embodies a central social-ethical dimension of urban consolidation: consolidation can be seen as a policy of facilitation which seeks to guarantee all people and groups equal access to the services and quality functions of the community and living environment that are vital to them. When complemented with an ecological/ethical dimension which emphasises the reduction of energy consumption and negative environmental impacts, it results in developing communities and living

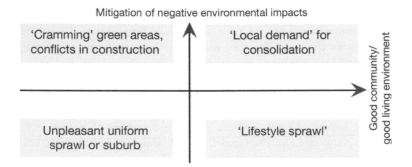

Fig. 1.
Consolidation
policy and the
dimensions of
eco-efficiency

Source: Maijala
and Sairinen

environments that facilitate access to functions and individual quality factors, in particular, by walking, bicycle or public transportation, i.e. living environments with 'affordances for an ecological lifestyle' (Kyttä and Kahila 2005 and 2006). This is above all egalitarian facilitation, supporting environmentally and socially sustainable choices. Structure, the end result of planning, has a probabilistic role in this process. How people ultimately act and choose is something that cannot be dictated by community planning.

Policy instruments for urban consolidation policy

Nuissl (2007), complementing and modifying the basic categorisation employed by Bengston *et al.* (2004), has divided policy instruments for 'smart sprawl' into four categories:

- Regulation – 'hard' legal instruments
- Incentives
- Public management – 'soft' instruments such as regional co-operation
- Information / education

The same classification scheme can also be used for an analysis of the instruments of an urban consolidation policy.

Regulation

The lack of legal regulatory instruments is not a problem, at least in Western Europe (Nuissl, 2007). The main problem seems to be the lack of willingness to use these instruments. There are many actors on different levels who oppose any restrictions on urban growth and supplementary development. Nuissl distinguishes between three main types of actors as 'users of anti-sprawl planning':

1. Core cities tend to be the natural allies for anti-sprawl policies.
2. Urban sprawl is often exacerbated by edge cities, which usually attract businesses and inhabitants with a total disregard for future costs. The growth of such cities is often perceived as sprawl from the perspective of the region, or at least from that of the core city.
3. Regional actors should be able to implement policies that mitigate urban sprawl.

There are two ways that regional land-use planning can be used to efficiently control suburbanisation: vertical and horizontal co-ordination. Vertical co-ordination is the harmonisation of municipal planning with more general goals. Horizontal co-operation balances conflicts between municipalities. In the Finnish context, the former refers to the traditional planning hierarchy between municipal and regional planning, the latter to the currently topical development of urban regional planning. According to Nuissl, one of the potential obstacles to successful regional co-operation is that local disinterest in anti-sprawl policy also operates on the regional level. Moreover, the regional level is easily beset with conflicts.

Incentives

One possible way of implementing an anti-sprawl policy is to try to influence the behaviour of actors that lead to urban sprawl. In practice, suburbanisation must be made to appear less attractive by altering incentives and the dynamics of development.

Public policy has two main instruments with which to alter the attractiveness of urban sprawl: by making sprawl-inducing behaviour more expensive and by reducing the attractiveness of some environments *vis-à-vis* others. Features that make suburbs attractive can be imported to inner cities, and perceived inner city disadvantages can be mitigated. There is an interesting contradiction in the case of public transportation policy: new or faster connections to suburbs tend to increase the attractiveness of decentralisation. On the other hand, new connections help to diminish the volume of private car traffic in inner city areas, thus making inner cities more attractive.

Public management

Nuissl's third group of policy instruments consists of a variety of 'soft' measures, i.e. non-regulatory and informal actions in which the public sector occupies a central role. In practice, such instruments are used to try to find cooperation-based solutions to specific problems that concern conflicting demands on land use. In reality, this comprises municipal collaboration and collaboration between municipalities and the state, where the key aspect is commitment to a common goal. As Nuissl points out, it is no surprise that such soft planning instruments are fairly successful in situations where there is a perceivable 'win–win' situation. Examples of this in Finland include the unofficial co-operation for master plans and structural modelling that has been implemented in certain urban regions.

Awareness and information

Policy instruments to increase awareness and dissemination of information about urban sprawl apply to both public and private actors, whose solutions have real and considerable impacts on the development of community structure. One key factor in this is the requirement that decision-makers must have better access to information about the consequences of their actions in order to be able to reliably assess the development of urban structure. The information will also promote public discussion about the matter and increase understanding and awareness of related problems. For this purpose, instruments for monitoring changes in land use and urban structure have been developed both in Europe (e.g. EEA 2006, Kasanko *et al.*, 2006) and in Finland (YKR) (Ristimäki *et al.*, 2003), especially for experts.

General consolidation policy in Finland

The concept of consolidation (or coherence) appears in the Finnish National Land-Use Guidelines accepted by the government in 2000 (Ministry of the Environment, 2002). Under the special guidelines it is also mentioned that possibilities for integrating the community structure should be included in regional planning and local master planning, and steps required for integration should be presented. It also notes that 'especially in urban regions, a functioning traffic system and a centre system improving the availability of services should be studied, and in this connection, also the placing of major retail trade units' (ibid.).

The progress of a consolidation policy in urban planning has been slow and difficult. Forces promoting the fragmentation of the urban structure are very strong. In the follow-up report of the new Land-Use and Building Act from 2000 (Ministry of the Environment. 2005), it is stated that the planning instruments provided by the Act are functional, but they are not being used sufficiently or in a way envisaged by the legislator. In its programme proposal, the Working Group for the Development of Steering of Urban Structure (YOKO) appointed by the Ministry of the Environment in 2003 arrives at similar conclusions (Ministry of the Environment, 2004). According to the group, the problem seems not to be the lack of guiding instruments, but insufficient use of existing ones.

Municipalities in Finland have a so-called planning monopoly; they are the only bodies that have the right to draw up land use plans that are legally valid and can be used as the basis for issuing building permits. This places municipalities in a key position as the implementers of a consolidation policy. In addition, regions (provinces) also have a distinct role in regional planning.

Growing and sprawling Helsinki Metropolitan Area

The Helsinki Metropolitan Area (Fig. 2) is located in the region of Uusimaa in southern Finland. It is home to approximately 1.3 million inhabitants, more than a quarter of the country's total population. The nation's capital, Helsinki, is also located in the region.

The Helsinki Metropolitan Area is one of the fastest growing urban areas in Europe at the moment. The effects of economic globalisation and new technology jobs demanding highly skilled labour have been felt mainly in the Uusimaa Region. The population of the Uusimaa Region is currently growing by about 15,000 people annually, and the proportion of young adults, i.e. under the age of 40, is particularly high. The population base in the region is estimated to grow by 160,000 people between 2000 and 2010. By 2030, it is estimated that the growth will be 200,000–300,000 people (Uusimaa Regional Council 2005; 2006).

As a result of migration into the region demand for housing has increased, pushing up the price of both dwellings and land. Coupled with increased affluence

Fig. 2. Helsinki Metropolitan Area 2005
Source: SYKE / Monitoring System of Urban and Spatial Structure 2007, Grid data Statistics of Finland. MML lisence num. 7 / MYY/07

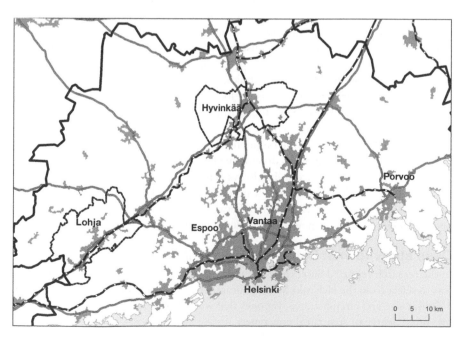

☐ Helsinki commuting region
▨ Densely built-up area 2005
☐ Lohja and Hyvinkää municipalities

–•– Railway
—— Main road network

and a scarcity of land for housing projects in the region, this has led to a situation where the pressure for housing has spread out into an expanding area around the edges of the urban region, partly also to suburbanised areas and even beyond the region. The gap between demand and supply has been particularly wide in the case of detached houses. Because of the lack of supply, many people have sought alternatives in satellite municipalities around Helsinki and even further away. Lower prices, certain quality factors and improved connections have also increased the demand for detached housing in the secondary zone of the metropolitan region, which also includes the towns of Lohja and Hyvinkää. The polycentric development of the whole region has been increasing.

Escalating growth and the demands of the new network economy have altered ideas about the structure of the Helsinki region. The centre of Helsinki continues to be the main nexus of the area. On the other hand, a considerable part of growth in jobs and traffic is taking place in other areas of the metropolitan region, and also on its periphery. The commuting area is also a fairly uniform market area for housing. As mobility increases, the entire metropolitan region becomes increasingly a common but dispersed area, where people search for services and leisure, even from a long way off.

Such development means that the urban sprawl is an increasingly serious problem in the Helsinki region (see also EEA, 2006), although in the European context, the area is very sparsely populated. Growth in traffic is strongest in the so-called ring zones around Helsinki. Increased cross-traffic in the region has mainly been of private cars; public transport has traditionally been based on a radial route network.

In terms of consolidation, the situation in the Helsinki Metropolitan Area is challenging. The municipal structure of the region is fragmented. The municipalities are financially challenged and are competing with each other for taxpayers. Regional co-operation in land use is very weak. Regional land-use planning is carried out through legislation. The effect of regional plans on growth is very small. However, the plans are often drawn rather loosely so as to include the margins for expansion that the municipalities require. Because of municipal autonomy and competition, there is hardly any prioritising about where growth is to be directed. The real choices in directing growth – if such choices are consciously made at all – are made in individual municipalities in the case of local master plans.

The regional land use plan for Uusimaa clearly aims at compaction and consolidation of the urban settlement structure. The structural goal for the area is a decentralised model depending on public transportation. In the regional land use plan, the zones of most vigorous development are the urban region of Helsinki and the development corridors around railways. The idea is to use the towns themselves as the basis for the development of the outer zone of Hanko–Lohja–Hyvinkää. One future possibility is a new railway line between Helsinki and Lohja, which would open up opportunities for new zones of development.

Case study towns: Lohja and Hyvinkää

In the research reported here, the basic idea behind the selection of the case studies was to select from the Helsinki Metropolitan Area (economic region) small towns of a similar size and circumstance, yet ones that were clearly different in terms of their measured degree of structural coherence.

Lohja and Hyvinkää are both on the very edge of the metropolitan economic region, both approximately 55 kilometres from the centre of Helsinki. The measured degree of structural coherence was based on the indicators and analyses of the monitoring system for the spatial structure of urban regions (YKR) developed by the Finnish Environment Institute. Monitoring has taken place of the 33 largest urban regions of Finland between 1980 and 2000. According to the first monitoring report (Ristimäki *et al.*, 2003), Hyvinkää (Fig. 3) seems well consolidated in terms of its spatial structure, whereas Lohja (Fig. 4) for its size is among the more fragmented urban areas. The area of new suburbs around Hyvinkää is exceptionally small for a town of its size. In Lohja, by contrast, the suburban area is larger than average, and it also grew proportionally more than elsewhere during the monitoring period. The average distance from the marginal and suburban areas to the functional centre of the urban region is about two kilometres in Hyvinkää, while in Lohja it is almost twice as far.

The populations in Hyvinkää and Lohja are about the same size, and a similar proportion of inhabitants live in densely built-up areas. The towns are equidistant from Helsinki and both their populations are growing at the same rate, with an average annual increase of approximately 0.8–0.9 per cent. There are similarities also in the history of the towns. Neither of them is an established old town, which would have entailed development according to a town plan. Hyvinkää and Lohja are both the result of industrialisation, and both retain strong industrial traditions. Yet their urban traditions are weak, and they lack the charm of small historical towns.

Typical of Finnish urban development, both Hyvinkää and Lohja obtained their administrative independence in 1926, becoming very small municipalities surrounded by a rural commune of the same name. In Hyvinkää, the municipality and the rural commune were merged in 1969, but in Lohja this did not take place until 1997.

Fig. 3. Hyvinkää
Source: YKR/
SYKE 2005;
Genimap Oy

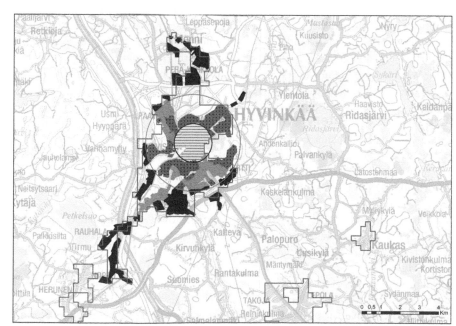

**Transit town
zones**

☐ Densely
built up
area

◯ 2.5 km zone

◯ 1 km zone

■ Car dependent
zone

≡ Walking zone

▦ Public transport
zone

▦ Cycling zone

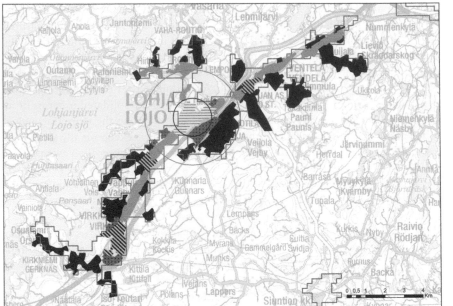

Fig. 4. Lohja
Source: YKR/
SYKE 2005;
Genimap Oy

Transit town zones

☐ Densely built up area
◯ 2.5 km zone
◯ 1 km zone
■ Car dependent zone
≣ Walking zone
▨ Public transport zone
▨ Cycling zone
\\\ Subcentre

There are also important basic differences between the two towns. The history of habitation in Lohja is very long, with many rural villages and scattered settlements, and the region is full of waterways. The centre of the town lies between dominating landscape elements, a very long high ridge and a big lake. Hyvinkää by contrast was for a long time a sparsely populated forested area, characterised by a relative lack of waterways and a few large manors. Spatially, Hyvinkää has developed basically around one centre and its railway station. In Lohja, industrial plants were established in a scattered pattern, and the town developed spatially into an urban region spread around a string of built-up areas with multiple centres. An important underlying factor in this development is the administrative division of the area. When a growing urban region extends over the area of several municipalities, this often leads to competition between the municipalities.

Themes of the interviews

There are no documents concerning the implementation of local consolidation policies. Our original research material comprised strategies, development programmes and actual land-use plans. Another important source of material was the data from the national YKR system (the monitoring system for the spatial structure of urban regions) and adjusted local analyses.

The main body of material for the research, however, consisted of thematic interviews conducted with leading local administrators and elected officials in August–November 2005. In total, twelve interviews were conducted in Lohja and ten in Hyvinkää.[1]

The main topics of the interviews included: how consolidation and urban sprawl were addressed in policy and planning, what the goals were and how consolidation was promoted; what the problems were; what the knowledge base for consolidation was; who were the relevant actors and the roles of elected officials and the various groups of municipal actors; how consolidation was manifested in regional co-operation; and how the success of a consolidation policy could be monitored and measured.

Successful consolidation policy in Hyvinkää

Using the indicators for community structure, Hyvinkää appears as a prime example of an urban community that has developed in an integrated manner. This raises some key questions: what role have local policies had in this development? What procedures and policy instruments have contributed to the result? Our key observations are summarised below.

The steering policy in Hyvinkää has been based on the following general principles:

- Systematic and consistent approach, foresight and equality
- Commitment and trust (within and between administrative branches and also *vis-à-vis* politicians and all other actors)
- Strong use of master plans for consolidation policies

The policy instruments have included:

- Acquisition of undeveloped land for development well in advance (using expropriation purchase if necessary). The pattern has a long tradition, going back to the beginning of the 20th century, and has been aided by the concentrated ownership of land
- A policy of deviation decisions that supports development: avoiding the creation of obstacles to future land-use needs
- Approved local master plans for both urban and rural areas that include:
  - Policy discussions and decisions made during the local master planning process (at 10- to 15-year intervals)
  - An approval procedure that makes for a more rigid process, but which gives continuity to the policy
- A 10-year construction programme, with an annual review that includes:
  - A monitoring tool of progress for politicians and others
  - Keeping matters involving urban structure to the forefront
  - Concrete and clear markers about the areas to be developed next, with cost calculations, and the number of plots to be released
- Contiguous areas that are developed sequentially, ensuring:
  - Efficient use of services and public utilities
  - Efficiency and sound economy to facilitate political support
  - Plots for detached houses are made available (but only on a modest scale), and the price of land has been kept in check
- Approved local master plans for villages (using the so-called sizing principles), ensuring:
  - Equal treatment of landowners
  - Emphasis on equal building rights for all (provided the threshold of densely populated area is not exceeded)
  - Concentrated land ownership to help steer development in a

Fig. 5. Hyvinkää from air
Source: City of Hyvinkää

centralised manner to structurally favourable sites
– Demand remains moderate while the number of sites has been adequate

The process instruments used include:

- A functional procedure and systematic approach (easy to continue; more difficult to change direction)
- Emphasis on the character of consolidation with clear aims and direction
- Commitment to consistency (it is demonstrated that development is under control and gets results: there are plots on offer, development is efficient and panic has been avoided)
- Systematic implementation of equality (all actors can rely on it)
- Effort, time and commitment, listening to and accommodating different parties, long processes. Consolidation is often more arduous than decentralising *laissez-faire*

Lohja – towards a new consolidation?

Many of the above features found in Hyvinkää are lacking in Lohja. Until very recently, Lohja has not had in place any systematic policy for urban structure to which the actors could have committed. This has prevented the emergence of a culture that would promote consolidation. The situation in Lohja could be described as follows:

- There have been very few resources to purchase undeveloped land, and purchases have not always been of structurally important sites
- Exemption policy has been lax, favouring decentralised construction
- Previously (before 1997) Lohja had no approved local master plan at all

- There are no plans for implementation; inter-sectoral co-ordination has been lacking until very recently
- Local detailed plans have included a lot of private land, which for different reasons has remained partly or entirely undeveloped; municipal investments in infrastructure are underused as a consequence
- Planning resources have been scarce, and planning for village areas has not been carried through

Our study indicates that Lohja is already well on the way to taking appropriate steps. Work for the preparation of a master plan for high-density areas is under way. This provides an opportunity to discuss issues and make strategic key decisions. The land policy programme will soon be completed. In the case of rural areas, the consolidation policy still needs adjustment and prioritising. Lohja is undergoing a critical period of investigation in the field of structural urban development. One difficult challenge for structural policy is posed by the new motorways adjoining the town.

What should be done in Lohja to allow a new kind of consolidation policy and planning culture to emerge? Possible measures might be:

- Making the concept of consolidation concrete: clear definition of aims, adopting a common-sense approach
- Outlining the benefits of consolidation to different actors
  - Municipal finances (savings)
  - Municipal economy (success, jobs)
  - Habitation (pleasant, equitably facilitating and functional environment)
- Preparation of an overall plan or strategy (setting goals and defining strategy), including:
  - Open and broad discussion, clearly and sensibly presented

alternatives, imported best practice and good examples
  – A binding framework plan that must be approved
• Preparation of plans for implementation, covering different contingencies
  including: organising monitoring tools; setting intermediate goals; and reminding
  people of the consolidation goals and strategy.

Results of the interviews

Consolidation is something that must be talked about in tangible terms and in a down-to-earth way

The concept of urban consolidation first emerged in urban land-use planning, and it is only natural that it is best known among circles responsible for land-use planning in urban areas. As a term, the concept is familiar to environmental and social authorities, but they are uncertain of its precise meaning. The town managers were familiar with the concept and its basic content. The level of familiarity among elected officials varied – natural considering the variation in the background, experience and interests of people in local politics. Most of them knew the concept as a term, but even some people who had been a long time in managerial positions in the town said outright that they did not know what consolidation was really about. This is a clear indication that *the concept of urban consolidation is hardly ever used as a slogan or title in local community or other policy.*

Many of the interviewed elected officials and also administrators said that they were more familiar with the term compaction, and that it had been used often in the context of structural development. Nor had any policy documents been prepared under the title of 'consolidation' or 'consolidation policy'. One difference between the two towns that emerged in the interviews was that politicians in Hyvinkää knew about the concept and related aims, whereas those in Lohja were very uncertain about the implications of the term. The director of the local Chamber of Commerce, who represented extra-municipal and, thus, also non-public-sector interests, was unfamiliar with the concept of urban consolidation and knew next to nothing about its meaning in the context of municipal urban planning.

It is obvious that a consolidation policy can only be created successfully if the concept is explained to the relevant actors, and also to those who are not experts in community planning. Elected officials, other administrative sectors or inhabitants cannot commit to the aims of a consolidation policy that they do not understand.

One thing that emerged in the interviews was that consolidation is most commonly spoken about in terms of the *economy of public utilities* and *economy of services*. These were quite clearly the strongest incentives for consolidation among the municipal management and elected officials. The economy and efficiency of services was also spoken about in terms of *integrating functions* to maximise the *accessibility of services and jobs to different groups.*

Another important issue in making a consolidation policy appear more down-to-earth is to stress that it is a process. Consolidation is not about individual measures to save costs. It is a continuous process where one looks hard at the direction in which one is going and the preconditions of future development. Regarding this, there were differences between planners in the two towns. The planners in Hyvinkää emphasised a process approach, whereas in Lohja they saw consolidation mainly in terms of its content, 'stopping gaps' in the urban structure.

There were interesting differences in how urban sprawl was perceived by individual respondents. The most common interpretation of urban sprawl was a growth in scattered settlements in rural areas. This notion is linked to the specifically Finnish context. Finland is one of the few remaining European countries that still maintains the so-called basic building right, i.e. the owner of a sufficiently large plot of land has the right to build a detached house on his own land. Another interpretation of sprawl was to see it in terms of scattered conurbations. The reason cited for this was bad planning. A third interpretation of urban sprawl was to see it as a weakening of interconnections between all functions, both physically and functionally.

Consolidation is often difficult – there is need for support

Four different types of major problems emerged in the interviews concerning the implementation of consolidation policy. Firstly, the most common problem was land ownership and land policy in general. This had to do especially with difficulties in obtaining undeveloped land in areas advantageous in terms of community structure. However, the difficulty also occurred in connection with areas already covered by land-use plans. The difference between the two towns in this respect was marked. Few interviewees in Hyvinkää made any reference to this, whereas in Lohja it was mentioned by all interviewees as a particularly problematic area. Land policy in Hyvinkää has been systematic and consistent for decades and all measures allowed by law (including expropriation) have been used there. Opportunities to purchase land in Lohja have been poor, and there has been no political support for expropriation.

Secondly, the interviewees referred to the fact that the areas offered by the town that are particularly advantageous in terms of consolidation *do not correspond to the wishes of the people* – their housing preferences; demand is for other areas or other types of housing. Differences between the two towns were again marked. In Hyvinkää, this was presented as a possible threat, but the consistent land policy has worked quite well up to the present. The town has developed its own lands one area at a time, finishing one area before developing another. By contrast, the problem was quite severe in Lohja, where there is a lot of planned undeveloped or underdeveloped private land for which demand is quite poor.

The third type of problem was the *cost of implementation*. Areas situated advantageously in terms of consolidation may be badly polluted or the ground poor for foundation construction. This problem was mentioned in both towns. Similarly, planning solutions for a high-density area may be demanding, complex and expensive. This is closely linked to the fourth group of problems, the relationship between consolidation and other interests. Operating within existing complex structures and environments often entails time-consuming negotiations and searches for compromise. There are *difficult discussions* and *difficult issues* to be resolved. Often people have to give up something. For example, locating a supermarket outside the urban structure along a highway on a disused field instead of on a partly developed town block with its layered history can seem like a tempting and easier way for many parties in the process – 'might as well let it sprawl a bit then', and powerful special interests may easily put a stop to the consolidation process by their inflexibility. Anti-sprawl advocates are few and far between in both municipalities and on the national level. In Hyvinkää, which had had a consistent consolidation policy in place for a long time, these problems emerged very clearly, and were evident also in attempts to develop the town centre in Lohja.

Local master plan as instrument and process: a foundation for strategic guidelines

How should consolidation policies be implemented in practice? Most of the interviewees specifically mentioned statutory land-use planning, local master plans in particular, as a central policy instrument for promoting urban consolidation. Master planning in Lohja has been in its infancy for a long time, and now they want to use it to create ground rules. Hyvinkää, on the other hand, has a long tradition where the local master plan is used as an instrument to make key strategic decisions that affect the urban structure. A local master plan is the local legislative framework instrument for spatial planning in Finland. It is a kind of 'multi-instrument of consolidation policy': it is not only a legal land-use instrument, it is also a tool that supports the creation, implementation and even monitoring of a consolidation policy. Being a legal instrument, the local master plan is a fairly rigid tool for planning and for that very reason it also ensures continuity for broad strategies. The local master plan calls for a more holistic perspective and offers a framework for creating strategic guidelines.

The interviews suggest that *at least as important* as the role of the local master plan as a regulatory instrument is the role of the local master plan *process as a forum* that offers time for development. Planners as well as other actors tend to be increasingly harried these days and they do not generally have the opportunity to engage in this kind of development work. A collective process provides an opportunity to get the various parties to commit to key strategies. The rigid local master plan can be made more flexible, if necessary by developing tools for a 'more sensitive master plan'. This might take the form of an assessment of significant structural impacts of new ventures and up-to-date databases containing basic facts.

As an instrument for a consolidation policy, the local master plan needs to be supported and complemented by other suitable planning, co-ordinating and

Fig. 8. Lohja's street
Source: City of Lohja

monitoring instruments (such as implementation programmes). Inter-sectoral co-operation and co-ordination are especially important in a consolidation policy. Finland has no legal instruments for the co-ordination of physical and functional planning between different sectors. Some instruments may have been developed at local level, however. A good example of co-ordination is found in the 10-year plans for the construction of new residential areas; such plans have been used in Hyvinkää for a long time. Prepared annually and subject to political approval, the plans have an important role as instruments for translating the goals of the consolidation policy (strategic decisions on the local master plan level) into concrete terms and monitoring their implementation.

Success is secured by trust and image

The interviews, especially those in Hyvinkää, indicate that, in addition to policy instruments, one important factor in successful consolidation policy is *trust* between actors. Important factors for creating trust include consideration of matters from different perspectives, equality and fairness in processing and discussing matters, and clear reasons for decisions and solutions. The ultimate test for trust is of course successful implementation of the policy.

The interviews suggest that the concept of consolidation has a fairly negative connotation for many people. The advantages of consolidation for everyday life and municipal economy are often ignored almost completely. The mental image of the end result of consolidation may be unpleasant: densely built areas plagued by social problems, restlessness, noise and hazardous traffic conditions. Communications, possibilities for compensation and marketing are areas that need to be developed in the context of consolidation.

The notion of consolidation that emerged in the interviews was not a very encouraging one; the quality of the living environment was hardly addressed at all except in terms of the interrelatedness and accessibility of functions. This applied equally to both towns. Such a quality as the pleasantness of the environment was only mentioned by the interviewees in reference to low-density detached rural housing, hardly ever to describe an urban environment. One reason for this is probably that the interviewees perceived consolidation above all as a structural concept rather than one that has to do with the quality of the living environment.

There is clearly a need also to consider factors contributing to the attractiveness of different areas as well as compensation in connection with consolidation. For instance, a dense spatial structure in residential areas should be offset with good local services or a high-quality living environment. Good examples are vital for this. The quality of an emerging environment must be ensured – through examples, consolidation acquires a positive sound to it. Otherwise it can become labelled as a policy that results in unpleasant environments.

The weak role of the state and local inhabitants

The interviews suggest that the key actors in consolidation in both towns are land-use planners, high municipal management and key elected officials. The municipal administrators emphasise their role as experts, to raise issues and prepare matters. The elected officials, on the other hand, emphasise their role as decision-makers and strategists. Generally speaking, the interviewees stressed the overwhelming importance of local politics in the promotion of urban consolidation. The interest of the state is often divided between strong sectoral perspectives, whereas consolidation would require a more comprehensive view of the whole. There is a need for the state to act in a more coherent way to give consolidation policy more rigour at both general and local levels.

According to the interviews, the role of local inhabitants in consolidation would seem slight, despite the conflicts and sensitivity of consolidation projects and infill development. A consolidation policy that stems from local needs and local demand was not considered particularly necessary. Heikkinen (2007) has in his research come to the opposite conclusion. According to him, local demand is actually an opportunity that should be actively exploited in a consolidation policy if we want to avoid unnecessary conflicts.

Regional consolidation is still in its infancy

Regional co-operation in the area of a consolidation policy seems to be very undeveloped, judged by our interview responses. Neither town prioritises regional co-operation in land-use planning. In both Hyvinkää and Lohja, their inclusion in the Helsinki Metropolitan Area was primarily perceived as a source of growth – something that provides both the means and the need for urban development. A regional consolidation policy hardly seemed to interest the local actors at all.

Conclusions

This chapter has assessed the steering of the development of urban structure in the margins of the Helsinki Metropolitan Area. The research focused on two satellite towns, Hyvinkää and Lohja, whose development is very much connected to the development of the Helsinki region.

Helsinki Metropolitan Area belongs not to the group of big global cities, but the dynamics of major changes in the urban structure are certainly very similar both in terms of urban sprawl, regional fragmentation and polycentrism. Helsinki Metropolitan Area has same challenges as many bigger global cities: How to improve urban sustainability? How to avoid fragmentation? How to support reasonable polycentric development?

The term 'urban consolidation policy' has been used here to refer to an urban development policy where both urban structure and the quality of the living environment are improved. Thus, the target of an urban consolidation policy is to improve the eco-efficiency of the urban structure (e.g. through densification), to diminish negative environmental impacts and to improve the quality of services. The research has shown that, in Finland at least, there are wide differences in consolidation policy. In Hyvinkää it was based on a systematic and consistent approach where various policy instruments have been used. In this context, local planning culture takes into account consolidation policy targets, and the consolidation-oriented planning culture has kept the urban structure compact and coherent. By contrast, Lohja in the recent past had no systematic policy for urban structure and this inhibited the emergence of a culture that would promote consolidation: however, the preparation of a master plan for high-density areas is under way. These two cases provide some lessons in moving towards more consolidated, and sustainable, urban forms.

Communication between all those involved in the process is vital, and the concept of consolidation needs to be explained clearly and the long term benefits articulated. It is a difficult process, as either it may not always coincide with the wishes of local people, or the land quality may not be the best (e.g. polluted or difficult to develop), and so support is needed to assist the process and the time-consuming negotiations needed to reach compromise. Drawing up a statutory local master plan is critical, and it is necessary to allow sufficient time to act as a forum for debate. Indeed, involving local inhabitants is necessary if conflicts are to be avoided. The same is true of local politics where narrow sectoral interests can dominate – a holistic view of urban consolidation is needed to help them arrive at a more comprehensive understanding. Many of the actors in the process of urban consolidation tend to view the process as negative, and so engendering trust between all parties is important and a necessary part of the negotiation and participation processes. A forward-looking approach would help. The advantages of consolidation for everyday life and municipal economy are often ignored and it seems that good communication, possibilities for compensation and marketing, are issues that need to be addressed in the context of consolidation. The lack of development of regional co-operation to achieve consolidation, identified in the research, reinforces the need for local and regional collaboration in land-use planning.

Finally, the development of consolidation policy for the whole metropolitan region means there is a need for two-dimensional policies. First, there needs to be 'regional wisdom of location policy' (how to allocate growth to various centres and towns around metropolitan region). This requires effective regional land-use planning. Second, the culture of local policy within the towns and sub-centres should favour (and support) urban consolidation policy in its various forms. Only if these two approaches are taken together do they provide potential for sustainable polycentric metropolitan development.

Note

1. The interviewees from administration included people responsible for management, land use, land policy, traffic and transportation, environment and social security. The elected

officials were people from the largest groups in the town council (four in Lohja and three in Hyvinkää). To cover economic policy, the managing directors of the local Chambers of Commerce also were interviewed.

References

Bengston, D., Fletcher, J. and Nelson, K. (2004) Public policies for managing urban growth and protecting open space: policy instruments and lessons learned in the United States. *Landscape and Urban Planning*, 69(2-3): 271–286.

European Environment Agency (EEA) (2006) *Urban Sprawl in Europe – the ignored challenge*, European Environment Agency, EEA Report No 10/2006, Copenhagen.

Heikkinen, T. (2007) Paikallinen tilaus eheyttämisen lähtökohtana (Local demand as a precondition for consolidation policy). In *Yhdyskuntarakenteen eheyttämisen toimivuus ja elinympäristön laatu*. Toimintapolitiikat, mittarit ja ihmisten arjen käytännöt, Ministry of the Environment, The Finnish Environment, Helsinki.

Kasanko, M., Barredo, J., Lavalle, C., McCormick, N., Demicheli, L., Sagris, V. and Brezger, A. (2006) Are European cities becoming dispersed? A comparative analysis of 15 European urban areas. *Landscape and Urban Planning*, 77(1–2): 111–130.

Kyttä, M. and Kahila, M. (2005) The Perceived Quality Factors of the Environment and Their Ecoefficient Accessibility, in *Forests, Trees and Human Health and Wellbeing*; *Proceedings from the 1st European COST E-39 Conference* (ed. C. Th. Gallis), Medical & Scientific Publishers, Thessaloniki.

Kyttä, M. and Kahila, M. (2006) *PehmoGIS elinympäristön koetun laadun kartoittaja (SoftGIS – Localizing the Perceived Environmental Quality)*, Teknillinen korkeakoulu, Yhdyskuntasuunnittelun tutkimus- ja koulutuskeskus, julkaisu B 90, Espoo.

Ministry of the Environment (2002) *Finland's National Land Use Guidelines*; Issued by the Council of State on November 30, 2000, Ministry of the Environment, Environment Guide 93, Helsinki.

Ministry of the Environment (2004) *Yhdyskuntarakenteen ohjauksen kehittämisohjelma (Development Programme for the Guidance of the Urban Structure)*, Ministry of the Environment, Helsinki.

Ministry of the Environment (2005) *Maankäyttö- ja rakennuslain toimivuus*, Arvio laista saaduista kokemuksista (Follow-up report of the new Land-Use and Building Act), Suomen ympäristö 781, Alueiden käyttö, Helsinki. Retrieved from: http://www.ymparisto.fi/default.asp?contentid=148830&lan=fi

Nuissl, H. (2007) Policy Responses on Urban Sprawl in *Europe, in Urban Sprawl in Europe* (eds C. Couch, G. Petschel-Held and L. Leontidou), Blackwell Publishing.

Ristimäki, M., Oinonen, K., Pitkäranta, H. and Harju, K. (2003) *Kaupunkiseutujen väestömuutos ja alueellinen kasvu (Population Changes in Urban Regions and Urban Growth)*, Ministry of the Environment, The Finnish Environment – series 657, Helsinki.

Uusimaa Regional Council (2005) *UTU35. Uusimaa 2035 scenario project*, Helsinki. Retrieved from: http://www.uudenmaanliitto.fi/files/512/UTUenglanti.pdf

Uusimaa Regional Council (2006) *Uudenmaan maakuntasuunnitelma 2030*, Visio ja strategia (Regional Strategic Plan for Uusimaa 2030), Helsinki.

7

Oto Hudec and Nataša Urbančíková

Spatial Disparities based on Human and Social Capital

Introduction: the centre of the heart of Europe

The 'heart of Europe' is an often used, but ambiguous term to characterise a central place in the continent of Europe. There are several cities that could claim to be at the heart of Europe, for example, Prague, Vienna, Bratislava, Munich, Krakow, Dresden. Often the term is used (or misused) for marketing purposes, as other cities such as Paris or Brussels also make the claim, although this may come from defining the heart as an economic, historical, cultural or political core of Europe. However, even when considered from a geographical point of view, there is an ongoing discussion about the geographic centre of Europe. The candidates, depending on how it is measured or defined, range from Dresden at the western side of Germany to Lithuania, Estonia, Belarus or Romania on the Eastern side. Self-proclaimed centres, for example, lie somewhere close to the cities of Vienna and Bratislava, and with some logic claim the term Central Europe to characterise their geographical position. A new initiative, 'Centrope', provides some support for that claim, declaring itself as being strategically located at the heart of the 'New Europe'. The 'Centrope' region brings together neighbouring areas of Austria, the Czech Republic, Slovakia and Hungary (Fig. 1), arguing that it is a heterogeneous structural region

Fig. 1. The Centrope Region – spatial scheme.
Source: Hudec and Urbančíková

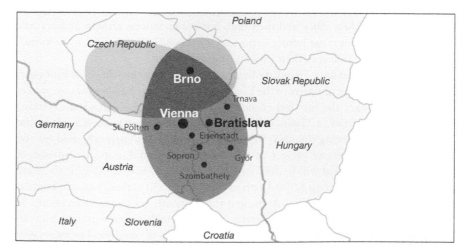

(Palme and Feldkircher, 2006; Schuh and Schuster, 2006). Although the idea seems promising at first sight, only Bratislava, Lower Austria and Vienna are prosperous parts of 'Centrope'; the other areas, Burgenland, West Transdanubia and South Moravia have a much lower economic performance (see Table 1). The first part of this chapter, within the context of the 'Centrope' region, focuses on the cities of Vienna and Bratislava, and concentrates on Bratislava, where the most interesting developments may be observed.

Centrope area	GDP per capita as a percentage of the EU 25*
Vienna	172.9
Bratislava	119.7
Lower Austria	97.5
Burgenland	81.5
West Transdanubia	59.8
South Moravia	56.9

Table. 1. GDP per capita in the Sub-Regions of Centrope (EU-25=100)
Source: Eurostat and calculations in Palme, Feldkircher (2006)

*Gross Domestic Product (GDP) per capita in Purchasing Power Standards (PPS): The volume index of GDP per capita in PPS is expressed in relation to the European Union (EU) average set equal to 100.

Bratislava and Vienna in a cross-border framework

The economic and political changes in Central and Eastern Europe in the late 1980s have changed the economic and political map of Europe. In the Danube Valley there are two capital cities within 60 kilometres of each other – the closest pair of capitals in the world. Such a short distance would immediately suggest their convergence, coexistence, and tendency to collaborate in forming a conurbation. But, although the cities have had many commonalities in their history, the 20th century did not dispose them to cooperate. Administrative, political and economic separation, mostly during the Iron Curtain period, cut off cooperation and development and prevented any formation of an integrated agglomeration of the two cities. Consequently, the picture today remains a two cities model comprising the 'Vienna–Bratislava-region'. The proximity of these cities and their high level of economic growth and activity raise questions of how cross-border issues and the revolution/evolution in the former communist Europe influence their process of development.

The Vienna–Bratislava Twin City Region, supported by the ARGE 28 European Programme, is a recent initiative of the Vienna and Bratislava chambers of commerce, aimed to interweave the life and economy of the world's two adjacent capitals.[1] Between the two cities, a 'Twin City Liner' provides a shuttle service along the Danube River, promoting the idea. The name 'Twin City' is currently used for the most expensive real estate project of all Central and East Europe, comprising a complex of buildings that will be dominated by a 42-storey tower. Already there are tangible results as more than 1,700 Austrian companies, 750 of them based in Vienna, have risked moving to Slovakia. At present this is mostly a one-way movement, as Slovak companies rarely move their activities to the neighbouring region of Vienna. This process is being encouraged by the gradual removal of barriers. The motorway connecting Vienna and Bratislava will be opened in 2007, and the two airports in Vienna and Bratislava cooperate and compete at the same time. The Slovak economy is open to foreign capital, offering low tax, investment incentives and a predictable tax policy.

A *Twin City Journal*, published by the city of Bratislava in cooperation with the city of Vienna, offers tourist information about both cities as a single destination, Sky Europe, a low-cost flight company is located and operates in the territory, and a Vienna–Bratislava TV station is in preparation. There are huge possibilities in transportation and logistics – the cooperation and potential merger of the airports; the Danube ports cooperation (Vienna's container port is overbooked and handling costs in Bratislava are half those of Vienna). Two cities and two capitals, located far from their respective countries' centres of gravity, with previously separate development caused by the Iron Curtain boundaries, are now showing genuine signs of cooperation and the possibility of joining in a polycentric cross-border vision. This cross-border polycentric framework for the Vienna–Bratislava region may be considered as a test site or laboratory of intertwining the old and new Europe.

Bratislava – transition, changeover and city re-branding

Bratislava's central location on the Danube River placed it at the crossroads of ancient trading routes and it was thus a melting pot of various cultures.[2] It has seen many changes over its long history from its emergence in 907, its status as the Hungarian capital city in 1536 and subsequently for 300 years as the coronation town for Austrian and Hungarian emperors, followed by 40 years of communist rule in the 20th century.

At the beginning of the 1990s, Bratislava was one of the least known new capital cities in Central Europe, but after becoming the capital of the newly-formed Slovak Republic in 1993, the city emerged from the shadow of Prague and Vienna. The geographical proximity to Vienna, Prague and Budapest has proved advantageous for investment.[3] The city, from the mid-1950s to the end 1990s, had one of the highest growth rates of residential, industrial and commercial areas (Uhel, 2006, Fig. 2), and traffic growth has reached an unsustainable level during the last 15 years.

EU membership in 2004 fully opened the gate to international business, producing a flow of investment, and influx of foreign managers and professionals. The relative cheapness of the market makes it highly attractive for foreign investment. There are, however, barriers in its infrastructure and services, mostly due to the heritage of the period behind the Iron Curtain as a centrally planned society and economy with inadequate, incoherent, low quality regional, city and environmental planning. Nevertheless, once the new motorway is complete in November 2007, the journey from Vienna will take less than 40 minutes. That, with the possible relaxation of the border régime, makes the position of the city very strong, enabling cross-border labour migration.

Many Slovaks have found jobs in neighbouring Austria, commuting to work every day. This is mainly true of the services, construction and agriculture sectors. The cross-border cooperation between the Vienna and Bratislava regions has meant the formal administrative, economic and transportation barriers are diminishing, although less formal yet important barriers maintain a confused image of disparate cities and neighbouring regions. Nevertheless, the prospects for a joint Vienna–

Fig. 2. The new look of the city of Bratislava Source: P. Funtal – SME

Bratislava mega-region as an economic area are alive as a vision of successful development on both sides of the border. This has resulted in Bratislava becoming more self-confident and willing to become a more equal partner to Vienna, Budapest and Prague.

At the same time as collaboration is growing, Vienna, Prague, Budapest and Bratislava are in competition because of their proximity. These, and other cities, may be considered as 'products' in the sense that they provide labour, land, premises and industrial infrastructures to businesses and at the same time offer housing, shopping, leisure and other amenities, and a social milieu to residents (Stewart, 1996; Barke and Harrop, 1994.) Where cities (as 'products') are in strong competition to attract new investments, tourists and visitors, they have tried re-branding themselves, for example, from old industrial cities to places of culture and leisure, such as Glasgow or Manchester in the United Kingdom.

From the outside, Bratislava is perceived as one of the most dynamic places in Europe, and efforts have been made to re-brand the city as a Central European place – and this is being realised through existing and planned investment. It is a city of growth and a city of the future. Cooperation with Vienna has afforded new possibilities, offering both the 'Twin City' or Centrope region as a destination. Foundations of a new Central European Region have been established. Bratislava, with its current population of 450,000, may grow to a limit of 750,000 and it will stay the economic, knowledge and administrative centre of Slovakia and one of the potential mega cities of Europe at the same time.

A polycentric vision of two cities?

Bratislava lacks a clear vision regarding Vienna and how closely the cities should converge and share common land and strategic plans. Should they form a polycentric city region, a conurbation, become twin cities or remain two cities? Should there be one region or two regions? Current cross-border relaxation is influencing the future of the territory, but still it is not clear either how to continue the process, or who will be in charge of managing it – the city or regional governments, and/or with help from chambers of commerce or academicians? Different cross-border visions lead to different strategic and land planning, with different communication and marketing plans, awareness-raising, and cross-border institutional and administrative support.

The retreat of communist ideology opened up unexpected and surprising perspectives for new development on both sides of the Austrian and Czechoslovakian (at that time) border. The extension of the European Union reinforced possibilities for a common strategy for development of the two cities of Vienna and Bratislava. Nevertheless a mental, cultural and historical border is still present and may remain for a long time. The heritage of the past Austro-Hungarian Empire and communist Czechoslovakia and of Slovak and Austrian populations living so close but in very different socio-political conditions over 40 years is hard to erase within the lifetime of one generation. Scepticism and suspicion on both sides is a very natural result of the recent history.

The application of European Union economic regulations may result in closer collaboration, as the attractiveness of the inter-border conurbation in Central Europe, connecting two large converging territories, becomes evident, as well as highly attractive for foreign investors.

A range of factors influence location investment decisions (e.g. Barrell and Pain, 1997) including supply side factors such as labour costs and corporate taxation, the skills level of the labour force; and demand factors, such as market size, growth and geographical location. Economic attractiveness is important, and three basic managerial questions are faced by the city fathers of Bratislava:

- How to manage smartly the actual and future high-speed development of the city,
- How to manage cross-border cooperation/competition aspect of development in relation to such a strong city partner/competitor, Vienna,
- How to manage the functions of Bratislava as the capital of Slovakia.

Managerial aspects of the strategic planning of the city, region, country and cross-border territory are crucial for the future. Bratislava considers itself part of the broader settlement region of western Slovakia, with close relations to the neighbouring Austrian, Czech and Hungarian territories, admittedly with accent on the city of Vienna. Both Bratislava and Vienna have an eccentric geographical position within the territory of their countries that is far from efficient for governing those respective countries. Although the economic proximity of the Vienna–Bratislava region or the even larger Central European region of Budapest, Gyor, Vienna, Trnava, Bratislava and Brno is evident, each of the cities has its natural community within its national state. Non-economic activities stay relatively independent due to language differences and, in the case of younger countries or nations, patriotism and national competitiveness.

Thus the emerging polycentric development in the Vienna–Bratislava region has many positive aspects, but has negative side effects by deflecting the centre of

Fig. 3. Images of Vienna
Source: Nikita Rogul, www. fotolia.com

economic gravity to Slovakia's geographic west. In most of the economic, social and transportation indicators, the urban region of Bratislava is drawing apart from the rest of Slovakia.

Strategic planning and regional disparities in Slovakia

The problems of regional and local development for post-communist countries were heavily influenced by the transition from central planning to a market system. The centrally planned system of resource allocation was characterized by hierarchically organized national, regional, and city planning (Sýkora 1999). Regional and city plans were strongly determined by priorities and decisions at the national level, using a top–down approach. National economic planning focused predominantly on massive industrialization in the first decades after World War II. The allocation of investment usually reflected regional requirements and centralised decision-makers often established new industrial plants in backward rural areas. This industrial enthusiasm often caused problems for the cities, their coherence and environment. When the centrally directed economic and urban planning régime was ended, the responsibility for city development and physical planning was passed to newly established municipal governments who developed systems of local self-government, with largely devolved powers given to regional and local governments.

The new local government system also created an institutional framework for local and regional planning and policy, delegating rights and responsibilities to municipalities. These included: the right to own property and exercise property rights; the collection of special taxes and fees, to manage their financial resources and formulate and promote municipal development; and the right to establish legal entities and participate in businesses. The main duties of municipal governments now include the maintenance of the local road system and public areas, public transport, water supply and sewage systems management, public order and safety, primary education and municipal housing policy, including provision of council housing.

At the beginning of the 1990s the city governments did not have enough power and maturity to protect their local and regional interests, and this has only very gradually been changing over the last decade. There is a consequent risk of global pressure influencing weak and less professional governance at all levels, from the national to the city level, to the detriment of plans for the public good. Recent laws[4] have strengthened the process of reform of public administration and strategic planning, and shifted rights and duties for development and planning to self-governing cities and regions. The independence granted to cities has led also to a shift toward strategic planning and the promotion of local economic development. The negative residuum of central planning contributes to weakly developed strategic planning skills, structures and knowledge notably at the city level.

Since the transition to a market economy, the countries of Central and Eastern Europe, including the Slovak Republic, have suffered from serious economic and employment-related problems. Market forces have tended to overwhelm the emerging strategic planning processes and the results are evidenced in many spatial disparities. The western part of Slovakia performs better economically than its eastern part. There are many reasons for regional disparities in economic development – for Slovakia the essential factors are proximity to western borders, urbanisation, diversification, quality of infrastructure, the level of human and social capital, entrepreneurial tradition, and their historic-cultural background. It is not surprising that during the transition period, capital cities and western regions have been generally much more successful, while the more eastern and rural regions are lagging behind. Regions with higher innovation and proximity to poles of growth have a very good chance to adapt to the new circumstances of the EU market.

The main influence on the economic structure and development of Slovak regional policy lies in its centre–periphery relationship. Bratislava assumed a more central position in the country after the split-up of Czechoslovakia, reducing the former dominance of Prague. There is only one serious economic competitor to Bratislava – the city of Kosice in the eastern part of Slovakia. However, its peripheral location, unskilled labour market, relative transportation inaccessibility, close proximity of its boundary with the Hungarian, Ukrainian and Polish poorest regions makes the future of Kosice, the second largest city of Slovakia, hard. Moreover, the accelerating development of Bratislava and its location close to successful foreign capitals reinforces the west of Slovakia at the expense of the east (Fig. 4). Further evidence of this dominance is seen in the unequal spatial distribution of the biggest companies in Slovakia. The economic journal *Trend* published a list of the top 200 of the biggest non-financial companies in Slovakia (*Trend*, 2006). Only 23 of the top companies are situated in Eastern Slovakia and more than one fourth – 59 companies – are located in the city of Bratislava. The situation is even worse when considering the top 20 companies: 13 are located in Bratislava, and only two in Eastern Slovakia. Considering that the top companies are leading innovation, the competitive advantage of the capital city is obvious.

While the unemployment rate is 2.13% in the Bratislava region, the average for Slovakia is 8.89% but in the south-eastern parts it is higher than 20% (in March 2007). The regional gross capital formation per capita, consisting of acquisitions less disposals of new or second-hand tangible fixed assets is four times higher than the average for Slovakia (Statistical Yearbook of the Slovak Republic, 2006). Market forces, and resulting spatial disparities, and its geography, are inexorably moving Slovakia towards a polycentric, cross border structure, which potentially leaves Slovakia fragmented.

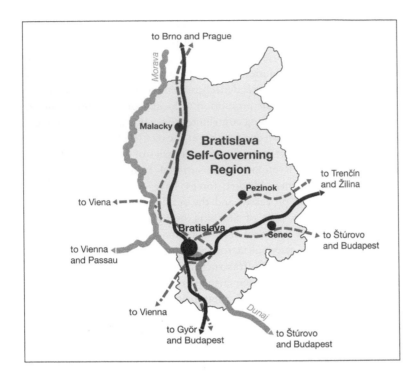

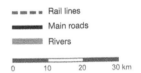

Human and Social Capital theory

An increasing interest has emerged in literature which focuses on the importance of knowledge, learning and innovation to the economic success not only of firms, but also of regions and nation-states (Lundvall, 1988; Forrant, 2001). The theory of capital differentiates physical capital (referring to physical objects), financial capital (referring to money and paper assets) human capital (referring to the properties of individuals) and social capital (referring to connections among individuals – social networks and the norms of reciprocity and trustworthiness that arise from them). The value of social capital is obvious: a society of many virtuous but isolated individuals is not necessarily rich in social capital.

The theory of social capital and human capital has assumed considerable importance in recent decades, although the variety of definitions and problems measuring their extent and quality limits the use of the concepts in practical implementation. However, human and social capital can be explained as follows: 'whereas economic capital is in people's bank accounts and human capital is inside their heads, social capital inheres in the structure of their relationships' (Portes, 1998). Some argue that theories of social capital remain underdeveloped and fragmented in specific disciplines (e.g., sociology, psychology, social psychology, political economy) (Storberg, 2002). Others argue that social capital shares commonalities with other forms of capital, notably human capital, in opposition to the opinion that social capital is the result of altruism and therefore not capital at all (Lin *et al.*, 2001; Hofferth *et al.*, 1999). The influence of education and training on the city/regional economy is also hard to uncover, yet its importance is not in doubt (Urbančíková, 2006).

The core thesis of human capital theory is that people's learning capacities are comparable to other natural resources involved in the production process; when the resource is effectively exploited the results are profitable both for the enterprise and for society as a whole (Livingstone, 1997). Economists such as Nobel Prize winners Gary Becker and Theodore Schultz established the concept in the early 1960s when they recognised the importance of including human knowledge and skills in models explaining economic development (Becker, 1964; Schultz, 1961). The most frequently used definition of human capital comes from the OECD report *The knowledge, skills, competencies and attributes embodied in individuals that facilitate the creation of personal, social and economic well-being* (OECD, 2001). The concept is intertwined with the social and collective ownership and development of knowledge, so although closely related, conceptually they are distinct and the various features need to be distinguished:

- Human capital resides in individuals,
- Social capital resides in social relations,
- Political, institutional and legal arrangements describe the rules and institutions in which human and social capital work.

The incorporation of technology, human capital and social capital into economic models as an endogenous variable (endogenous growth theory or new growth theory) opened the way to new growth models (Martin and Sunley, 1996). The importance of human capital is highlighted and it is argued that highly skilled workers tend to be more productive and innovative and are therefore of crucial importance to both companies and economies. It follows that governments have an incentive to invest in training, subsidies for research, development and innovation, and support of education; these are necessary for growth.

If there is a considerable debate and controversy about defining human and social capital in an acceptable way, this is particularly true when dealing with the possibility of measuring them. Although there is a common belief and understanding of their role in growth, comparisons of countries, regions or cities are usually based on a very small number of empirical surveys with methodological problems and uncertainties over the quality of their data.

Regional competitiveness: Slovakia and Austria – Bratislava and Slovakia

Human and social capital theory helps explain the recent developments in Slovakia, such as the position of Bratislava region in Slovakia, and the disparities between Austria and Slovakia. So what are the main factors of city, regional and country competitiveness? Most studies and literature put forward the following factors as main drivers of regional competitiveness (Sepic, 2004):

- Clusters (networks of firms, organisations, and suppliers)
- Human capital and social capital
- Enterprise environment and networks
- Innovation/Regional innovation systems
- Governance and institutional capacity
- Sectoral structure and type of enterprises
- Infrastructure (broad understanding)
- Typology of regions and level of integration of firms

- Internationalisation and nature of foreign direct investment (FDI)
- Geographical location
- Attractiveness for investments

There is not enough agreement on these rather broad definitions of factors; however, some different sources can be used as indicators when comparing Austria and Slovakia, with a focus on the Vienna–Bratislava region on the one hand and Bratislava and the rest of Slovakia on the other. Considering the factors noted above, these can be grouped into four key indicators relevant to the comparison as follows:

1. *Geographical location, attractiveness for investment and internationalisation and nature of foreign direct investment (FDI), enterprise environment and networks*

The attractiveness of the Vienna–Bratislava for investment has been highlighted in many studies and within this chapter. The region, known also as the 'Golden Triangle', together with the neighbouring Hungarian region of Gyor, has been marked out as the most promising investment region in Europe in 1993. The Vienna and Bratislava regions show a similar quality in their national frameworks – a concentration of a highly qualified labour force, knowledge, research and development, high tech companies and services. Austria and Slovakia also show a number of similar results when their enterprise environment is measured (World Bank, 2006; Table 2). In a number of surveys Vienna and Bratislava as regions are considered more attractive than the cities themselves.

Country	Ease of doing business rank	Starting a business	Dealing with licenses	Employing workers	Registering property	Getting credit
Austria	30	74	50	103	28	21
Slovakia	36	63	47	72	5	13

Table. 2. Ease of doing Business Ranking
Source: World Bank (2006)

2. *Infrastructure and sectoral structure and type of enterprises and typology of regions*

The European Competitiveness Index, benchmarking 118 regions of Europe, finds that only three regions within the European Union's new member states are more competitive than the European average. Prague enters the regional rankings in 7th position and Bratislava in 10th (European Competitiveness Index, 2006). The Infrastructure Index can be used as a measure of infrastructure development, combining three quantities: total length of roads in a country compared with the expected length of roads; the percentage of the population with access to improved sanitation facilities; and the percentage of the population with access to improved drinking water. An index value of '0' represents the country with the least infrastructure while the value of '100' is for the country with the highest infrastructure provision. The last survey in 2001 shows values of 80.61 for Austria and 72.71 for Slovakia.

The Vienna region has a well elaborated highway system, high-speed trains, and Vienna International Airport in Schwechat. Still, there are some infrastructural deficits in the transport connection to the neighbouring Czech Republic and Slovakia. The situation of Bratislava is worse; the future development of the Bratislava airport is still not clear and the highway system in Slovakia is underdeveloped, although proximity to Vienna makes the infrastructural problems of Bratislava less problematic. Despite these disparities, the best sectoral areas for cooperation include: information and communication technologies; the automotive industry; biotechnology and environmental technologies; and the creative industries.

3. Innovation/regional innovation systems and level of integration of firms and clusters (considered as geographic concentration of firms and supporting organisations that trust one another and exchange knowledge)

Based on a database produced by the Austrian Business Agency, seven clusters located in the Vienna region, including a strong automotive cluster can be identified. The special feature of the clusters in the Vienna Region is an effort at cooperation beyond national borders, including with the Slovak Republic. In Slovakia, no known clusters exist, but with the highest concentration of multinational automotive firms in Central Europe, cross-border cluster formation looks promising. Both regions have developed regional innovation systems involving the interaction of a range of actors and resources. The Austrian system of innovation is much more developed. At the same time, the Vienna and Bratislava regions are working much better than the average for their respective countries. For Slovakia, some figures are noteworthy: 54.8% of the R&D workers of Slovakia are located in the Bratislava region (45% of Austrian R&D workers in Vienna). However, there is a difference in the distribution of R&D researchers between the countries, as in Austria 66.3% work in the business sector and only 19.9% do so in Slovakia. This trend results in a lower application of R&D and innovation knowledge in Slovakia than in Austria, and the same is true of the Bratislava region and the rest of Slovakia (Table 3). Existing and intensified cross/border R&D cooperation is expected to improve the positioning and competitiveness of the region within Europe.

Table. 3. Regional Innovation performance

Source: The European Innovation Scoreboard (2006)

Country/Region	Regional Innovation Performance 2005
Austria	0.51
Slovakia	0.26
Vienna Region (AT)	0.68
Burgenland (AT)	0.29
Bratislava Region (SK)	0.66
Eastern Slovakia (SK)	0.19

4. Human and social capital, and governance and institutional capacity

How far does the level and quality of human and social capital affect the spatial disparities in Slovakia, Bratislava and its Region? Human and social capital has been measured by indicators such as: civic participation (e.g. membership in voluntary organizations, used by by Putman (2001): trust (the most commonly used empirical measure of social capital, popularised by Fukuyama (1997): density of networks (density of ties between individuals), and philanthropic generosity (i.e. altruism).

The Eurobarometer surveys commissioned by the European Commission and carried out by Gallup Europe (2006) and other sources are used to show the development of the social capital (Eurobarometer, 2004; Fidrmuc and Gërxhani, 2005). Regarding civic participation, turnout figures for European parliamentary elections ranged from as low as 17% in Slovakia to 42% in Austria; the average rate was 45.5% in the EU25, with marked variations between member states. In Austria 33% of people trust the political institutions, whereas Slovak political institutions are only trusted by 15% of Slovaks, reflected in low voter turnouts (calculated as the average of trust in three political institutions – national government, parliament and political parties). For example, in the Slovakian municipal elections in 2006 turnouts for Bratislava were 32.75%, and in Kosice, 26.54% (Statistical Yearbook, 2006). In fact the majority are dissatisfied with the way democracy works in Slovakia (80% dissatisfied), in Austria, 64% were satisfied.

Civic participation, trust, density of networks and philanthropic generosity are very low, as an inheritance of communist central decision-making. Local Slovak society either tends to accept decisions made at the national level, or attempts to solve social problems by migration to work in Prague, Vienna, or the USA, as well as by a flow of skilled labour to Bratislava. It can be seen that the east of Slovakia is typical in its low civic participation, low level of social capital and of paternalism (Krivý, 2000; see Fig. 5).

Human capital in the Central and Eastern European regions has been measured using an indicator based on formal education and qualifications (Schmid and Hafner, 2005). The results of these formal human capital indicators for Austria and Slovakia are *de facto* the same and the regional differences within the countries are not large. Regional differences in human capital in its broader definition follow from the concentration of universities, knowledge workers, financial and governmental institutions, large firms, and media in capital regions. Migration from less developed regions implies regional human and social capital disparities.

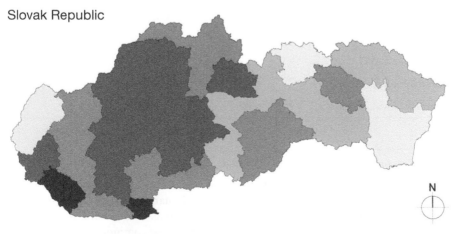

Slovak Republic

Fig. 5. Election to the Parliament 2006, turnout in Slovak regions

Source: based on Statistical Office of the Slovak Republic, 2006

Voter turnout in %

45,40 - 49,37
49,38 - 52,77
52,78 - 55,76
55,77 - 59,33
59,34 - 62,62

Table. 4. Human, Social Capital and Governance indicators
Source: Lattin and Young, 2005

World comparisons of human and social capital and governance, which are related, can be found in the paper of Lattin and Young (2005). Social capital achievement (SCA) is an aggregate of three subsystems: economic variables, social/cultural variables, and legal and political institutions. In Table 4 the results for Austria and Slovakia are shown, and the particular measures of human capital and governance.

Country	SCA	PPP	HD	FREE	EF	S&P	VA	GE	RQ	RL	CC	CP
Austria	0.93	0.91	0.91	0.99	0.87	0.99	0.89	0.92	0.95	0.96	0.93	0.91
Slovakia	0.71	0.77	0.76	0.99	0.58	0.64	0.76	0.67	0.73	0.65	0.65	0.61

Note: PPP – Gross Domestic Product Per Capita in Purchasing Power Parity: HD – United Nations Development Programme's Human Development Index: FREE – Freedom House's Freedom Index: EF – Fraser Institute's Economic Freedom of the World: S&P – Standard & Poor Foreign Currency Issuer Ratings: VA – World Bank Governance Indicator – Voice and Accountability: GE – World Bank Governance Indicator – Government Effectiveness: RQ – World Bank Governance Indicator – Regulatory Quality: RL – World Bank Governance Indicator – Rule of Law: CC – World Bank Governance Indicator – Control of Corruption: CP – Transparency International's Corruption Perceptions Index

Social capital has been measured in other studies (e.g. Van Schaik, 2002), and Svendsen (2003) published the following values related to social capital:

Table. 5. Social Capital Indicators
Source: based on Van Schaik, 2002, and Svendsen, 2003

Country	General Trust Index	Civic participation Population
Austria	31.82	5.97
Slovakia	17.38	3.38

There is a high correlation between the level of social capital and regional development. Generally both values are lower for the former communist countries, and the significant differences between Austria and Slovakia explain the different levels of economic development.

Conclusion: a Vienna–Bratislava mega region?

A common and agreed vision for the two cities of Vienna and Bratislava has not yet been fully realised, but the trend appears to be moving in the direction of a well-connected polycentric city region. This is particularly true if the definition of polycentric development, as a spatial and functional form of development in which there are many centres and not just one large city that is dominating all the others is used (Hague and Kirk, 2003). Both cities believe there is a benefit from regional polycentric development, and this is highlighted in a number of programmes, documents and initiatives. Polycentrism is expected to increase competitiveness of the region, and enhance infrastructure and access to knowledge. The main attractive driving forces of the polycentric vision are the economic interests of both cities, expected prosperity of the region and its European centrality, and common membership of the EU.

The economic differences can be a positive aspect when looking for a balanced development, supporting exchange of capital, goods and labour force. Nevertheless, there are strong non-economic barriers resulting from different histories, languages, population, attitudes and political culture. These barriers may be strong enough to keep both cities at a distance for at least a generation. Vienna and Bratislava lived for many decades in relative isolation and these still exist in relics of the transition from communism to capitalism, mutual mistrust, communication problems, mis-understandings, and intercultural differences.

Willingness to cooperate in urban planning and implementation of an effective strategic plan is important. There are many plans and links, at different levels, but although there exist a number of strategic documents, their implementation is still underdeveloped in Slovakia. Given these initiatives, however, Vienna assuming a dominant central position in a monocentric region is now less probable. That makes the vision of two very close but autonomous regional centres of a polycentric region more likely.

The polycentric vision as a spatial and functional form of development with two centres is thus on the way to being realised for the Vienna and Bratislava region; the positive externalities of the polycentric process are vivid, pulling institutions and people to work together and utilise the advantages offered by a cross-border region. The door is open for polycentrism, and economic development will certainly bring close cooperation and ideas for a common future.

Is there another chance of polycentric mega cities or functional areas in Europe crossing the former West–East European borders? According to ESPON (2005), there exist the following potentially polycentric transnational areas:

- Vienna/Bratislava/Györ Area
- Copenhagen/Malmö Area and the Øresund region
- Krakow/Katowice/Ostrava Area
- Lyon/Grenoble/Geneva/Lausanne Area

But within these mega regions there is no case as special as Vienna–Bratislava, fighting relics of historical development in the 20th century. Economic, spatial, transportation and strategic planning has started and it is bringing success. Possibly with Györ in the future, there is a chance to have three countries within one functional polycentric mega region in the centre of Europe.

Notes

1. ARGE 28 was part of an EU co-funded programme in 2001/2002 about 'Border regions: Initiatives on the Impact of Enlargement in the EU'. It was an SME project 'Growing Together with Europe' involving a network of 28 Chambers of Commerce located in EU border regions.
2. The city was also known as Pressburg, and Poszony, but has been known only as Bratislava since 1919.
3. Slovakia, driven by its capital Bratislava, achieved the highest economic growth in Europe among members of the OECD in the 4th quarter of 2006 (OECD, Quarterly National Accounts database).
4. In particular Act No. 416/2002 Coll. on the Transmission of Certain Competencies of State Administration Authorities to Municipalities and Higher Territorial Unit.

References

Barke, M. and Harrop, K. (1994) Selling the Industrial Town: Identity, Image and Illusion, in *Place Promotion: The Use of Publicity and Marketing to Sell Towns and Regions*, J. Gold and S. Ward (eds), Wiley, Chichester, pp, 79–94.

Barrell, R. and Pain, N. (1997) The growth of foreign direct investment in Europe. *National Institute Economic Review*, 160: 63–75.

Becker, G. (1964) *Human Capital: A Theoretical and Empirical Analysis, with Special Reference to Education*, National Bureau of Economic Research, New York.

ESPON (2005) *Potentials for polycentric development in Europe*, Planning Practice and Research, Report 1.1.1, 20.

Eurobarometer 61 (2004) *The European Union, Globalization, and the European Parliament*. Retrieved from: http://ec.europa.eu/public_opinion/archives/eb/eb61/eb61_en.htm

European Commission (2006) *The European Innovation Scoreboard 2006*, European Commission. Retrieved from: http://www.proinno-europe.eu/doc/EIS2006_final.pdf

European Competitiveness Index (2006) Robert Huggins Business & Economic Policy Press.

Fidrmuc J. and Gërxhani K. (2005) *Formation of social capital in Central and Eastern Europe: Understanding the gap vis-à-vis developed countries*, William Davidson Institute Working Papers Series, No. wp766.

Forrant, R. (2001) Pulling Together in Lowell: The University and the Regional Development Process. *European Planning Studies*, 9(5): 613–628.

Fukuyama, F. (1997) Social Capital and the Global Economy. *Foreign Affairs*, 74(5):89–103.

Gallup Europe (2006) *Innobarometer on Cluster's Role in Facilitating Innovation in Europe*, The Gallup Organization Hungary & Gallup Europe.

Hague, C. and Kirk, K. (2003) *Polycentricity Scoping Study*, Heriot-Watt University, School of the Built Environment, Edinburgh.

Hofferth, S. L., Boisjoly J. and Duncan G. (1999) The Development of Social Capital. *Rationality and Society*, 11: 79–110.

Krivý V. (2000) *Politické orientácie na Slovensku – skupinové profily*. Bratislava, IVO – Inštitút pre verejné otázky, p. 102.

Lattin R. and Young S. (2005) *Country Ranking: Social Capital Achievement*. Retrieved from: http://www.cauxroundtable.org/PDF/SocialCapitalAchievement.pdf

Lin, N., Cook K.S. and Burt R.S. (2001) *Social capital: theory and research*, Aldine de Gruyter, New York.

Livingstone D. (1997) The Limits of Human Capital Theory: Expanding Knowledge, Informal Learning and Underemployment, *Policy Options*, 18(6): 9–13.

Lundvall, B-Å. (1998) Innovation as an interactive process: from user–producer interaction to the national systems of innovation, in G. Dosi *et al.*, *Technical Change and Economic Theory*, Frances Pinter, London.

Martin, R. and Sunley, P. (1996) Paul Krugman's Geographical Economics and its Implications for Regional Development Theory: A Critical Assessmen. *Economic Geography*, 72(3): 259–92.

OECD (2001) *The Well-Being of Nations: the Role of Human and Social Capital*, OECD, Paris.

Palme G. and Feldkircher M. (2006) *Wirtschaftsregion 'CENTROPE Europaregion Mitte': Eine Bestandsaufnahme*, 2005 Study of WIFO-Österr. Inst. für Wirtschaftsforschung, p. 84.

Putman, R. D. (2001) *Bowling Alone*, Simon and Schuster, New York.

Portes, A. (1998) Social Capital: its Origins and Applications in Modern Sociology. *Annual Review of Sociology*, 24: 1–24.

Regional Operational Programme Bratislava Region (2006) Bratislava Self-governing Region.

Schmid K, and Hafner H. (2005) *Human Capital in the Central and Eastern European Countries. International Benchmarking on the Basis of the ibw's Human Resources Indicator*. Ibw Research Brief No 16.

Schuh E. and Schuster P. (ed.) (2006) *New Regional Economics in Central European Economies: The Future of CENTROPE*, Oesterreichische Nationalbank.

Schultz T. W. (1961) Investment in Human Capital. *American Economic Review*, LI, 1–17.

Sepic D. (2004) *The Regional Competitiveness: Some Notions*. RECEP'S Reports No 4.

Statistical Office of the Slovak Republic (2006). Retrieved from: http://www.statistics.sk/nrsr_2006/angl/obvod/results/kart1.jsp

Statistical Yearbook of the Slovak Republic (2006) Bratislava, VEDA.

Stewart, M. (1996) Competition and Competitiveness in Urban Policy. *Public Money and Management*, **16**(3): 21–27.

Storberg, J. (2002) The Evolution of Capital Theory: A Critique of a Theory of Social Capital and Implications for HRD. *Human Resource Development Review*, **1**(4): 468–499.

Svendsen G.T.(2003) *Social Capital, Corruption and Economic Growth: Eastern and Western Europe.* Working Papers 03-21, Aarhus School of Business.

Sýkora, L. (1999) Local and Regional Planning and Policy in East Central European Transitional Countries. In: Hampl *et al.*, *Geography of Societal Transformation in the Czech Republic*, pp.153–179. Prague, Charles University.

Trend (2006) *Top 200, ,Trend Holding, Bratislava,* November 2006, pp. 47–8

Urbančíková, N. (2006) *L'udské zdroje v regionalnom rozvoji (Human Resources in Regional Develoment)*, Kosice.

Uhel R. (2006) *Urban Sprawl in Europe: The Ignored Challenge*, European Environment Agency, Copenhagen, p. 60.

Van Schaik, T. (2002) *Social Capital in the European Values Study Surveys (EVS)*: Paper for the OECD-ONS International Conference on Social Capital Measurement, London, September 25–27.

World Bank (2006) *Doing Business in 2007: Creating jobs*, World Bank and International Finance Corporation, Washington.

8

Jaume Carné and Aleksandar Ivančić

The Barcelona Model:
1979–2004 and beyond

Introduction: Barcelona as a 'world class' city

Barcelona is well known as a successful urban model of the compact, Mediterranean city. It generates new ideas and allows for the reinterpretation of traditional ones already well established in the existing urban pattern of the city. After the 1992 Olympics, Barcelona finally joined the club of prestigious 'world class' cities. During those years, the popular expression was that the event 'has set the city on the world map'. This chapter reviews the last 30 years of urban development of the city, starting in the late 1970s with the first democratic local government, and analyzes the process up to the present day and beyond. What is important to note is that the 'Barcelona Model'[1] is in a continuous process of improvement, learning from the implementation of specific projects and the accumulation of experience over the years.

Despite having a small population (1.5 million habitants in 2005) and a small total area (98 km^2), it has a population density worthy of an Asian city (15,000 persons per km^2 and reaching 35,000 persons per km^2 in the central neighbourhood of the Eixample). Barcelona is the capital, and dominant city, of an important network of metropolitan cities, a Euro-region of more than 4 million inhabitants.

The city is well-known around the world because it is a principal city on the Mediterranean, a circumstance that implies good weather, hospitality and gastronomy. Its more than 2,000 years of history also provides an interesting cultural background, from Gothic to Modernism. Furthermore, its role as a meeting point at a local and global level has meant that the city has offered a very high standard of services and shopping. Within this framework, the range of possibilities for tourism and leisure activities is increasing strongly.

Barcelona, mainly an industrialised city during the 19th and 20th centuries, is rapidly changing: practically two thirds of its exports today are high or medium-high technology goods, and it has become a city of services. The increase in tourism (in 2007, more than 14 million overnight stays per year, with a constant 10% annual increase since 1990) helps other developing sectors to firmly establish themselves. The most highly developed sectors are those related to design, such as fashion and architecture, and those related to knowledge, such as the academic world, universities and new technologies (Ajuntament de Barcelona, 2006).

However, Barcelona, like other European cities, faces an important challenge: to remain competitive in the 'new geographies' of a more and more globalized world. It is making great efforts to improve its attractiveness for foreign and

Data		City	Metropolitan Area	Metropolitan Region
Area	km²	100	633	3,240
Population	million	1.6	3.1	4.2
Population density	inh/km²	~16.000	4,900	1,296

Indicator	variation per year	period	referent data (2005)	
Foreign population	81.9%	1996–2005	15.9	%
Jobs (soc.sec. affiliation)	3.1%	1998–2005	904,935	
GDP (constant 2000)	2.5%	2001–2004		
Cars/capita	-1.2%	2001–2005	387.5	/1000 inhab.
University students	1.3%	2001–2005	188,170	
Broadband internet connected homes	45.6%	2004–2005	26.2	%
Cellular phones	5.2%	2004–2005	84.4	%
Visitors	22.5%	2001–2006	7,180,000	(2006)
Overnight stays in hotels	15.1%	2001–2006	14,000,000	(2006)
Fairs/exibitions	6.9%	1992–2006		
Airport passangers	11.9%	1997–2006	30,000,000	(2006)
Cruises passangers	64.0%	1992–2005	1,230,000	
Congress/conventions assist.	26.0%	2001–2005		
Solid waste total	0.0%	2001–2005	1.51	kg/capita, day
Solid waste separated	36.5%	2001–2005	0.64	kg/capita, day
Water consumption /domestic	-2.2%	2001–2005	122	l/capita, day
Final energy consumption	2.1%	1999–2004	11.6	MWh/capita, year
Power generated within the city	93.2%	1999–2004		
Energy intensity	0.7%	1999–2004		
Solar thermal capacity	92.0%	1999–2004		
Solar PV capacity	107.0%	1999–2004		
Biogas capacity	151.0%	1999–2004		
GHG emmisions	0.7%	1999–2004		

Table 1: Significant indicators for the City of Barcelona
Source: Based on data from: Dept. d'Estadistica, Barcelona City Council; IDESCAT; Area Metropolitana de Barcelona; Agencia d'energia de Barcelona

national investment, which may bring in new technology-based industry and know-how. Nevertheless, the market trends point to tourism and construction/real estate activities as the main or emerging drivers of the local economy. This present positive economic trend may have the hidden danger of turning the city into a 'tourist resort' or retirement residence for the elderly of Europe, simply becoming the 'European Florida', leading to an undesirable lack of diversity, or an enforced monoculture.

One of the main characteristics of the Barcelona Model is the reinvention, or 'recycling', of existing space – a model opposite to that of urban sprawl. The heritage of the traditional compact and complex Mediterranean has constantly been reinterpreted within new, contemporary, social, economic and cultural parameters. This contemporary urban model needs to integrate new modern infrastructure into the urban space in order to support today's demands for quality and to be a backbone for good services. It is only in this way that regenerated urban space will remain efficient – architectural intervention on its own is simply not enough. Furthermore, sustainability and environmental requirements are becoming more and more important, not only as general concepts but on the daily agenda of the planning processes and their implementation in Barcelona. The challenge of sustainable growth for cities rests, among other factors, on both infrastructure development and environmental preservation.

The present day transformation of Barcelona cannot be described just as a physical urban transformation but as a conceptual or cultural transformation. There are obvious physical components bringing profound changes: industrial relocation that leaves behind obsolete space which is often unused; the gradual transformation into new residential/service sector neighbourhoods, never in a single-use way but by always striving to find the right mixture of uses for each territory. This transformation also requires new criteria including: the mobility of people and goods; the introduction of new service networks such as optical fibre or district heating and cooling (DHC) systems; a concern for environmental effects; and, the effort required to improve air quality. But above all, efforts have to be made, both to attain social cohesion in the face of a rapidly changing social environment, and to elaborate a city model that stimulates knowledge.

All of this means maintaining the necessary dynamics of a compact, complex, accessible and creative city; a future Barcelona as a 'liveable city'. It is evident that for the type of changes that Barcelona proposes, what is needed is not only political but intellectual leadership as well.

The evolution of the 'Barcelona Model'

Clearly recognisable techniques of planning intervention were adapted to meet the needs of the moment. These included: 1) 'acupuncture' as a precise intervention with spin-off effects; 2) the organisation of the metropolitan centre; 3) the urban transformation; and 4) the metropolitan region's structuring and cohesion. The above techniques have gradually been incorporated into urban strategies and any new ones have simply been overlaid on previous ones, without any one of them becoming obsolete. In short, intervention techniques used on the city have been enriched.

Fig. 1. Plaça Reial: One of the first interventions of so-called 'urban acupuncture'. In the mid-1980s this emblematic neoclassical plaza was re-urbanized including new pavement, streetlights, urban furniture and landscaping
Source: Aleksandar Ivančić

Regaining public pride in Barcelona: *Re-equipping and rebuilding the city or 'urban acupuncture' (1979–87)*

The city had been left to its fate after 40 years of dictatorship.[2] During the 1960s and 1970s the city experienced massive growth without rigorous planning. The first actions of the new democratic municipal government, elected in the very first democratic local elections in 1979, aimed to provide new spaces for the citizens: squares, parks and amenities in all parts of the city.

The first elected mayor, Narcís Serra (socialist) requested Oriol Bohigas,[3] a prestigious local architect seriously involved in the political/cultural life of the city,

to take charge of the Town Planning Department of Barcelona. Bohigas gave up his post as head of the Faculty of Architecture of Barcelona University and together with Josep Acebillo set up the Municipal Department of Urban Projects at the heart of the ancient municipal bureaucracy. Hand in hand with a few young architects he instigated changes to the municipal administration's decision-making processes on investment in public spaces and facilities. From these beginnings, the architectural quality of the urbanisation projects has been acknowledged both by numerous national and international awards, and by the way in which the real demands of the citizens have been fulfilled.[4]

It is worth mentioning two key decisions taken by the City Council. The first was in 1984 at the beginning of the second term of office of the democratic era. The municipal administration was reorganised into 10 municipal districts. The aim was to decentralise civil government and transform Barcelona into a 'city of cities', each district being self-governed, like a city of 100,000 to 150,000 inhabitants (Ajuntament de Barcelona, 1999, pp. 117–133).

The second defined the way in which the City Council would act in the creation of public companies to be commissioned to solve specific problems. *Promoció de Ciutat Vella SA* (PROCIVESA) was set up in the district of 'Ciutat Vella', which took in the area of the medieval town and the seafaring quarter of Baceloneta (begun in the 18th century). The company was founded in 1988, at the start of the third term of office of the local council, with the aim of solving the problems of degradation of the oldest part of the city. Over time, the company, and its successor 'Foment de Ciutat Vella', transformed 'La Rambla del Raval', the 'Mercat de Santa Caterina' and surrounding area, and initiated a recent project in 'La Plaza de Gardunya'.[5]

In the first stage of 'urban acupuncture' there were those who thought that the city had to go even further. In the mid-1980s work began on the Olympic project which bore fruit when Barcelona was elected to hold the 1992 games.

Setting the city on the map: *Forming the metropolitan centre by way of the construction of the Olympic infrastructures (1987–1992)*

The nomination of the city of Barcelona (in 1986) to hold the Olympic Games in the summer of 1992 was a challenge for the new-born sense of pride of Barcelona citizens.

Fig. 2. Rambla del Raval: More complex urban intervention methods in the mid-1990s – poor condition blocks demolished making way for the construction of the public space that became the landmark of the collective identity of the neighbourhood

Source: Aleksandar Ivančić

It was agreed that the best impression of the city should be given for the event. The City Council built the sports facilities for the Olympic competition (Olympic Ring on the Montjuïc Hill, and other sports facilities in Diagonal and Vall d'Hebron areas), a new residential neighbourhood was constructed in a run down area of the city (Olympic Vila, widely recognized as an exemplar of urban redevelopment), not to mention the new ring road infrastructure for inner metropolitan connections (Ronda Litoral and Ronda de Dalt).[6]

The two objectives were clear: First, as Mayor Pasqual Maragall pointed out, 'the city had to be put on the map'. The success of the games, remembered as one of the best ever, served as a launching pad for the city in the international community. Second, the objective of greatest value to the citizens was transformation of the city, modernizing it so that it could act as an efficient centre for the metropolitan conurbation.[7]

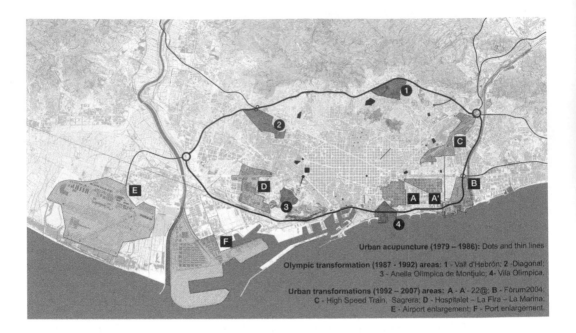

Urban acupuncture (1979 – 1986): Dots and thin lines

Olympic transformation (1987 - 1992) areas: **1** - Vall d'Hebron; **2** -Diagonal;
3 - Anella Olímpica de Montjuïc; **4**- Vila Olímpica.

Urban transformations (1992 – 2007) areas: **A** - **A'**- 22@; **B** - Fòrum2004;
C - High Speed Train, Sagrera; **D** - Hospitalet – La Fira – La Marina;
E - Airport enlargement; **F** - Port enlargement.

Fig. 5. Concept of urban interventions in Barcelona: growing in scale Drawing: G. Delbene

The Olympic investment was the excuse to transform the city. Many other cities that have been the seat of the Olympic Games have invested heavily in sports facilities, with varying degrees of positive impact on the city concerned. In Barcelona the key was the construction of the ring roads that were deemed to be an essential factor for the rapid connection between the four Olympic areas. Indeed the 'Rondas' were the key to structuring and modernizing the central core of metropolitan Barcelona.

Four Olympic areas were chosen. Firstly, the Olympic Ring in Montjuic was a choice made by the City Council to explain past historical events; the Montjuic Olympic Stadium was originally constructed to present the city's candidature for the 1936 Olympic Games. The stadium, which was practically in ruins, was remodelled and a sports pavilion was added (Palau San Jordi), an outdoor swimming pool (the Picornell) was also remodelled, as was a diving pool in Poble Sec. The site for the gymnastic events was constructed and is today the centre of the National Institute for Physical Education. Secondly, the 'Diagonal site' was an obvious choice. Sports facilities in this area were already in existence, the Barcelona Football Club being of particular importance because it was the sports facility with the city's greatest seating capacity.

However, the choice of the other two areas was not so obvious. Vall d'Hebron only had a cycle track that had been constructed in 1984, a short time beforehand, for the world cycling championships. In this area one can sense the underlying logic behind the Olympic gamble; to take advantage of the event to transform the city. In this particular instance, an area of the city that had been developed in the 1960s and 1970s as a disjointed collection of housing estates was reorganised. The Vall d'Hebron Park, designed by Eduard Bru, was given a number of facilities: a sports pavilion, a rifle range (today reconverted into football pitches), a tennis club, and a residence for journalists. The fourth and final area, the Olympic Village, designed by Oriol Bohigas' studio (MBM), was where the most novel aspects were introduced.

The scheme offered not only a residence for athletes during the Olympic Games, but also the regeneration of an obsolete industrial area, including the waterfront. The result was a neighbourhood including 2,000 dwellings, a pleasure harbour (used for yachting events during the 1992 Games), and two skyscrapers (a hotel and office premises). It was the first project designed to transform an urban area while opening up the city to the sea. It represented the beginning of the transformation of an obsolete industrial area close to the heart of the city and thus of crucial importance to Barcelona. Subsequently, this regeneration would become one of the most powerful initiatives in the city.

Staying there: *a Metropolitan Capital of Southern Europe (1992– 2004)*

Cities compete with each other to achieve the best possible ranking in the club of 'world class' cities. For this reason the next step was to create the elements to compete in this global context. As the metropolitan centre had already been reorganised, the urban strategy focused on the transformation of obsolete urban fabric.

The 22@ Project for the transformation of Poble Nou

Poble Nou covers practically the entire eastern quadrant of the Eixample, which was the plan for the extension of medieval Barcelona, designed in 1859 by the civil engineer Ildelfons Cerdà. During the second half of the 19th century and the first half of the 20th century it was precisely in this quadrant that the majority of industrial complexes in the city were set up. The type of manufacturing corresponded to that of the first industrial revolution. Poble Nou specialized in textile manufacturing and the metallurgical industry. From the last quarter of the 20th century onwards, this industrial fabric became obsolete.

Poble Nou was characterized by three important features: firstly, by the imprint of the street layout, comprising the typical grid-iron pattern suggested by Cerdà of 113m × 113m street blocks with canted corners, forming an octagonal square at every street crossing; secondly by its size, not only in total surface area (the 22@ Project covers almost 200 hectares) but also in its position and proportion with respect to the whole of the city of Barcelona (Fig. 5, area A-A'); and, thirdly, by the challenge of reusing an abandoned industrial heritage made up of important works of industrial architecture.

In the current General Metropolitan Plan (PGM) of Barcelona, originating from 1976 (Area Metropolitana de Barcelona, 2005), the key '22a' identifies those areas considered to be industrial zones. This is where the name 22@ comes from. The 22@ Project[8] aimed to set up a new régime to transform land that was designated for industrial use into one of mixed use (industrial, services, residential and commercial) while moving away from the strict specialization that the '22a' classification proposed in the PGM. The proposed change highlights the introduction of activities related to the knowledge economy: telecommunications, information, education and culture. It is worth stressing that the 22@ plan respects the power of Cerdà's street layout and the general rules concerning land use and intensity.

A summary of the data gives an idea of the magnitude of specific modifications to the PGM passed in 2000.[9] The development covered a total of 198 hectares, corresponding to 115 street blocks in the Eixample. This means that the transformation, involving either new construction or the renovation of what exists at present, totals up to 3,200,000m², of which 2,700,000 were designated for economic activities. The total number of dwellings is close to 8,000, of which 3,500–4,000 are new construction with special attention paid to public housing development. Finally, the total surface area dedicated to land transfer for public facilities and green spaces comes to 200,000m². Together with the PGM modification for the 22@ area, a parallel Special Infrastructures Plan (PEI) was designed to set the basis for the investment to be carried out in public spaces and urban development of the streets and public green spaces. The PEI is a flagship for innovative planning in Spain.[10] It introduces up-to-date criteria and formulates specific proposals for traditional public service networks (e.g. power grid, cleaning, water supply) and also introduces some new networks (e.g. district heating and cooling (DHC), groundwater and the latest generation in communications). For the smooth running of these new activities the PEI suggests the following:

- A model of urban planning, overlaying the existing Cerdà layout, to structure the use of street space.
- Improvement of the infrastructures of internal and external mobility, road networks, public transport, and car parking, and especially the extension of the metro system and the introduction of trams, vital given the location of commercial premises.
- An integrated overview of infrastructure at both ground and underground level and building rooflines
- Regulation of rights of way of infrastructure on private property and public space, allowing the networks optimization and continuity.

As for the infrastructure, a more innovative plan has been designed for:

- The energy system (power network with fall-back systems, integration of renewable energy installations, district heating and cooling, profiting from the generation of heat in the urban solid waste treatment plant)
- Communication infrastructures (cabling the sector, dark fibre,[11] mobile telecommunication infrastructures).
- Infrastructure related to the automated waste collection with a separation system for selected refuse collection and waste recovery.

To manage a project of this magnitude, Barcelona City Council, in addition to its administrative and managerial instruments, created a limited liability company

*Fig. 7. Glorias
Square and
a part of
22@ district;
aerial view
photomontage
Source: courtesy
of 22@Barcelona*

Fig. 7. Glorias Square and a part of 22@ district; aerial view photomontage Source: courtesy of 22@Barcelona

'22@bcn S.A.' wholly owned by the council. The creation of this company has meant that more than half of the retail projects foreseen in the global project are already in the development stage.

It is still too soon to evaluate the results of a plan whose development is planned over a 10- to 15-year period. 22@ has stimulated a rapid and predictable growth in public investment and in the provision of facilities (e.g. building the new departments for the Pompeu Fabra University for New Technologies for Image and Information Studies). The project has also had an invigorating response from the private sector (e.g. the administrative district 22@ has the fastest increase in real-estate developments in the city).

The Waterfront – River Besòs and FORUM

The next area selected to undergo major reform was the mouth of the River Besòs. An area measuring more than 200 hectares at the north-eastern end of the city between Barcelona and Sant Adrià de Besòs was to be transformed from a highly deteriorated, marginal area to be one of the outstanding centres of the city's development (Fig. 5, area B). It was chosen by the Barcelona City Council as the principal venue for the celebration of the *2004 Barcelona Universal Forum of Cultures*, an international event focused around three main subjects: cultural diversity, sustainable development and conditions for peace.[12]

The urban transformation had six fundamental development objectives:

- Restore the coastline (parks, facilities, marina, areas for bathing and areas for coastal ecosystems).
- Integrate pre-existing technical installations.
- Break down barriers to gain access to the shoreline and create a climate of interchange between the city and the sea.
- Create a new central urban area (including a Convention Centre (15,000 people), Auditorium (3,000 people), hotels, university campus and offices).

- Develop new residential areas and renovate existing ones.[13]
- Bring about the environmental recovery both of the land and the marine environments (new energy criteria, restoration of the river and the marine biotope).

The Besòs coastline accommodates the technical plant essential for the day to day running of the city and its metropolitan area: sewage treatment plant, solid waste treatment plant and the Besòs power plant complex.

Several schemes were included in the rehabilitation and redevelopment project of the Forum 2004 area to improve environmental parameters. All of the infrastructure and technical plants were maintained and none have been relocated outside the area which was to be transformed into a new focal point of the city. Thus the Forum project implements a fundamental concept: the acceptance of typically undesirable, but necessary, infrastructures for the running of the city. As observed by Ramon Folch, it is a process similar to the integration of toilets inside dwellings at the end of the 19th century, but now on an urban scale. It seems obvious that a city that strives towards sustainability cannot simply go on externalizing its effects on the environment by dumping undesirable functions in areas further and further away as the city grows.

Remodelling and expansion of the Besòs sewage treatment plant was complex due to its location, limited space and tight building schedule at the time. This was further aggravated by the need to undergo works without halting sewage treatment, and without the possible reduction in treatment levels during the most critical stages of the work, which would have affected the quality of the water in bathing areas. The new installations are compact, covered, and use powerful treatment systems for obnoxious smells and gases. The total area of the sewage treatment complex is some 11.5 hectares with the capacity of $600,000m^3$, the equivalent to serving a population of 3,250,000 inhabitants. The plant is located under the main square of the Forum in the urban section of the city, freeing some 15 hectares. Above this, and without invading public space, is a solar plant for generating electricity. The remodelling of the sewage plant, together with other interventions, has contributed to the recuperation and improvement in quality of the Barcelona waterfront.

Fig. 8. Forum – aerial view. This intervention re-values what used to be the city edge brownfields. By breaching the access barriers and integrating the infrastructure, a new tertiary and public space is created on the waterfront.

Source: courtesy of BIMSA (Barcelona d'Infraestructures Municipals)

Various projects related to energy production have been developed in the same way as the Forum, such as: renewal of the power generating complex by substituting steam cycles for combined cycles in the Besòs power plants; laying electricity cables along the mouth of the river; development of a photovoltic solar plant; and the introduction of a silt drying process in the sewage water plant. Along with these measures, a plan was developed for the integration of facilities to promote the use of renewable and residual energy, including the introduction of district cooling and heating infrastructures using local energy sources and implementing the efficient use of energy. Promotion of the use of renewable energies is another step forward towards a less contaminated and healthier environment.

New challenges for the future: 'world class' cities

London has extended its leading position over Paris, but these two cities are well ahead of their nearest rival Frankfurt. Barcelona overtakes Brussels in fourth place ... Barcelona, Madrid and Prague are seen as the cities doing the most to improve themselves as business locations. These were the top three locations in 2005(Cushman and Wakefield, 2006).

The *European Cities Monitor*, cited above, suggests that Barcelona is part of the select club of 'world class' cities. What is important, however, is to know what its role is in this group. It is evident that the internationalization of the economy that has grown with globalization has meant that cities, or more precisely, metropolitan agglomerations, directly compete between themselves.

Viewed generally from the outside, it would not be too difficult to identify the role that Paris, London, Berlin or Milan play on the global chessboard and identify the differences between them. Looking from the inside, it is somewhat more complicated because, to a certain extent, a similar process is happening in Barcelona today. While the Spanish economy is facing some problems, Barcelona's economy is changing rapidly, leaving its industrial production to one side while promoting its commercial and service sectors and tourism. Barcelona has to continue its dependence on tourism and the service sector while it sows the seeds in other sectors such as bio-technology and medicine, in which it can become competitive in the

Fig. 9. Forum: One of the most outstanding points of the Forum project is to dare to put together the prime uses with the generally undesirable infrastructure, something the opposite to the well-known NIMBY attitude

Source: Aleksandar Ivančić

future. There are a number of aspects that define the Barcelona of globalization: territorial structure; mobility; environment and new technologies; and housing.

Territorial structure

The central location of the core of metropolitan Barcelona is an area enclosed by the coastal strip extending from Castelldefels to Mongat and from the Tibidabo to the sea – an area that is geographically understandable. With this size of area, town planning can still be undertaken in detail, sketched in almost architectural detail, far removed from abstract methodologies using zoning as a working model. However, there was a fundamental problem: the lack of a political organ that could govern in the metropolitan area. This has been the case since 1986 when the right-wing government of the Generalitat de Catalunya dissolved the Barcelona Metropolitan Corporation which had been created precisely to develop the General Metropolitan Plan of Barcelona (1976) and 27 neighbouring municipalities. It does not seem that the creation of a new metropolitan governmental organisation is going to be easy, given the lack of agreement between the majority of political parties.

Mobility

Mobility is the key to the organisation of the metropolitan region. Although the institutional framework may make the situation seem forlorn, there are certain aspects of global planning where there is general consensus. One example of special importance is the Master Plan of Collective Public Transport Infrastructures 2001–2010 (MPI), in which diverse local, regional and state administrations agreed on a development plan and the construction and modernization of the public transport infrastructure (Prat, 2002; Casas, 2002). This is basically a railway system including, metro, suburban metro, commuter trains and even a new tramway[14] network, totalling some 100 km. This plan is expected to radically improve mobility in Metropolitan Barcelona.

Mobility statistics within the city of Barcelona indicate that individual journeys are distributed among the different means of transport in the following way: 47% on foot or by bicycle, 30% by public transport and 23% by private transport. The picture regarding connecting journeys in the metropolitan area is quite different: 2% on foot or by bicycle, 46% by public transport and 52% by private transport. Nevertheless, this situation could be improved within the city itself and indeed it is essential to do so in the context of Barcelona as a city and within its metropolitan area.

The city has a policy of promoting public transport while discouraging the use of the private car. This includes: integrated public transport fares within the metropolitan region, implemented at the beginning of 2002; new neighbourhoods designed to change the traditional dominance of the private vehicle in public areas through giving precedence to pedestrians and cyclists; 30km per hour speed-restricted areas; and future projects such as the separation of the BUS–VAO[15] lane. To promote the use of bicycles, the network of cycle lanes is continually being extended (128km at present) and the bicycle is beginning to be seen more on the street and left in underground parking facilities. Throughout 2007 the 'Bicing' scheme is being implemented, making 3,000 council-owned bicycles available to the public with the aim of introducing the bicycle as a form of individual transport, to be used together with public transport.

One of the cornerstones of the MPI is the new L9 metro line visualized as a circle line servicing the outlying neighbourhoods, and restructuring the radial network in place at the moment. The L9 is made up of a total of 47 new stations over a length of 44km, resulting in 17 interconnections with metro and other rail networks.

Presently, there are three physically independent rail systems within the city (two networks controlled by the Catalan Government (FGC) and one operated by the Spanish national railroad (RENFE)). The totally radial structure of the commuter trains operated by RENFE and the two unconnected networks of the FGC, both with terminal stations inside the city, are nothing but disadvantageous. There is a need to convert the rail networks into an authentic regional metro (similar to the RER in Paris), perhaps using some of the RENFE tunnels under the city serving its goods trains and intercity services. This is planned, and pending the construction of a new tunnel for the forthcoming high-speed train service.

The high-speed train, which will connect Barcelona to Zaragoza and Madrid as well as to the French border, is one of the most urgent needs of the city. Initially scheduled to arrive in 2004, the project is now expected to reach Barcelona (Sants station) at the end of 2007. The completion of a line across whole the city in combination with a new mainline station (Sagrera[16]) is scheduled to take place between 2009 and 2011.

The Sants and Sagrera railway stations are planned as major intermodal intersections where practically all the metropolitan and national/international public transport networks will converge. However, to be able to set up a truly operative regional rail network it will be necessary to construct an orbital railway line, to avoid having to pass through the city centre.

Environment and new technologies

Environmental criteria are becoming more and more important in the planning process, including issues of water consumption, waste, greenhouse gas emissions, and efficient use of energy.

Considerable effort has been made to reduce the consumption of drinking water and is beginning to bear fruit: at present it is around 121 litres per inhabitant per day which represents an 8% decrease in the last 6 years (Area Metropolitana de Barcelona, 2006). This figure is substantially below that of other large European cities.

Domestic refuse decreased for the first time in 2005 and today amounts to 1.51 kilos per inhabitant per day. At the same time, the percentage of recyclable refuse disposal is increasing year after year. As for the treatment of solid urban waste, four 'ecoparcs' (fully integrated waste treatment plants) have been constructed, allowing for the final closure of the city's major refuse dump. This has resulted in the improvement in the treatment and assessment of waste, a highly important reduction in greenhouse gases (GHG) emissions (due to the collection of methane produced in the process of the biological breakdown of organic waste) and the use of the gas as a fuel for electric power generation.

The city's concern for GHG emissions and local energy prospects were identified in the 'Energy Improvement Plan for Barcelona', and adopted by the City Council at the beginning of 2002 (Ajuntament de Barcelona, 2002). The plan sets the goal of not increasing GHG emissions between 1999 and 2010, although there is expected to be a 26% increase in energy consumption. The Energy Plan is a tool that structures interventions, places them into a wider context of the city and facilitates scheduling of priorities from a cost-benefit point of view, as well as for their social and environmental impact. The Plan develops various concepts, one of particular interest being the re-centralization of the energy system, incorporating renewable sources (Ivančić *et al.*, 2004).

In compact urban fabrics the density of energy demands, both electric and thermal, is high. This gives rise to the opportunity for local power-generating systems to supply electrical and thermal energy to the final consumer. The joint generation of electric and calorific energy means the efficiency of fuel use is considerably greater than for independent systems, and as a consequence the environmental impact is reduced.

The strategy of locally generated energy is promoted when new construction or renovation is planned. Energy generation *in situ* is conditioned by the planning and development of urban infrastructures. The linking and compatibility of electricity, gas and heating/cooling systems is crucial in achieving a system that is both highly reliable and offers good energy performance. This strategy has resulted in the important development of DHC systems in the north-eastern part of the city (Forum, 22@ and Sagrera in the near future) and intensive planning in the south-western section of the city, where it is expected to use residual cooling from an important industrial plant to provide cooling to adjacent areas.

The evaluation of the Energy Plan, undertaken in 2004, indicates good performance in the mobility sector but an undesirable performance in the residential and service sectors. Results at the city level continue to be negative for the time being: an average annual growth of GHG emissions of 0.66%; an average annual growth of 0.69% in energy intensity. However, the use of renewable energy systems saw important growth and reached around 100% annually in the case of solar power.

Housing

At first sight, with moderate population growth and a large property market, the problem could seem to be non-existent. However, two factors which are common to other European metropolitan areas have to be taken into account. The population is not increasing but there is a reduction in the number of people living in each dwelling: those living alone, the young and the elderly. More dwellings are needed for the same total population and, of course, new construction is needed for recent arrivals. The overall buoyant economic climate in Spain has led to considerable immigration from North African countries and South America.[17]

A seemingly contradictory comment must be made on the housing problem. How is it possible that there is a housing problem in an economy that has based its expansion on the property boom? Effectively, the Spanish economy is growing faster than other European economies thanks to growth in the construction industry. Barcelona, as in the rest of Spain, has noticed that as demand for housing increases, speculative capital plays its part in inflating prices. The percentage of public housing in the total market is very low and a lot of the new housing built is destined for weekend homes and investment properties.

To alleviate this situation, the council adopted the Barcelona Housing Plan (2004–2010) in 2004 with the objective of increasing the city housing stock by 10% not actually using any new land, but through urban renewal. The plan includes the City Council's commitment to promote at least 19,000 protected dwellings in this period. To judge by the number of housing licences granted in Barcelona in the period 2000–2006, the plan is starting to work: from less than 10% of subsidised dwellings, within the total dwelling number in 2000/01, it reached over 30% in 2005/06 (Ajuntament de Barcelona, 2007).

Instead of conclusions …

Over the past 30 years Barcelona has developed strategies that have lifted the city into the world's consciousness. The transformation of Barcelona is not the result of a random process nor a set of unlinked strategies, but an evolutionary path of inclusion and superposition of different techniques of planning intervention, adapted to meet the needs of every new challenge resulting from the growing scale of its territory: from small public urban spaces (squares) up to the Metropolitan Region.

The city's transformation is closely related to the urban culture of Barcelona. First, it has its origins in the cultural and geographical framework rooted in the culture of the compact Mediterranean city. Second, it has happened in a socio-political context when urban development was guided by successive local democratic governments. Finally, the urban development approach is characterized by its sharp architectural character.

The Barcelona Model continues to evolve and faces new and important challenges. It is necessary to consider a new conceptualization for the next future stage of development. There is general agreement between professionals and politicians that the city has to face problems related to its density, compactness and sustainability. The debates about low density and the anti-urban tendencies of the 1970s and early 1980s have been left behind. In developing a new conceptualization, the geography of Barcelona helps, in the sense that vacant land for lateral expansion has run out. Effectively, Barcelona is contained between the Tibidabo mountain, the sea and the River Besòs and River Llobregat. The only available territory of any significant size belongs to the lower area of the Llobregat Valley and its delta, recently marked out as an agricultural park. Consequently, one is faced with an urban area which cannot expand and as such, can only grow in on itself, by imploding. It is here that there are two aspects of fundamental importance: firstly, the operations of its urban restoration (which makes the 22@ project for Poble Nou exemplary as its criteria and planning techniques extend into other areas of town planning), and secondly, strengthening the relationship with the territory that makes up the functional unit itself, that being the whole Metropolitan Region.

Finally, in future Barcelona needs to demonstrate that it is capable of continuing as a model city, paying special attention to the Metropolitan Region. It needs to extend its innovative thinking and planning to the region and maybe beyond. Its influence on the world stage may stem from its innovative planning and policies that focus on public housing and environmental issues, building public facilities and improving the infrastructure to provide efficient and sustainable mobility at the Metropolitan Regional level.

Notes

1. The idea of the 'Barcelona Model' evolved at the beginning of the 1990s with the awareness of the changes that had taken place in Barcelona during the previous decade. The idea is so wide-ranging that it has been used not only by urban planners but also by politicians, economists, sociologists, geographers and even management. It represents a transformation of urban life special to the citizens of Barcelona – the process being led by the city council

and closely controlled by its citizens. Some authors question the existence of a 'Barcelona Model' and the use of the term 'model'. The term may not be totally adequate if referring only to the physical urban transformation, but if 'model' is understood as 'a thing used as an example to follow' (*Concise Oxford English Dictionary*), then the transformation of the city in its widest sense has indeed become a 'model'.

2. During the Spanish Civil War, 1936–39, the city remained loyal to the republican government who lost the war; the war was followed by the General Franco dictatorship, 1939–1975.
3. Oriol Bohigas' intellectual leadership is evidenced in his contributions in the media, his writing and books.
4. For example: the urbanisation of the 'Moll de la Fusta' by Manuel de Solá-Morales (1981), 'El Paseo de la Via Julia', and the re-urbanisation of squares in the Gracia neighbourhood.
5. Just as PROCIVESA PROEIXAMPLE was set up in 1996 to ensure quality standards for the Eixample district of Barcelona, PROEIXAMPLE converted the interior courtyards of 28 street blocks into public gardens. One effective mechanism was the creation of *ad hoc* companies for specific aspects that complemented the municipal machinery.
6. Growth and modernization of the city through international events is not new to Barcelona; important spin-offs occurred in the 1888 Universal Exhibition and in the 1929 International Exhibition.
7. For in depth explanation of the 1st phase of the Barcelona urban transformation see Bohigas (1985). For a wider explanation see Busquets (2005).
8. See: www.22barcelona.com/documentacio/22@_Barcelona_ENG.pdf
9. See: www.22barcelona.com/content/blogcategory/62/126/lang,en/
10. See: www.22barcelona.com/content/view/101/141/#3
11. Dark fibre refers to unused fiber-optic cables – ones that have been already laid but have still to be used.
12. See: www.barcelona2004.org
13. The neighbourhoods La Mina and La Catalana were two of the most degraded in the area.
14. The tramway operated in Barcelona between 1872 and 1971; it was reintroduced in 2004.
15. VAO – high occupation private vehicle/car with more than 2 passengers – this permits private vehicles with more than 2 passengers to use separate lanes.
16. Shown in Figure 5 as area C.
17. See Inmigrantes. El continente móvil, *Vanguardia dossier* vol. 22, Jan. 2007.

References

Ajuntament de Barcelona (1999) *Barcelona: gestión y gobierno*, Diaz de Santos, Barcelona.
Ajuntament de Barcelona (2002) *Pla de Millora Energètica de Barcelona*, Ajuntament de Barcelona, Barcelona.
Ajuntament de Barcelona (2006) *Barcelona ciudad del conocimiento: Economía del conocimiento, tecnologías de la información y la comunicación y nuevas estrategias urbanas*, Ajuntament de Barcelona, Barcelona.
Ajuntament de Barcelona (2007) *Habitatge asequible i qualitat de vida a Barcelona (2004–2010)*, Ajuntament de Barcelona, Barcelona.
Area Metropolitana de Barcelona (2005) *Normativa urbanística metropolitana; Reedició actualitzada 2004*, AMB, Barcelona.
Area Metropolitana de Barcelona – Entitat del Medi Ambient (2006) *Dades ambientals metropolitanes 2005*, EMSHTR, Barcelona.
Bohigas, O. (1985) *Reconstrucció de Barcelona*, Edicions 62, Barcelona.
Busquets, J. (2005) *Barcelona: The Urban Evolution of a Compact City*, Harvard University Graduate School of Design, Cambridge.
Casas, X. (2002) A fundamental element in urban transformation. The Present and Future of Metropolitan Transport. *Barcelona Metròpolis Mediterrània*, 2: 80–92.
Cushman and Wakefield (2006) *European Cities Monitor 2006*, Cushman and Wakefield Global Real State Solutions, London.
Ivančić, A., J. Lao, J. Salom, J. and Pascual, J. (2004) Local Energy Plans – a way to improve the energy balance and the environmental impact of the cities: Case study of Barcelona. *ASHRAE Transactions*, 110 (1): 583–591.
Prat, J. (2002) The MPI 2001–2010, a consensus-based response to the infrastructure needs of public transport; The Present and Future of Metropolitan Transport. *Barcelona Metròpolis Mediterrània*, 2: 32–44.

9

Sidh Sintusingha

Sustainable 'World Class' Cities and Glocal Sprawl in Southeast Asian Metropolitans

'World class' and sustainability: conflicting objectives

Cities have always been brands, in the truest sense of the word. Unless you've lived in a particular city or have a good reason to know a lot about it, the chances are that you think about it in terms of a handful of qualities or attributes, a promise, some kind of story. That simple brand narrative can have a major impact on your decision to visit the city, to buy its products or services, to do business there, or even to relocate there (Anholt City Brand Index, 2005, p. 2).

'World class' and sustainability have become ubiquitous catchwords in the competitive branding of cities – applied to the city as a whole or in parts e.g. 'world class' cities, 'world class' architecture, 'world class' transportation, 'world class' education', 'world class' workforce.[1] Through globalization's commercialization processes, they have become plastic words – more often superficially self-proclaimed than 'earned' – employed towards the main ends of attracting skilled migration, investment and tourism. In practice, they often follow prescribed patterns of the development of transportation mega-structures, inner cities, brown fields, green fields and waterfronts, resulting in the gentrification and homogenization of cities' forms, images and experiences.

Moreover, the notion of 'world class' is problematic for sustainability as it sets and imposes specific urban practices for the assumed 'lesser' cities to emulate through the 'modern religion' of 'development' (Rist, 1997, p. 21), the *modus operandi*. Additionally, the objectivity of the 'world class' assessments, often framed by Western urban culture, should be questioned and one wonders how long it would take, if ever, for cities at the bottom of various world-city ranking lists such as Anholt's City Brand to make it to the upper rankings monopolized by Western cities. Furthermore, the city's often high socioeconomic and physical fragmentation and hence the significant disparities within each city itself problematize where the label 'world class' applies and does not. This is particularly apparent in the sprawling suburbs, characterized as neither 'world class' nor sustainable, where an increasingly large proportion of the city's population reside.

1st world / 3rd world cities: globalization or neo-colonization

Development consists of a set of practices, sometimes appearing to conflict with one another, which require – for the reproduction of society – the general transformation and destruction of the natural environment and of social relations. Its aim is to increase the production of commodities (goods and services) geared, by way of exchange, to effective demand ... (Rist, 1997, p. 13).

Indicators considered fundamental to both 'world class' and sustainability are those of material growth and rising living standards, based on indicators including per capita GDP and consumption levels. According to environmental proponents such as Rees (1999), Girardet (1999), and Yencken and Wilkinson (2000), these dominant neoclassical economic[2] indicators of growth, by concentrating on market efficiency, neglect the environmental and social 'costs' of development. Sustainability is often associated with lessened or zero economic growth, giving weight to the view that environmental and social issues are frequently considered by economists as 'valueless'. Environmentalists, in contrast, prescribe 'practicing sustainability' by 'dematerializing' society, or by calling for decreased, rather than increased consumption,[3] which clearly conflicts with the objectives of economic growth and vitality crucial to 'world class' aspirations. In the narratives of sustainability and 'world class', developed economies are thus faced with the contradictions of the precedents they set and messages they pass on to developing economies:

The spread of international best practice on these issues is not a management fad, nor a conspiracy for world domination from certain industries or advanced nations – it is essential for the future of the world. Thus a global orientation is a precursor to understanding sustainability (Newman and Kenworthy, 1999, p. 3).

Moreover, the notion of 'world class' city is consistent with and reinforces the hegemonic neoclassical worldview that has classified the world into the developed and developing nations[4] (as measured by economic productivity or GDP per capita). This division is problematic for global sustainability as it supports the unequal relationship that lets cities of developed countries exploit resources from, and unload wastes onto, developing countries in the name of lowest prices.[5] Across this divide, the developed countries also provide unsustainable, increasingly materialistic examples and particularly urban lifestyles, as targets for others to emulate – manifested clearly in the rapid urbanization and suburbanization of cities in developing countries.[6] The terminology reinforces this inherent bias, and developing countries are left chasing and aspiring after the ever-shifting goal of 'developed', 'First World' status[7] while their cities chase 'world class' status.

Unfortunately, this 'homogenizing', urbanization process also often comes with a loss of indigenous cultures and know-how ' ... derived from a highly developed understanding of their local environments ... ' (Girardet, 1999, p. 25). As the differentiation between rural and urban lifestyles becomes blurred, the local becomes assimilated into the global neoclassical economics notion of the 'city-as-market' (Bridge and Watson, 2002, p. 107). This rejection of indigenous, vernacular culture and its knowledge of local ecology, perpetuated by globalization, is one of the root causes of current environmental ills worldwide. Old mistakes are merely being replicated at an unprecedented scale and, according to Campbell (1996, p. 302) while the indigenous communities had no choice but to be sustainable, ' ... we must *voluntarily choose* sustainable practices, since there is no immediate survival or market imperative to do so. Although the long-term effects of a nonsustainable economy are certainly dangerous, the feedback mechanisms are too long-term to prod us in the right direction'.

Human settlements of the past, out of necessity, generally maintained a constant, sustainable relationship with their immediate rural/natural surrounds. In contrast, contemporary cities/settlements aided by the advance in transportation technologies, are able to draw resources from anywhere across the globe. Lacking and shielded from the immediate feedback mechanisms of past settlements, modern cities seem to have the choice to exploit and over-utilize their (and other's) surroundings.

The past, however, should not be seen as an 'ideal model' to be uncritically emulated. What is argued here is not for the abandonment of the values and practices of those indigenous cultures – but rather studied modifications: to maintain a continuum with the past rather than reject it outright for imported urban sensibilities, or to treat the past as an unfashionable *tabula rasa* as is often the case in developing countries, often suffering cultural theft and insecurity as a result of colonization and globalization, while eyeing the modern Western city as the model to follow.

Technology and the unsustainable alliance between urban sprawl and cyberspace

'Nothing endures but change.' That observation by Heraclitus often seems lost on modern environmental thinkers. Many invoke scary scenarios assuming that resources – both natural ones, like oil, and man-made ones, like knowledge – are fixed. Yet in real life man and nature are entwined in a dynamic dance of development, scarcity, degradation, innovation and substitution (Vaitheeswaran, 2002, p. 11).

Technology, as in any era, has the potential to alter our cultural practices and, thus, could transform the ways we define and practice sustainability. It will always remain the unknown variable in future developments – for example, improvements in systems of communication technology as recently as in the late 1960s led to the modern globalization phenomenon (Giddens, 1999, p. 10). Left to its own devices, technology can be less than beneficial, such as in the processes in which technology has transformed the city from a centre of 'civilization' into a city of 'mobilization' (Girardet, 1999, p. 11).

The visionary promises of automobile and highway technologies have led to the nightmares of urban sprawl, further augmented by cyberspace technologies:

Current thinking on the future of the city tends to be split between materialists, who concern themselves with sustainable urban forms, whether compact, polynucleated or decentralized, and non-materialists, who proclaim the supplanting of urban space by cyberspace, as if it were a *fait accompli* ... (Hagan, 2001, p. 167).

The Internet – the electronic flow of information, a rhizome-like mesh covering the planet's surface – reduces and hinders the need for face-to-face contact. It is highly compatible with and, thus, further encourages spatially dissipated settlements. One can be amongst pristine nature in a gated resort and still 'engage' with the world through the multiple media windows of cyberspace. Cyberspace is arguably the delayed realization of 1930s visions when ' ... better futures were imagined in positive technological terms ... ' (Rowe, 1991, p. 192), then referring particularly to transportation technology[8] which 'enabled widely dispersed citizens to live and work wherever they might choose ... [effectively] eliminating distance' (ibid., p. 192, quoting Segal, 1986, p. 127).

Cyberspace challenges and further complicates a materialist notion of form – adding another intangible (yet, becoming more tangible with advancement in simulation technology) spatial dimension that seems to compete with all pre-existing forms and the 'real' spaces of everyday life.[9] In a dispersed suburbia, the 'nonplace urban realm' (Rowe, p. 58 from Webber, 1964), cyberspace may be less of a challenge than an enhancement. It possesses the potential to threaten the environmentally sustainable compact settlement models, once the technology becomes more affordable and available to the masses – such as radios or TVs in every household of all economic classes. According to Hagan, this is the dilemma faced by designers/planners – the possibility of architecture and the city condemned to insignificance with the proliferation of the Internet (and 'world class' cyberspace):

> If we then counter this trend with compaction, how do we do it? By bulldozing the environmentally wasteful and architecturally deprived 'in between' to create, or recreate, 'compact cities'? Or do we perform a patch-up job on the old centers and new peripheries that retrofits some degree of environmentally sustainability into the existing mess? (Hagan, 2001, p. 175).

The last few questions posed by Hagan are critical points for the city's sustainability – how do we counter the sprawling tendencies of the city's forms, begun first by successive improvements in road and automobile technologies,[10] now reinforced by cyberspace? Do we treat these 'mistakes' as another *tabula rasa* or, more practically, through 'patch-ups' and 'retrofits'?

It must also be noted that these are issues more directly relevant to the wealthier northern hemisphere cities – developing cities often face different problems. Essentially, developing cities, late adaptors of industrialization, are facing issues that Western cities have confronted earlier (and resolved with decentralization via highways and cars) but at a compressed time-scale. Thus, developing cities often have to tackle urban congestion, poverty, environmental degradation, concurrently with suburban sprawl – combining the compounded problems of both the industrializing and post-industrial city.

Contextual and cultural differentiation: the glocal[11] suburbs

> It is not surprising therefore that after decades of independence, many ex-colonial countries are still struggling to relink to their own traditions, and to redefine their cultural identities, while encountering the influences of Western modernity and, more recently, of aggressive globality ... (Lim and Jaikawang, 2003, p. 72).

The dichotomies – global vs. local; modern vs. traditional; homogeneity vs. differentiation; dispersed vs. compact settlements; or active vs. passive technologies – epitomized the history of urban evolution from the pre-Industrial Revolution age to its present fossil fuel dependent state, amplified by technologies that dispersed people ever more spatially apart. This phenomenon is manifested in the sprawl of urban activities further away from its traditional centres and ever deeper into the hinterlands to the extent that many would proclaim that there are no more hinterlands. The 'machine' is no longer in the 'garden' to twist Leo Marx's (1964) phrase but has displaced, or even besieged the garden. This is the globalized urban form. Had there been no social/environmental impacts, this economically driven form would never be critiqued, questioned, or resisted (for we would have achieved the modern 'utopia').

The message here is that there are choices, enhanced by the global communication network, critically to pick and appropriate lessons and practices from far and wide – and drawing deep into the historical evolution of various cities in contrast to the common practice of copying forms and pasting them undifferentiated into varying cultural, geographical contexts. Despite competition from cyberspace (challenging the notion of 'reality'), priority must be given to the biosphere where all lives hinge. Here Bull posits:

> Achieving a more sustainable balance between the natural and made worlds requires us to be more careful about how and where we choose to expand developed areas and to improve the environmental quality of the development matrix itself. Rather than relying as we have in the past, on sustainability being achieved by balancing urban areas with non-urban areas beyond, we must address the balance between the built and natural worlds within the matrix so that the developed landscape is, of itself, more sustainable (Bull, 2002, p. 1).

At the broader scale, in particular for the many cities characterized by sprawl, this is consistent with the calls for mediating the relationship of the city to its rural and natural surroundings from being exploitative and parasitic to mutually beneficial – such as: Spirn's call to view ' ... the city, the suburbs, and the countryside ... as a single, evolving system within nature, as must every individual park and building within that whole ... ' (p. 5); Nicholson-Lord's ' ... injection of ecology into the formal planning of cities ... ' (p. 184), or; Ebenezer Howard's vision of ' ... a garden city perhaps, but the garden is wild' (p. 185); Rowe's (1991) 'Modern Pastoralism' that reconciles the machine with the garden in the middle landscape of the suburbs. The sustainable city must be based on that critical dialectical relationship regardless of actual physical forms that emerge from that process. It is about the balance of the built and natural worlds, and also the reconstructed natural world, within the city.

Sprawl is a common global phenomena tied in closely with economic and population growth. For any city it is a manifestation of the ' ... many non-local forces that are shaping the city – forces that frequently do not even know the specific cities in which they are at work' (Frey, 1999, p. 15); however, there are also local factors in interplay. Each city's particular economic, social, environmental and cultural traits must be identified and understood in order for effective measures to be custom-planned and designed towards more sustainability.

The following sections present preliminary observations from Southeast Asia, of the specific cultural landscapes production/re-production and evolutionary trajectories (including socio-economic and political environments), from three city edge suburbs: Roxburgh Park in Melbourne (geographically, politically and economically in the region), Muang Thong Thani in Bangkok, and Linh Dam in Hanoi. The suburbs were selected as they demonstrate and physically manifest the acculturation processes, however imperfect, of those global and local forces that shape cities. They manifest the multiple dialectical relationships reflecting the varied cultural and ecological contexts in which 'glocal' suburbs operate with implications for both 'sustainability' and 'world class'.

Roxburgh Park: regional picturesque

Seen broadly, Roxburgh Park housing estate, developed by Urban Land Corporation (Victorian Government's urban development agency – today incorporated into VicUrban), is a typical example of the global highway induced development following the path of suburban precedents set in post-World War II America. The housing estate, established in 1992, is located twenty-six kilometres to the north of the CBD towards the northernmost edge of Melbourne's urban growth boundary[12] (Figs 1 and 2). Built on former agricultural land, the 651 hectares estate is to contain 6,200 houses for 20,000 people. The grid road network, consisting of six primary and 350 secondary roads, was imposed and adjusted to the slightly undulating topography, yielding a picturesque landscape composition. The development is linked and integrated with Melbourne's green wedges – an interconnected non-urban, open space system that defines the urban growth boundary.

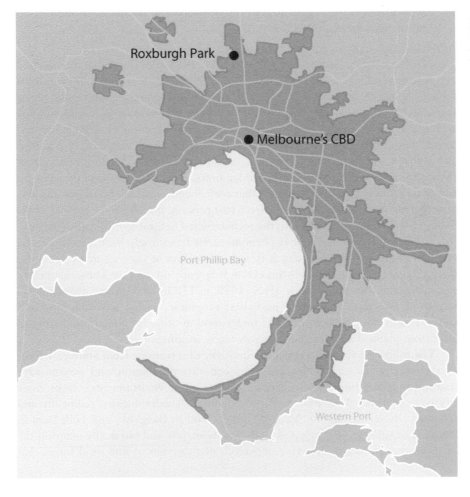

Fig. 1.
Melbourne and
Roxburgh Park
Source:
Adapted from
Department of
Infrastructure,
2002, p. 3

Residents are highly reliant on private cars whether for basic chores or to get to employment, mostly not within the estate (there are industrial estates lining Hume Highway to the east). The estate is serviced by bus routes which are infrequent and patronage is low, limited to the young and old without other travel alternatives. A rail station to link to the CBD has been planned at the suburb's commercial centre on Somerton Road to the southeast, however, poor overall integration and the dispersed and low density housing would not significantly alter travel patterns.[13] The low density planning also resulted in the under-utilization of the extensive pedestrian and cycling networks.

What differentiates Roxburgh Park from previously developed suburbs is the utilization of indigenous flora species in the picturesque landscape of the public realm (Fig. 3). This is not the 'voluntary' contract with nature but regulated by the government in response to the dry conditions – within the fences of individual house compounds water-intensive exotic species still dominate private gardens and many of the houses are heavily influenced by an English architectural heritage ill-suited to Australian climes (both now being redressed with the issues of climate change coming to the fore, prompted by the 'feedback mechanism' of the extended drought).

Fig. 2. Aerial view of Melbourne and Roxburgh Park

Source: Adapted from Google Earth

As in any contemporary suburban development in Melbourne, houses modelled on the Victorian and Georgian styles common in England during its colonization of Australia from the 18th century are popular. Developers market the houses highlighting this colonial heritage to capitalize on the prestige associated with this style with models such as the 'Majestic', 'Sovereign', 'Embassy Regency' and 'Carrington' from which to select. Consumers can also purchase models such as the 'Nevada', 'Sierra', 'Sacramento' and 'Californian', associated with American suburban culture.

Fig. 3. Roxburgh Park landscape with native plants and lakes Source: Sidh Sintusingha

Muang Thong Thani: palimpsest of global forms

Muang Thong Thani (literally 'Golden City'), a privately developed satellite city located forty kilometres north of Bangkok in the province of Nonthaburi, has been envisioned as a partially self-contained city that at its full development will house a population of 360,000 in an area of 640 hectares, combining work, residential and recreational activities (Figs 4 and 5). As a wholly private development begun in 1987, all public infrastructures within the project are delivered by Bangkok Land, the project's developer. Muang Thong Thani is well serviced by public bus routes and highways, including the Chaeng Wattana road branching off the major north bound highway from Bangkok, and the Bang Pa-In Pak Kred expressway, which links the project directly to downtown Bangkok.

Muang Thong Thani in its current form, partially due to its expanse and decades-long time span of development (compounded by Thailand's weak urban planning infrastructure), is more a collection of disparate parts than a comprehensive, coordinated whole. There is a huge diversity of housing types in the project, catering to a wide range of clientele of varying economic classes. The accumulated built forms lining the main axis that runs from the entry at Chaeng Wattana Road extending into 'Bond Street', the designated main business district, reveals this diversity in forms and functions. Generic shop-houses mark the entry while stylish offices line one side of the palm-lined Bond Street and residential towers on commercial podiums line the other. Beyond this main axis is a variety of house types, from the detached houses built and occupied during the earlier phases of the project, to the apartment blocks that are the residences for the less well off (Fig. 6).

The project is the product of the highly intertwined global-local aspirations and processes, from its finance to its potential inhabitants – once speculating Hong Kong

Fig. 4. (above–left) Bangkok and Muang Thong Thani
Source: Adapted from www. sawadee.com

Fig. 5. (above–right) Aerial view of Muang Thong Thani
Source: Adapted from Google Earth

Fig. 6. Muang Thong Thani Puae Kahrachakarn.
Source: Sidh Sintusingha

migrants from the 1997 reunification with China and as recently as 2001, a 'mini-Tokyo' for '30,000 households of Japanese retirees' (Intarakomalyasut, 2001).

Regardless of the eclectic appropriation of foreign forms, the liveliest part of Muang Thong Thani, in terms of the local inhabitant's patronage, appears to be the commercial node at the suburb's entry on Chaeng Wattana Road. This street is lined with the ubiquitous shop-houses that cater for a wide variety of needs and services for the residential areas inside. This type of development is the generic pattern of development throughout Bangkok and is in contrast to the relatively quiet, subdued precinct of the European styling of Bond Street. The 'global' and 'local' forms co-exist awkwardly on the same axis, in particular the typically animated shop-house-lined entry juxtaposed with the vacant condominium towers (remnants of the 1997 economic bust).

This is in stark contrast with the vast mosaic of rice-paddies irrigated by networks of natural and man-made canals on the floodplains of the Chao Phraya River that characterized this area only three decades ago. There is little evidence that the developers of Muang Thong Thani have made attempts to reconcile notions of tradition and modernity, agriculture and the city. Nor is there any clear formal attempt at integrating the various introduced parts/phases of the project together. The process appears to be of one erasing the local in pursuit of the primary goal, one of global and local marketability. Muang Thong Thani is not unlike Bangkok, a 'palimpsest' where forms are imported with great enthusiasm and assimilated with varying successes.

Linh Dam: conforming to global trends

Linh Dam-Lake project, located eight kilometres from the centre in the southern part of Hanoi, is one of the largest and most significant urban housing developments in Hanoi developed by the state-owned Housing and Urban Development Corporation (Figs 7 and 8). Like the previous two examples this estate was developed in conjunction with the city's highway construction projects and is on the main north–south highway connecting the capital to Ho Chi Minh City and Vietnam's southern provinces. The proposed Ring Road Number 3 will run through the Linh Dam estate and onto Highway Number 1 to the east and Thang Long Bridge and Noi Bai International Airport to the north and northwest. Located at the junction of these highway networks, Linh Dam is a gateway feature, signifying development and modernity, at the southern entry to Hanoi. The project is also well served by public bus routes that connect it to various parts of Hanoi.

The Linh Dam housing estate, planned to be a modern commercial service and housing district, has a total area of 184 hectares (including 74 hectares of an expanded natural lake). The Linh Dam project has been divided into two stages. The 24 hectares first stage begun in 1997 included detached houses and 3 high-rise buildings containing 368 apartment units. The second stage includes many detached

Fig. 7. (below–left) Hanoi and Linh Dam
Source: Adapted from photograph of street side tourist map

Fig. 8. (below–right) Linh Dam aerial
Source: Adapted from Google Earth

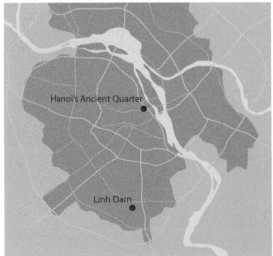

Fig. 9. 'French style' villas with high-rise apartments in the background at Linh Dam

Source: Sidh Sintusingha

houses and approximately 2,000 apartments in 16 eleven-storey buildings (HUD 2001) on an area of 35 hectares. In both stages the high-rise buildings line the main roadways and create a symmetrical and formal urban planning typology.

In contrast to Hanoi's older housing stocks, road access was given high priority, as were the separate drainage and sewage systems, underground power supply, parklands and facilities for businesses. The objective, according to the developers, is to achieve modern international standards of housing and these facilities have seldom been part of previous settlements.

There are two types of housing in Linh Dam: the villa, which caters to the high end residents, and the apartment, which accommodates the middle class residents. Negotiating between Vietnam's socialist foundations and the new capitalist outlook (as part of *Doi Moi*), Linh Dam brings the élite into contact with the middle as well as lower classes in surrounding rural villages (and thus does not exactly conform to the gated community typology). The absence of the ubiquitous row-house with commercial space at the street frontage, the most common residential typology in Hanoi, suggests that the developers deliberately distanced themselves from this indigenous housing type at Linh Dam, favouring global typologies.[14] Moreover, the villas are decorated in the 'French' style associated with higher status despite Vietnam's violent colonial history (Fig. 9). Global trends in apartment designs are emulated and here is one of the earliest examples in Hanoi of open-plan principles where the kitchen, dining and lounge room spaces are combined. The one concession to local needs is the location of the kitchen close to the balcony to suit Asian cooking. It will be interesting to see whether the inhabitants eventually adapt the villas and apartments – or even the public realm as has occurred at Muang Thong Thani – to suit their local lifestyles.

Discussion

These three suburbs demonstrate the tensions between the prevalence of regional suburban characteristics and the forces of globalization. Shared construction technologies, automobile ownership (not common yet in Hanoi, but inevitable with increasing economic affluence) and the use of air-conditioning enable the universal global suburb to be discussed. However, an attempt at comprehensive comparisons across cases (which is beyond the intent of this chapter) would need to consider regional particularities – some of which have been touched upon.

An interesting point is the huge differences in population density whereby Roxburgh Park and Muang Thong Thani, with similar land areas, have densities of approximately 31 persons/hectare and 563 persons/hectare respectively. Linh Dam, dominated by high-rise apartments, will have a population density closer to Muang Thong Thani, and higher than the surrounding peri-urban settlements. Although Muang Thong Thani and Linh Dam were also conceived as new urban centres, and Roxburgh Park as a residential suburb, their densities are more or less consistent with urbanized areas in their respective cities[15] (Table 1). The much lower density and the lack of high-rise in Roxburgh clearly differentiates it from Muang Thong Thani and Linh Dam, which could be attributed to the underlying cultural differences in settlement habits as well as the relative abundance of land resources in Australia. On the other hand, it could also be attributed to the relatively more affluent Australian economy, where citizens can more easily realize their aspirations of owning detached houses – as wealthier residents of Muang Thong Thani and Linh Dam also could do.

The disparity in suburban density and form between Bangkok/Hanoi and Melbourne can also be accounted for what McGee (1991) termed as 'desakota' (literary coining two Bahasa Indonesia words for 'village' and 'city') which characterizes ASEAN urbanization/suburbanization patterns whereby, through direct foreign investments, factories are located in densely populated rural areas to take advantage of the local work force – in contrast to Europe, America and Australia where a large majority of the workforce are employed in urban-related jobs and where the rural areas, relying on mechanical tools rather than manual labour, are much less densely populated.

Unfortunately, both city administrators and developers ignore the morphological processes of these villages and they are often displaced with and transformed by modern suburbs as is the experience at both Muang Thong Thani and Linh Dam. The imported forms and images of those global suburbs combined with the post-colonial conditions in Melbourne and Hanoi, and to a differing extent Bangkok's 'un-colonized' condition, influence the design and planning of the suburbs. Roxburgh Park's curvilinear streets and its houses demonstrate the Australian obsession with Englishness and the picturesque. The detached houses at Linh Dam demonstrate the continued aspiration for French style. Both the villas and high-rise apartments are arranged symmetrically around a wide linear central axial avenue with precedents in Hanoi's French Quarters. The lack of an obvious colonial model makes it more difficult to suggest that suburbs like Muang Thong Thani are beholden to one particular style – it in fact, draws unabashedly upon the most diverse of styles. On the other hand, the planning on closer inspection resembles the indigenous transportation pattern of main roads with branching off, often dead-end, *sois'* (distribution roads), descendant of the *khlong*'s (canals) and *khlongsoi*'s (branching waterways) typology.

At another level is the planning response to local climates and landscape. Out of necessity rather than preference, Roxburgh Park re-establishes the indigenous Australian flora that is well suited to the drought conditions, in that process achieving a distinct synthesis in the use of indigenous flora with picturesque landscape style. In doing so, Roxburgh Park arguably came closer to fabricating the ideals of the early suburbs of living close to indigenous nature and not its modern artifice. Linh Dam adapts and integrates a large pre-existing lake to its planning – a tradition going back

Suburb	Roxburgh Park (from 1992)	Muang Thong Thani (from 1987)	Linh Dam (from 1997)
Type of development	Public–private; mainly residential with services serving suburb	Private; commercial, residential – 'self contained community' to create 120,000 jobs	Public; commercial, residential when completed
Area (hectares)	651	640	184 (59 for stages 1 and 2)
Projected population	20,000	360,000	–
Density (persons/ hectare)	30.7	562.5	>172.6*
Distance from city centre (kilometres); transportation	26; highways, public bus, new train station to link to CBD	40; highways, expressway, public bus and other local modes of transportation (motorcycle taxis, vans etc.)	8; (poor road network – means more time travelled – being addressed with new highways); highways, public bus and other local modes of transportation
Planning	Grid and cul-de-sac	Thanon and Soi	Axial
Housing typologies and styles	Detached houses; townhouses mainly 'English' styles	Mixed; varied	'French' style detached villas; multi-storey apartments
Social mix	No	Yes	Yes
Landscape integration	Yes	No	Yes
Planning controls	Yes (developed consistent with original plan in 1992)	Yes (weak – urban development mostly led by private sector)	Yes (increasingly compromised by rapid development related to economic growth)

* Stage 1 and 2 – based on the combined 2,368 apartment units times 4.3 the number of persons per household for Tranh Tri district (for the year 2000 from Thi *et al.*, 2005, p.7) where Linh Dam is located.

*Table 1.
Comparison of
the features of
the three glocal
suburbs*

*Source: Sidh
Sintusingha*

to the pre-colonial settlements of Hanoi Ancient Quarter. Muang Thong Thani's insensitive response can be attributed to the wholly capitalist nature of the venture (there is no government involvement at all) but also to the hegemonic narrative of 'development' and 'progress' that runs deep across all sectors of Thai society and that seemingly override any consideration for the environmental contexts. On that note, Linh Dam's combination of villas and high-rise apartments owned by differing economic classes also retains some semblance to the socialist ideals and intentions of the Vietnamese government through centralized housing policy – but that is increasingly under threat from the rapid economic development since *Doi Moi*.

Conclusion

As it is practiced, the notion of 'world class' is a homogenizing force that favours the economically wealthy, technological advanced 'first world' and encourages unmitigated economic growth for the 'third world' to catch up, and is arguably unsustainable. To remedy that, a conciliation between the concepts of 'world class' and 'sustainable' is critical. This can be done by practically incorporating sustainability benchmarks in the assessment of 'world class' or, going further, making the two concepts synonymous which, however, would require a radical global cultural shift.

For the city, lessons from and characteristics of pre-industrial settlements have frequently been quoted as part of the solution for living in and designing sustainably – especially the notion of the 'organic' self-sufficient, mixed-use, compact, walking city where food is produced locally in the surrounding open spaces. This, it has been suggested, should be complemented by a highly efficient mass-transit system, which caters for inter-nodal/centre travel (Girardet, 1999; Yencken and Wilkinson, 2000). In essence, this calls for the synthesis of applicable lessons from a much less materialistic past with the best of today's technologies. This implies the need to revisit the fossil-fuel dependent transit-oriented cities of the late 19th, early 20th century before the private car superseded rail travel (e.g. Melbourne's CBD and inner suburbs' effective integration by train and tram networks) and the 21st century reconfigurations of Ebenezer Howard's garden city model,[16] and consider higher population densities in the centres and the satellite suburbs. While cyberspace poses a challenge for a city's spaces and forms, the compression life/work/shopping (etc.) electronically could also reduce the need for travel and could be utilised to benefit sustainability.

This is the broad global framework towards the sustainable city which will have to be planned and acculturated on the ground with the particularities of each city and its parts taken into consideration. Arguably the most problematic part is the city's sprawl and the three projects featured the phenomenon in Melbourne, Bangkok and Hanoi. While the three projects cannot be considered 'sustainable' in the strictest sense, each displaying some characteristics but not others, they are examples of a complex, incremental process. The three projects may resemble any global automobile-friendly suburb and look to Europe for stylistic inspiration. That is merely the exterior 'shell' – whether English, French, modern, postmodern – that multiple local forces, whether cultural, social, environmental, assimilate into unique global settlement forms. And it is these 'glocalizations', it is argued, that the narratives of sustainability and 'world class' need to assimilate and hybridize to form the basis for policies, planning and design towards more 'genuine' sustainable, 'world class' cities, both in quality and diversity. In other words, the current divergent paths towards sustainable and 'world class' must be reconciled; yet they must also be diversifying and open-ended, not homogenizing, as is currently practiced.

Notes

1. The Age columnist Terry Lane compared the proliferation of world class advertisements with 'propaganda art' (2006). This chapter appropriates the term 'world class' semantically in the context of its everyday, widely propagated usage in the media to brand and advertise the city and its parts and thus has arguably greater societal impact than the academically defined concepts of 'world city' and 'global city'. As the concept of world city is also problematic for sustainability in the same ways, 'world class' can be seen as the popular, secular synonym of 'world city'. Its common usage arguably makes it 'easier' to achieve world class, in contrast to the academic exercise of achieving 'world city' status.
2. ' ... an analytical framework in which emphasis is placed on easy substitution of labour and capital in the production function to generate a steady-state of growth, and where all variables are growing at a constant, proportionate rate ... Neo-classical growth models identify the sources of growth as technical progress and population increases, with capital accumulation determining the capital-to-labour ratio in the steady state' (Bothamley, 1993, p. 367).

3. Although mentioning that the reduction would be partially offset by growth in new sustainable industries.
4. *The New Lexicon Webster's Dictionary of the English Language* bluntly defined 'developed nations' as 'countries with a per capita annual income of more than $2,000 and consequent higher standards of living than in so-called developing, or underdeveloped, nations' and 'developing nations' as 'countries with a per capita annual income of less than $2,000 and a commensurate poor standard of living among most of the population' (Cayne and Lechner, 1992, p. 261). Subsequent definitions have tended to be vaguer.
5. Exacerbated further by subsidies rich countries pay their farmers, such as to cotton farmers in the US (Touré and Compaoré, 2003) and Europe's Common Agricultural Policy (CAP).
6. The neoclassical economics' 'bid-rent model' combined with land as goods ' ... results in the famous concentric ring banding of land uses around the city ... ' with variants of the model having the effect of ' ... rapid suburbanization of many cities in the second half of the twentieth century' (Bridge and Watson, 2002, p. 108).
7. At a seminar of Thai Rak Thai Party MPs, Thailand's Deputy Prime Minister, quoting various economic indicators, proclaimed that ' ... the government ... had the potential to transform Thailand from a developing into a developed country ... ' (Lim and Jaikawang, 2003).
8. John Perry Barlow in 1996 gallantly declared ' ... We will create a civilisation of the mind in cyberspace. May it be more humane and fair than the world your governments have made before' (Manasian, 2003, p. 3).
9. Marshall McLuhan, taking a positive view, sees the globe 'electrically contracted' reverting into 'a village' – releasing ' ... men from the mechanical and specialist servitude of the preceding machine age ... ' (in Nicholson-Lord, 1987, p. 23).
10. That not only enhanced sprawl, but shifted scale-relations between humans, manmade artefacts, and natural landscape – the combination of larger human constructs and speedier movements led to the increasing domestication of the wilderness in scale (Rowe, 1991, p. 32).
11. 'Glocal' is a concept synthesized from the terms 'global' and 'local' with wide application in the areas of business, technology, sociology, etc. In the context of this chapter, the concept applies to urban forms and practices. The term more accurately reflects the phenomenon of globalization, which is not wholly homogenizing and undifferentiated, but is also defined, to varying degrees, by dialectical localization processes.
12. Introduced in the Melbourne 2030 plan in 2003.
13. 'Train' and 'Train and other' patronage of one percent each (http://www.domain.com.au accessed 16 April 2007).
14. In fact, the expanding project threatens an adjacent ancient village (Thang 2005).
15. Muang Thong Thani's projected density is in fact higher than Bangkok's densest district of Pom Prap Sattru Phai with 389.5 persons/hectare (2000 number from BMA, 2001, p. 1).
16. Originally published as *Tomorrow – a peaceful path to real reform* in 1898, republished in 1902 as *Garden Cities of Tomorrow*, with many reprints by Faber, e.g. 1946, 1965.

References

Anh, M. T. P., Ali, M., Anh, H. L. and Ha, T. T. T. (2004) *Urban and Peri-urban Agriculture in Hanoi: Opportunities and Constraints for Safe and Sustainable Food Production*, AVRDC, Shanhua, Taiwan.

Anholt, S. (2005) *How The World Views Its Cities: The Anholt City Brands Index December 2005 Report*, Global Market Insite, Seattle.

BMA (2001) *Statistical Profile of BMA*, Department of Policy and Planning, Bangkok Metropolitan Administration, Bangkok.

Bothamley, J. (1993) *Dictionary of Theories*, Visible Ink Press, Detroit.

Bridge, G. and Watson, S., eds (2002) *The Blackwell City Reader*, Blackwell Publishing, Oxford.

Bull, C. (2002) Contributing to ESD through landscape planning, design and management, *Environment Design Guide*, August 2002, DES 48, RAIA Publications, pp. 1–5.

Campbell, S. (1996) Green Cities, Growing Cities, Just Cities? Urban Planning and the Contradictions of Sustainable Development. *Journal of the American Planning Association*, **62**(3): 296–312.

Cayne, B. and Lechner, D. E., eds (1992) *The New Lexicon Webster's Dictionary of the English Language* (Volume 1). London, Lexicon Publications, Inc.

Department of Infrastructure (2002) *Melbourne 2030 Implementation Plan 1: Urban Growth Boundary* (draft). Melbourne, Department of Infrastructure.

Domain (2007) *Suburb profiles*: Roxburgh Park 3064. Retrieved in April 2007 from: http://www.domain.com.au/Public/SuburbProfile.aspx?searchTerm=Roxburgh%20Park&mode=research.

Frey, H. (1999) *Designing the city: towards a more sustainable urban form*, E & FN Spon, London; New York.

Giddens, A. (1999) *Runaway world: how globalisation is reshaping our lives*, Profile, London.

Girardet, H. (1999) *Schumacher Briefing No.2 Creating Sustainable Cities*, Green Books, Devon, UK.

Hagan, S. (2001) *Taking shape: a new contract between architecture and nature*, Architectural Press, Oxford; Boston.

Housing and Urban Development Corporation (HUD), (2001) *The Official Information on Linh Dam Housing Projects*, The Ministry of Construction, Hanoi.

Intarakomalyasut, N. (2001) *Mini-Tokyo seen at Muang Thong Thani*. Retrieved from: http://www.siamfuture.com/ThaiNews/ThNewsTxt.asp?tid=1023

Lim, S. and Jaikawang, K. (2003) *DEVELOPED-NATION STATUS: TRT will 'lead the way'*. The Nation (internet edition issue 7th December, Bangkok. Retrieved from: http://www.nationmultimedia.com/page.news.php3?clid=5&id= 105729& usrsess=1).

Manasian, D. (2003) Digital dilemmas: A survey of internet society (supplementary). *The Economist*. **366**: 1–18.

Marx, L. (1964) *The machine in the garden; technology and the pastoral ideal in America*, Oxford University Press, New York.

McGee, T. G. (1991) The Emergence of Desakota Regions in *Asia: Expanding a Hypothesis. The Extended metropolis: settlement transition in Asia* (eds N. Ginsburg, B. Koppel and T. G. McGee), University of Hawaii Press, Honolulu, pp. 3–35.

Newman, P. and Kenworthy, J. (1999) *Sustainability and Cities*, Island Press, Washington, D.C.

Nicholson-Lord, D. (1987) *The greening of the cities*, Routledge & Kegan Paul, London; New York.

Rees, W. (1999) Scale, complexity and the conundrum of sustainability, in *Planning Sustainability* (eds M. Kenny and J. Meadowcroft), Routledge, London; New York, pp. 101–127.

Rist, G. (1997) *The History of Development: from western origins to global faith* (translated by Patrick Camiller), Zed Books, Atlantic Highlands, NJ.

Rowe, P. G. (1991) *Making a middle landscape*, MIT Press, Cambridge, Mass.

Spirn, A. W. (1984) *The granite garden: urban nature and human design*, Basic Books, New York.

Thang, V. (2005) *Development threatens Linh Dam village*. Vietnam News, Hanoi. Retrieved in February 2005 from: http://vietnamnews.vnagency.com.vn/showarticle.php?num=02SUN200205).

Touré, A. T. and B. Compaoré (2003) *Your Farm Subsidies Are Strangling Us. New York Times*, New York. Retrieved in July from: http://www.nytimes.com/2003/07/11/opinion/11CAMP.html?th).

Yencken, D. G. D. and Wilkinson, D. (2000) *Resetting the Compass: Australia's Journey Towards Sustainabilty*, CSIRO Publishing, Collingwood.

Vaitheeswaran, V. (2002) How many planets? A survey of global environment July 6th 2002 (supplementary). *The Economist*, **364**: 1–16.

UNFCCC (1994) *Understanding Climate Change: A Beginner's Guide to the UN Framework Convention*, UNEP/WMO Information Unit on Climate Change, Geneva.

Kiyonobu Kaido and Jeahyun Kwon

10

Quality of Life and Spatial Urban Forms of Mega City Regions in Japan

Introduction

This chapter examines the relationship between regional urban forms and quality of life in Japanese cities. Japan has three mega city regions: Tokyo with 30 million inhabitants; Osaka with 15 million; Nagoya with 8 million. Today, Japan is characterised by a declining population, and one that is rapidly ageing with a commensurate decline in birth-rates. Although the pattern is not uniform in every region, the changes may lead to a different form of urbanized society in Japan. The trend is for people to move from the countryside, small cities and suburbs into mega- or large-cities, and to inner city or central areas. Recently, Japanese central government has developed a new policy for urban regeneration aiming to achieve urban compaction and the revitalization of city centres. Many policy makers and planners are concerned with the relationship between urban form, sustainability and quality of life. It is clear that urban forms and regional spatial structures are transforming under new policies and conditions in Japan.

This chapter discusses the transformation of city regions and regional plans for sustainability. The past changes in, and prediction of, population changes in Japanese regions are examined, showing population decline in local areas, and increases in metropolitan areas, particularly Tokyo, Osaka and Nagoya. The quality of life related to urban density in the Nagoya metropolitan area is explored, testing the claim that higher density can bring higher accessibility to urban facilities. The impact of the process of urbanization in the Nagoya metropolitan area on traditional local cities around the central city is explored, showing a weakening of their centralities. By way of comparison, sustainable urban form and regional plans in Europe are reviewed, using the cases of the Stockholm, Copenhagen, Manchester regions. Certain commonalities of regional structure are explained, using the urban growth and decline cycle model in the city region; illustrated through Paelink and Klassens's theory. Finally, the chapter draws some lessons when planning systems to realize more sustainable mega city regions.

Population change in Japan

The changes in the Japanese population in the past and those predicted for the future are significant (see Fig. 1). After the Second World War, the population of Japan increased by between 5–7% over a period of five years. The rate of increase slowed down after the late 1970s. Since the 1990s the population decline accelerated, occurring in 12 prefectures between 1985 and 1990, and extending to

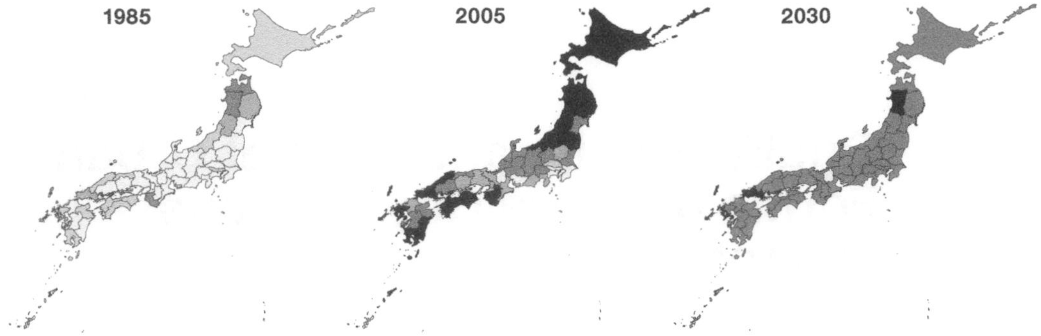
1985
2005
2030

Fig. 1. (above) Changes and predicted changes in the population of the Japanese prefectures

Source: Kaido and Kwon

Key

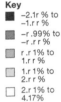

- –2.1r % to –1.r r %
- –r .99% to –r .r r %
- r .r 1% to 1.r r %
- 1.r 1% to 2.r r %
- 2.r 1% to 4.17%

23 prefectures over 1995–2000.[1] After the year 2000, only 15 out of 47 prefectures, most of them located in metropolitan areas, showed an increase in population. By 2005 the overall trend in Japan was of a reduction in population. The predictions show this trend set to continue with the 2005 population of 128 million reducing to 90 million by 2055 (National Institute of Population and Social Security, 2006) (see Fig. 2). While the total population is in decline, the percentage of ageing people over 65 (estimated at 40.5%) is increasing, meaning that Japan will have the most rapidly growing and ageing society in the developed countries worldwide. Now, one of the highest priorities in Japan is how to maintain a good quality of life with a falling birthrate and an ageing society. Recently central government has made changes to the planning system to take account of the reduction in population (described later in the chapter).

Three mega city regions in Japan

Measuring the 'compact city level'

Throughout the 20th century, particularly in the first half, Japan experienced a massive and rapid urbanization. In this process, three mega city regions were formed. To set these mega city regions in an overall context, urbanisation across Japan was measured in order to assess its relative intensity: this measurement is termed the 'compact city level'. One of the integral characteristics of a compact city is higher

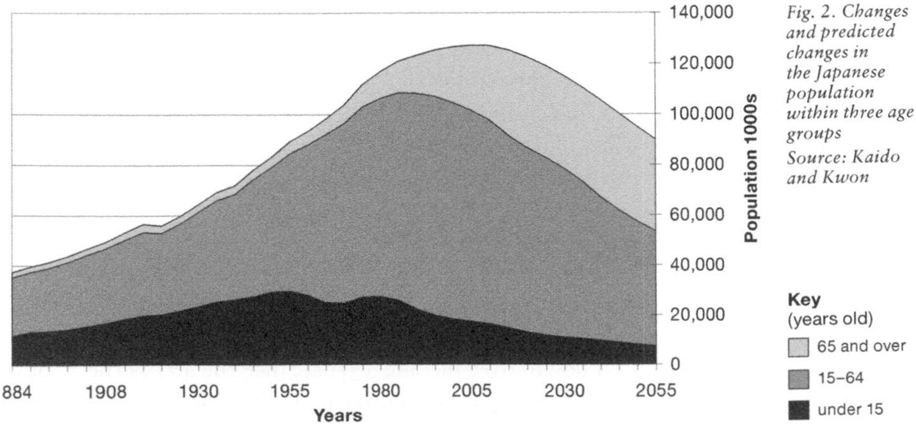
140,000
120,000
100,000
80,000
60,000
40,000
20,000
0
Population 1000s
1884
1908
1930
1955
1980
2005
2030
2055
Years
Key
(years old)
65 and over
15–64
under 15

Fig. 2. Changes and predicted changes in the Japanese population within three age groups

Source: Kaido and Kwon

Group	Population Density (people per km²)	Employment Density (people per km²)	Urbanization Level	DIDs' Population Density (people per km²)	Urban District Level	Compact City Level (integrated account)
1	8641	6166	0.98	11078	0.76	4.26
2	6258	2974	0.95	8936	0.66	2.89
3	1969	822	0.72	7074	0.20	0.66
4	991	463	0.51	5773	0.09	-0.10
5	692	314	0.40	4597	0.06	-0.50

Note: Population and employment density are calculated from the numbers of people per km² of the habitable areas of each group of prefectures. The Urbanisation Level is the ratio of population living in DIDs to the population in the habitable areas. The DIDs' population density is the number of persons per km² in the DID areas. The Urban District Level is the ratio of the DID areas to the habitable areas. The Compact City Level is a measure integrating all five factors.

Table 1. Compact city level

Source: Kaido and Kwon

density; accordingly this measure classifies all 47 prefectures into several groups, assessing statistically the 'compact city levels'. Five indicators are used for analysis: population density; employment density; level of urbanization; population density in DIDs (Densely Inhabited Districts (urbanized areas)); and the urban district level. All the prefectures are divided into five groups, showing the most to the least intense urbanisation level (see Table 1 and Fig. 3). Group 1 includes only the Tokyo prefecture as the most intense compact city area. Osaka and Kanagawa Prefectures are contained in Group 2 with the second highest level. Seven prefectures are in Group 3, including Nagoya and Fukuoka and some prefectures adjacent to Tokyo and Osaka. Seven prefectures are included in Group 4. The result of measuring compact city levels in all prefectures shows that the three mega city regions are the densest and most widespread urbanized areas in Japan – with Nagoya metropolitan area (included in Aichi prefecture) a degree less so than the Tokyo and Osaka metropolitan areas.

Fig. 3. Prefectures grouped according to compact city level scores

Source: Kaido and Kwon

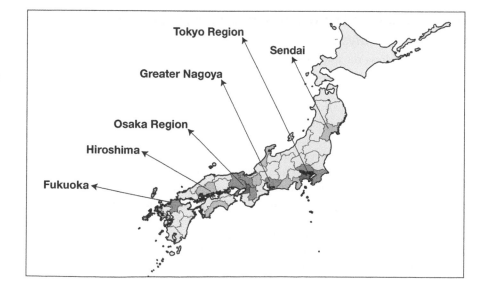

Key
- ■ Group 1
- ▨ Group 2
- ▨ Group 3
- ▨ Group 4
- □ Group 5

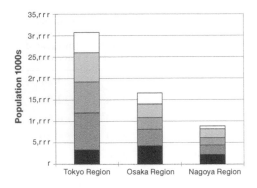

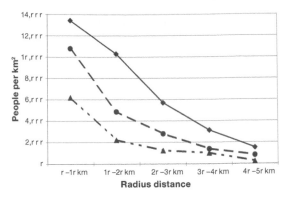

Spatial structure of three mega city regions

Three metropolitan areas, defined within a 50km radius from the regional centre, have different spatial structures. Tokyo metropolitan area has the population of some 30 million, and is acknowledged as a world city along with New York, Paris and London. The spatial structure of Tokyo is that of a continuous, high density built-up area. The inner area, comprising a 10km zone, is not very densely populated compared with other world cities, but is the densest inner city area in Japan, with about 13,000 people per km^2. The population of Osaka metropolitan area is about 15 million with a density in its inner area of 11,000 people per km^2. Nagoya metropolitan area, half the size of the Osaka region, has a population of 8 million and density of around 6,000 people per km^2 (see Figs 4 and 5). These densely populated mega city regions accommodate almost half the population of Japan, and so the question of whether or not they provide their citizens with a good quality of life is a significant one. It is hardly contentious to suggest that people who live in world class mega city regions should expect to enjoy a high, even world class, quality of urban life.

Quality of life and spatial urban form in Nagoya Region
Methods of analysis

In order to address the question about the quality of urban life in these large city regions, Nagoya metropolitan region[2] is examined in some detail. The concept is a complex one, but arguably one key indicator of a high quality of life is the ease of accessibility to urban facilities for every residence in every area, and it is generally considered that the higher the density the easier accessibility becomes. Across both the city and region, because each has its own spatial structure, population densities vary, the lowest being found in suburban areas. Nagoya metropolitan region's density is rather lower than Tokyo or Osaka region. The central city in the metropolitan region is Nagoya City which has a population of 2.2 million and 16 wards. Nagoya City is a part of Aichi prefecture which has a population of 5 million comprising a further 30 cities (not including Nagoya City). Gifu Prefecture adjacent to Aichi is also included in the analysis, and it has a population 2.1 million and includes 14 cities. The number of wards and cities analysed in Aichi and Gifu total 60. The research measured the accessibility of convenience stores, banks and/or post offices,

Fig. 4. (above-left) Population in every distance zones of three mega city regions
Source: Kaido and Kwon
Key
☐ 4r–5r km
▨ 3r–4r km
▨ 2r–3r km
▨ 1r–2r km
■ r–1r km

Fig. 5. (above-right) Spatial structure of mega city regions
Source: Kaido and Kwon

Key
→ Tokyo Region
–●– Osaka Region
–▲– Nagoya Region

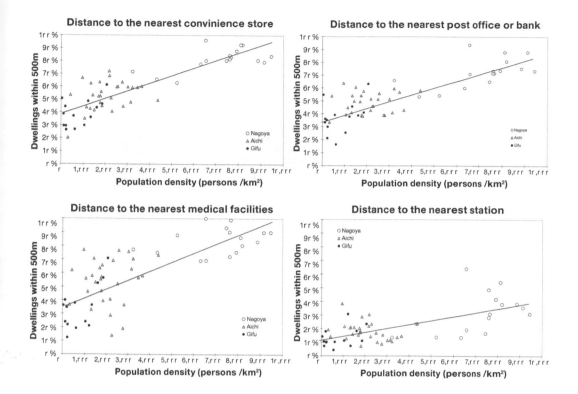

Fig. 6. (above–left) Accessibility to convenient stores
Source: Kaido and Kwon

Fig. 7. (above–right) Accessibility to banks or post offices
Source: Kaido and Kwon

Fig. 8. (below–left) Accessibility to medical facilities
Source: Kaido and Kwon

Fig. 9. (below–right) Accessibility to rail stations
Source: Kaido and Kwon

medical facilities and rail stations and related these to population densities in each of the sample wards and cities. The percentages of homes located within 500 metres of each of the nearest facilities were used for evaluating accessibility. The resulting scatter plot figures and best fit curves were drawn using statistical data based on these measurements.

Accessibility of facilities

Population densities in the 60 wards and cities range between some 10,000 to 300 persons per km², with the wards of Nagoya City having the highest densities, followed by Aichi, and with cities in Gifu having the lowest densities. Convenience stores are generally located in the higher density areas which tend to provide a more commercially viable environment. Convenience stores in the more densely populated wards of Nagoya City are most accessible to residents living there, but accessibility reduces as densities get lower (see Fig. 6). The multiple correlation coefficient of the best fit curve for convenience stores, R2, is a significant 0.725. The other facilities that are almost as significant in terms of accessibility are banks and/or post with a correlation coefficient of 0.6616 (see Fig. 7).

The policy for public medical facilities is to locate them for residential convenience. This occurs in many cases, but in some areas, especially in isolated areas or those with small populations, medical facilities are often lacking. Despite the good intentions, medical facilities are generally less accessible to local people than convenience stores and banks/post offices, as can be seen from the correlation between medical facilities accessibility and population density (see Fig. 8).

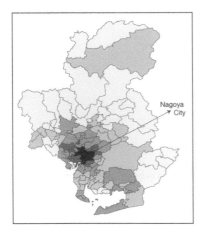
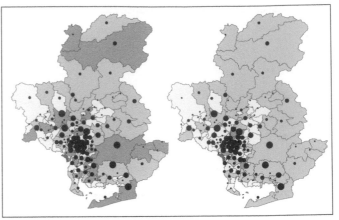

Accessibility to transport, especially to rail stations, is not so easy. The correlation between accessibility and density is lower than for other essential facilities, and absolute levels of accessibility are also lower (see Fig. 9). Even in Nagoya city, the proportion of homes within 500 metres of a rail station is between 60% to 15%, although this is to an extent counterbalanced by good bus service networks. However, in the outer Nagoya Region, it is very difficult to enjoy easy access to a rail station without reliance on buses or private cars.

Lifestyle and private car dependency

The reduction of private car use in daily life is one of the aims in making cities or city regions more compact and sustainable, and this has been of increasing interest to Japanese planners and policy-makers. However, it is very difficult to reach this objective in practice, as public transportation systems need certain levels of population density to be viable. There is inevitably an inconsistency between living in low density environments and enjoying access to urban facilities.

The use of the private car as the sole means of commuting from home[3] varies between the Nagoya and Tokyo Regions. The Tokyo Region has an advanced public transport system, and this has led to a low proportion of private car use of only 11.6 % in the year 2000. Outside Tokyo, the proportion of car use is higher – in the Tokyo Prefecture it is between 22% to 33%, in the Nagoya Region (in the Aichi Prefecture) 51.3%, reaching 64.6% in the neighbouring Gifu Prefecture.

Figure 10 shows car dependant commuting is lower in central areas but higher in suburban areas. This is clearly linked to lower densities, and so raising densities through urban compaction or intensification is one radical strategy to reduce private car use. Another closely associated strategy would be to encourage the development of polycentric-urban forms, including a mixture of uses and employment which could help to decrease the distance of commuting from the surrounding areas. The extent of the daytime population is a useful indicator when identifying polycentric 'centralities' of a city or region. If the proportion of a city's daytime population reduces, it means that the centralities of a city become weaker. Figure 11 illustrates this in the Nagoya Region. Comparing 1985 with 2000, several more densely populated areas have lost daytime population, showing that traditional cities, or

Fig. 10. (above-left) Average per cent of car use for commuting, 2000
Source: Kaido and Kwon
Key

☐ 84%

☐ 72%

▨ 63%

▨ 53%

■ 40%

Fig. 11. (above-right) Proportions of daytime to residential population in 1985 and 2000
Source: Kaido and Kwon

Key
Proportion of daytime population

☐ Lowest

▨

▨

▨

■ Highest

Population

• 1,739–23,444

● 23,445–53,077

● 53,078–96,459

● 96,460–193,004

● 293,005–411,743

polycentric centralities, in the outer Nagoya Region have declined since the 1960s. Traditional local industries, e.g. the ceramic ware industry, textile industry and clothing trade, have declined there. Instead, many housing areas have been developed in the outer Nagoya Region for families who commute to work in central Nagoya and other urbanized areas, a trend that has grown since the 1970s and 1980s. By the year 2000, the area of densest population had shifted to centre on the city of Nagoya and its immediate surroundings. The result is that many cities elsewhere in the region have lost their original identities.

Urban sprawl in Greater Nagoya

Economic growth, urbanization and motorization occurred rapidly during the 1960s and 1970s which, combined with a relaxed planning control system, encouraged the spread of urban sprawl. One of the dominant characteristics of the Japanese planning system has been land-use zoning, with permitted development for a broad range of mixed land use. Boundaries were designated to contain urban growth ('limit-to-growth') in cities with a population of 100,000 and over. However, urbanization inside 'limit-to-growth' boundaries has often proved too limited to contain the urban growth within some boundaries. Thus in the Nagoya metropolitan region spatial development has been characterised by an expansion of urbanization and low density sprawl in the suburbs, which accelerated rapidly after the 1960s. In the Aichi Prefecture, including Nagoya City, and neighbouring Gifu Prefecture, gross population rose by 186% from a baseline in 1960. Over the same period, the number of DID areas rose by 316% while their density reduced to 56% of the 1960 baseline (see Table. 2).

Table 2. DIDs in Aichi and Gifu Prefectures Source: Kaido and Kwon

Year	1960	1970	1980	1990	2000
Population	100	135	158	178	186
Areas	100	168	232	300	316
Density	100	77	65	57	56

Notes: DIDs are defined as urbanised areas with 4,000 persons per km² and over
index, 1960 = 100

At first, in the 1960s expansion, new housing areas were developed some 10–20km from the regional centre. Subsequently further housing areas were developed progressively in suburbs ever more distant from the centre (see Fig. 12). By contrast, the population of the central area (a 10km radius from centre) did not grow, and indeed the population of the centre of Nagoya City reduced from the 1960s to the end of 1990s. At the end of 1990s, a modest re-population of the central area occurred, and many multiple dwellings were developed – a process of re-urbanization and return to the city centre. This overall legacy has presented new urban problems for Japan, both in relation to the urban sprawl that is destroying good townscape and urban environment, and an unsustainable ageing housing stock in suburban areas.

The Japanese context of rapid growth, followed by shrinkage in population, the resultant spread of urbanisation at low densities into the surrounding city regions, and unsustainable commuting problems, raises the question about possible solutions. The aim of planners has been to look for 'compact' solutions, either through urban

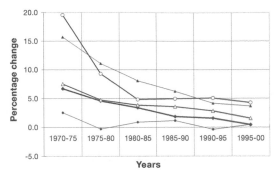

Fig. 12. Changes in population by distance zones
Source: Kaido and Kwon

Key
(km radius)
◆– 40–50
–△– 30–40
–▲– 20–30
–○– 10–20
–●– 0–10

intensification, or by strengthening inherently polycentric spatial structures to provide sub-centres in easy commuting distances of local populations. The next section explores three cases in Europe to assess whether the experience there can provide lessons from successful practice that could be applied, with caution, to Japan.

Urban forms in European city regions

Compact city and polycentric urban form strategies in the European Union (EU)

In 1999 the EU published its *European Spatial Development Perspective* (ESDP) recommending the development of polycentric urban forms as a key policy direction (see Fig. 13). In addition ESDP recommended five guidelines for the sustainable development of towns and cities:

- control of physical expansion;
- mixture of functions and social groups;
- wise use of resources – conservation management of the urban ecosystem;
- better accessibility by different types of transport; and
- conservation and development of natural and cultural heritage.

The ESDP also notes that member states and regional authorities should pursue the concept of the compact city (p. 22). The EU published its environmental political agenda in a document, *Thematic Strategies* (EU, 2002), in which sustainable urban design and construction are given due prominence. A working group on urban design published *Urban Design for Sustainability* in 2004 (EU, 2004). The report

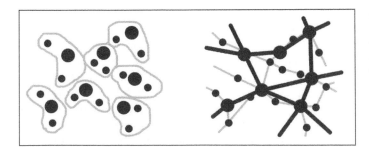

Fig. 13. Polycentric Urban pattern models
Source: Adapted from ESDP, EU (1998)

reinforces the validity of the Compact City Strategy advocated by the European Commission in its 1990 *Green Paper on the Urban Environment* as a basic model for sustainable urban design. It also promotes polycentric development, stating that '[i]t should draw on other approaches such as "decentralized concentration" at the urban regional scale' (EU, 2004). These key EU documents provide a policy context against which to examine three case studies of urban regions in Europe – Stockholm, Copenhagen and Manchester.

Population change and regional urban forms
The Stockholm Region

The Stockholm regional plan recommends a sustainable 'Green Compact City' strategy for the city as a whole (EU, 2004). Compared with Japan, Sweden is a small country with a total population of 10.5 million in 2005. Stockholm's city region expanded after the Second World War, the population increasing from just over 1 million in 1950 to 1.8 million in 2005. While the region was expanding the city was shrinking, and the population of the inner city areas halved from 440,000 in 1950 to 226,000 in 1980. Subsequently, the population of the inner city areas gradually recovered rising to 289,000 in 2005. The majority of the growth occurred in the suburbs with an almost quadrupling of population between 1950 (258,000) and 2005 (951,000). Much of the suburban development comprised housing areas built around suburban rail stations.

Copenhagen Region

Copenhagen exhibits a number of characteristics of growth and decline similar to Stockholm, although its country, Denmark, is smaller, with a population in 2004 of 5.3 million. The population of the Copenhagen Region is some 1.8 million (in 2002), almost the same as the Stockholm Region. This has been gradually increasing from 1.6 million in 1960. By contrast, the population of the central and inner city areas in Copenhagen decreased by half until about 2000, when the trend reversed, with a gradual increase in population moving back to the city (see Fig. 14, Illeris, 2004).

Figure 15 illustrates Copenhagen's four steps of development in its well known five-finger regional plan. The city centre was developed in medieval times and the population was around 130,000 until the mid 19th century. The second

Fig. 14. (left) Population changes in each area of Copenhagen Source: Based on Illeris (2004)

Key

- City centre
- Inner areas
- Outer city
- County of Copenhagen
- Rest of region

Fig. 15.(right) Copenhagen's development Source: Based on Jørgensen (2004)

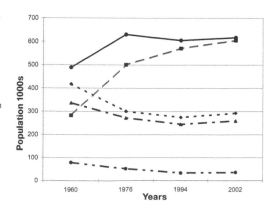

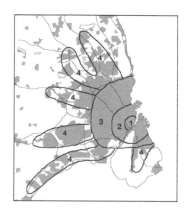

development occurring from the beginning of 20th century was known as the Tram City (for fairly obvious reasons). The third expansion was served by a train system. These three areas comprise the palm of the 'hand'. After the Second World War, suburban areas rapidly grew, making the five fingers of the plan. Today, the planning authorities want to locate urban functions around rail stations within distances of between 500m and 1km (Jørgensen, 2004).

Manchester Region (see Fig. 16)

Manchester is an old industrial city and its urban form is different from the two city regions of Stockholm and Copenhagen. Today, Greater Manchester Region has a population of some 2.5 million. Manchester's urbanization stemmed from the 18th century industrial revolution and rapid development of cotton industries, and at its peak the population of Greater Manchester reached 2.7 million (in 1931). The population then declined until the early 1990s. At the heart of the region, Manchester city's population also declined until early 2000s, and for a time it was known as a shrinking city (Ostwalt, 2005), but then after 2001 it began to recover. Industry declined again after the energy crisis in 1970s. The rate of unemployment (long term out of work over one year) increased from some 20% in late 1980s to 43.2% in 1994. Manchester had many unfit houses especially in its inner city areas, and urban regeneration based on neighbourhoods became the key strategy of Manchester City Council. The success of Manchester's regeneration, combined with the upturn of the UK economy and reduction of unemployment rates has been reflected in the re-population of the city – the 'shrinking city' of Manchester had entered a new era (Ferrari and Robert, 2004, p. 50). But the Manchester City Region, although it has geographically mixed areas with a vibrant economy, also contains areas with high levels of deprivation (AGMA, 2006, pp. 4–7). Even so, the city and its region appear to be moving towards 'world class' status.

Urban growth and decline cycle model in the city region

From the examination of cases in Japan and Europe, the question arises: how closely do population changes in these four city regions match theory? In 1979,

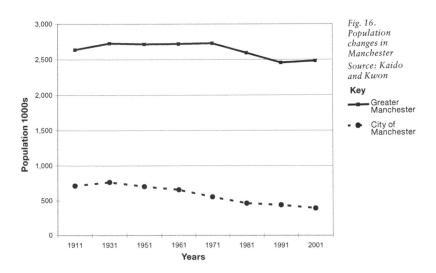

Fig. 16.
Population changes in Manchester

Source: Kaido and Kwon

Key

——■—— Greater Manchester

- ● City of Manchester

Paelink and Klassens published an influential theory about the transformation of spatial urban forms. They divided urban areas into the central city, suburban areas and the urbanized region. The stages of urbanization are illustrated at three levels: urbanization, suburbanization and anti-urbanization. They argue that: the population of the central city grows rapidly in the first stage and then decreases; the population of suburban areas first decreases and then grows in a second stage; the population of the region grows in both the first and second stages; at the last stage of urbanization, the population decreases in every area especially in the central city and region. The basis for this theory was through research into the transition of population in many European city regions in first half of the 1970s.

The four cases in this chapter show a different sort of transition of city regions. Figure 17[4] shows a general model of the development of a city region. The horizontal axis shows the growth or decline of the inner area or central city. The vertical axis shows the growth or decline of suburban areas. In the first stage of development the inner area grows (1) and then the city area expands widely into the suburban areas (2). After that, both the inner area and suburban areas grow (3). Next, the population of the central city or central regional zone declines, at the same time as suburbanization continues (4). After this stage, suburban growth reduces (5), and lastly both the inner areas and suburban areas decline (7). This stage is often termed the 'shrinking city' or reverse-urbanization. However, with effective policies, the city area is likely to be regenerated firstly in the inner area (8) and then in suburbs (6). Sometimes the stages of urban regeneration (or re-urbanization) may take different paths, and these are indicated in the dotted lines in the figure.

Fig. 17. Growth and decline cycle in city regions
Source: Kaido and Kwon

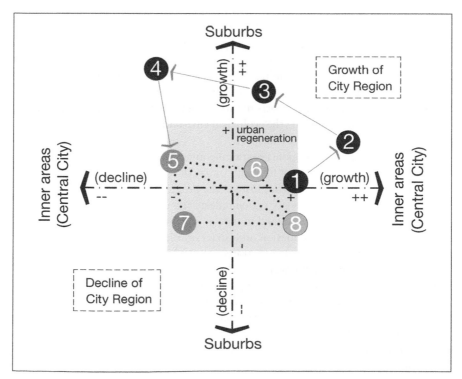

Growth and decline in the cycle of a city region is not a natural phenomenon, but rather its physical conditions are related to social and political influences and economic shifts and policies. The most advanced of today's economy is service industries based on culture, information, IT or creativity. That economy needs creative people who, generally speaking, seem to prefer working and living in places close to the city centre. Many city regions, especially for 'world class' cities, are now in competition. Many city regions are tackling the regeneration of their urban spaces. The economic structure has changed dramatically since the time when Paelink and Klassens first published their theory. Attractive central cities or central areas are essential for economic success. If the central city achieves a successful regeneration, the city region may also be able to grow. In other words, spatial strategies and urban forms need to be commensurate with achieving a good quality of life for all residents.

Types of spatial urban form in the city region and regional plans

While the general trends of population changes, growth and decline of urban areas are apparent, the outcomes are reflected in various identifiable spatial forms of mega city regions in both European and Japanese planning strategies.

The Office of Regional Planning and Urban Transportation (RTK) cooperated with 26 municipal authorities and regional offices of central government to draw up the Stockholm Region Plan 2000–2030 (RTK, 2003) which envisages one regional centre and seven regional cores. The plan has three basic objectives: international competitiveness; good and equal living conditions; and a long-term sustainable living environment.

The advanced five finger plan of the Copenhagen Region Plan 2005 shows an extension of the fingers. The principles of the five-finger plan include easy access to infrastructural facilities such as green spaces, bike paths, commuter trains and motorways. The finger plan has steered growth for almost 60 years (Cahasan and Clarke, 2004). The revised Planning Act sets regulations for planning governing the 34 new municipal councils in Greater Copenhagen in 2006. These two regional plan strategies show long term efforts to satisfy contradictory goals, e.g. permitting urban growth, while conserving the ecological environment and providing access to service facilities by using public transport.

The Manchester City Region Programme aims to achieve a 'world class' city region at the heart of the North of England. Unlike the two other European examples there is no comparable spatial plan; instead the programme states that it is critical that the spatial strategy for the city region reflects and underpins a dual approach to regenerate deprived communities and to exploit its existing assets.

The relocation of the population, industries or other functions in metropolitan regions to local areas was a thematic strategy in the National Development Plan of Japan after the Second World War until the late 1990s. New strategies have been drawn up from 2006 to cope with population reduction and the rapidly ageing society. They aim to control urban sprawl more rigidly and revitalize city centres. However, the Japanese planning system has no regional planning system. There is a similarity in that both shrinking Greater Manchester and Greater Nagoya have no clear spatial plan strategies; however, they aim to encourage a polycentric urban form.

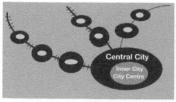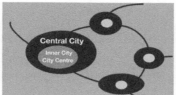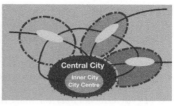

Conclusions

Fig. 18. Patterns of sustainable urban forms in city region Source: Kaido and Kwon

From the analysis in this chapter, three models of spatial forms of mega city regions are abstracted (see Fig. 18). 'World class' city regions should keep the balance between a high quality of life, characterized by the accessibility of essential facilities, and economic and spatial growth, characterized by sustainable urban forms. Of course, spatial growth does not simply mean the physical expansion of urban areas. A city region, whether 'world class' or not, should have appropriate spatial strategies with a regional governing system and sustainable urban forms. City regional planning and appropriate urban design are essential in supporting a good quality of life and social sustainability.

As for city regional planning, some important lessons emerge from the case studies. It is clear that higher densities and compact urban forms are essential to enjoy quality of life, especially with respect to ease of access to essential facilities. The central city is showing signs of becoming more important. In the sample of central cities considered here, it is evident that they have attracted inhabitants from suburbs. This is partly due to urban regeneration and vibrant central cities, which are also needed to support the sustainable growth of the city region.

The essential need to keep a balance between economic growth and urban containment does not rest on a single physical form. As seen here, the spatial forms of city regions studied are of various types, and these can be both successful and potentially sustainable. However, there also needs to be an effective regional planning body, or at the least effective and efficient cooperation of local authorities, if successful regional panning strategies are to be realised. Furthermore, a 'world class' city needs a spatial strategy based not only on the city as a whole, but also in the context of a 'world class' city region.

Notes

1. Japan is divided into 47 sub-national jurisdictions, known as prefectures. These are areas generally larger than cities or towns, and have governmental bodies led by a directly elected governor with a single chamber parliament (see http://en.wikipedia.org/wiki/Prefectures_of_Japan).
2. The Nagoya Metropolitan Area comprises a 50km radius distance from central Nagoya embracing three Prefectures, Aichi (including Nagoya City), Gifu and Mie. However the Nagoya Region used here, to simplify the characteristics of areas by using population densities, includes just the two prefectures, Aichi and Gifu. Commuting by employed persons and persons attending school, 15 years of age and over, based on place of usual residence
3. In Fig. 17 the numbers, surrounded by a circle, refer to stages of development. Numbers 1–4 mean a development stage, numbers 5 and 7 mean decline, and numbers 6 and 8 mean regeneration. The dotted lines between numbers 5, 6, 7 and 8 show the different paths to urban regeneration.

References

AGMA (2006) *The Manchester City Region development Programme 2006: Accelerating the Economic Growth of the North*, Association of The Greater Manchester Authorities (AGMA), Manchester.

Cahasan, P. and Clark, A. (2004) *Copenhagen, Denmark, 5 Finger Plan*. Retrieved from: http://depts.washington.edu/open2100/Resources/1_OpenSpaceSystems/Open_Space_Systems/copenhagen.pdf.

EU (1990) *Green Paper on the Urban Environment*. Retrieved from: http://ec.europa.eu/environment/urban/pdf/com90218final_en.pdf.

EU (1999) *The European Spatial Development Perspective (E.S.D.P.)*, The European Consultative Forum on Environment and Sustainable Development, Secretariat: European Commission DG.XI. Retrieved from: http://ec.europa.eu/environment/forum/spatreport_en.pdf.

EU (2002) *Seven Environmental Thematic Strategies*. Retrieved from: http://ec.europa.eu/environment/newprg/index.htm.

EU (2004) *Urban Design for Sustainability*, Final Report of the Working Group on Urban Design for Sustainability to the European Union Expert Group on the Urban Environment. Retrieved from: http://ec.europa.eu/environment/urban/pdf/0404final_report.pdf.

Ferrari, E. and Roberts, J. (2004) *Regrowth of a shrinking city*, Shrinking Cities – Manchester/Liverpool 2, Federal Cultural Foundation, Germany.

Illeris, S. (2004) *How did the population in the Copenhagen region change 1960-2002?*, Cities in Transition, Dela 21, Cities in transition, pp. 404–421. Retrieved from: http://www.ff.uni-lj.si/oddelki/geo/publikacije/dela/dela_21.htm.

Jørgensen J. (2004) *Copenhagen – Evolution of the Finger Structure, From Helsinki to Nicosia* (ed. Genevieve Dubois-Taine), p. 187-197, METL/PUCA, (C10 - COST Action).

Oswalt, P. (ed.) (2005) *Shrinking Cities – Volume 1 International Research*, Hatje Cantz Verlag, Ostfildern-Ruit, Germany. Retrieved from: http://www.shrinkingcities.com/publikationen.0.html?&L=1.

RTK (2002) *Regional Development Plan 2001 for the Stockholm Region (RUFS 2001)*, The Office of Regional Planning and Urban Transportation (RTK), Stockholm. Retrieved from: http://www.rtk.sll.se/english/About_us/about_rufs.htm.

11

Jiaping Wu

Global Integration, Growth Patterns and Sustainable Development:

A case study of the peri-urban area of Shanghai

Introduction

Economic globalisation has become 'a strong force' driving the urban development of Asian countries. The formation of megacities which have 10 million or more inhabitants, or a new class of urban agglomeration of Extended Metropolitan Regions (EMRs), has increased in number and scale (McGee, 1991, 1995; Douglass, 2000). It is estimated that the urban population in East Asia will increase by approximately 200 million people over the next twenty-five years, of which 40% will occur in the peri-urban area (PUA).[1] The environmental challenges of these megacities – high rates of land and energy consumption, severe pollution of air, water and soil, and decrease of quality of life – have become of increasing concern (Marcotullio, 2001; Yueng, 2001; Webster, 2002; Ng and Hills, 2003).

It has often been said that sustainable urban development means different things to different people (Gale and Cordray, 1994; Satterhwaite, 1997). The environmental view of sustainable development focuses on the stability of biological and physical systems. The economic approach to sustainability is to use resources efficiently in the pursuit of consistent economic growth. Very often, the exploitation of natural resources in some cities or regions goes beyond the capacities of their ecological systems. Economic globalisation has increasingly linked the sustainability of the world together, so that environmental problems cross national boundaries through international trade and global flows of manufacturing activities, in particular those with high level of pollution intensity. For instance, recent research shows that CO_2 embodied in goods exported from China to the US accounted for 14% of China's CO_2 emissions in 2003 (Shui and Harriss, 2006).

Attraction of foreign investment is an important strategy for the economic development of developing countries. To integrate with the world economy more effectively, many countries in Southeast Asia have created different types of targeted zones in the peri-urban areas of their large cities to accommodate foreign direct investment (FDI). Fuchs and Pernia (1989) noted a decade ago that the primate cities in Thailand, the Philippines, Malaysia and Indonesia, for example, dominated the concentration of foreign investment; each accounted for 80–90% of inward FDI of the nation. The unswerving concentration of FDI in these areas has created a ring of manufacturing plants in the peri-urban areas. This has been well demonstrated in Jabotabek,[2] Bangkok, Kuala Lumpur and other of Asia's megacities, placing great pressure on environmental resources available for sustainable development of these cities. In many places, these 'will affect the future shape of society, the sustain-ability of economic development, and the environment of these cities' (Webster, 2002, p. 6).

Foreign investment is a main vehicle for China's global integration, which has become significant in China's urban transition. The absorption of FDI has repositioned large extended coastal metropolitan regions in the global economy and restructured the metropolitan areas of large Chinese cities (Lin, 2001; McGee *et al.*, 2007). Shanghai has experienced the most rapid economic growth of any Chinese coastal city in the last ten years and the most furious pace of change in the built environment (Wu, W., 1999a; Wu, F., 2000). The social, economic and physical transformation of the peri-urban area in Shanghai, an important environmental resource in support of the city's sustainability, is no less significant, but has attracted little attention.

This case study focuses on the urban transformation in the urban–rural interface in Shanghai, aiming to provide some empirical information on the links between economic globalisation and sustainable development in this mega urban context. It examines the causes and consequences of Shanghai's urban expansion by addressing some principles of sustainable development, including equitably offering a good quality of life to its residents without compromising the welfare of its surrounding areas, increasing capacity to handle change, and minimising automobile dependence and environmental damage.

This chapter consists of five main sections. In the first section, the policy change about urban planning and development in the peri-urban area of Shanghai, which shifted from central planning to global integration, is reviewed. The growth of foreign direct investment and its impact on restructuring of production spaces in Shanghai is then charted and discussed. The third section focuses on population movement and social and spatial transformation of the peri-urban area, in particular the patterns of peri-urbanisation. Discussion of how this growth pattern has emerged to challenge the sustainable urban development of the city comprises section four. The environmental impact of FDI, links between growth patterns and sustainability and the vulnerability of urban health in peri-urban areas are explored. The discussion is followed by a conclusion.

Reconfiguration of urban space: towards global integration
Central planning and Chinese cities

Chinese cities were characterised by a sharp rural–urban division during the central planning period. This included China's policy of categorising urban population as either urban or rural, and structuring of individual cities into urban districts and suburban rural counties. In most cities, areas of suburban rural counties were much larger than those of urban districts (Zhou and Shi, 1993). The two parts of the metropolitan areas were assigned different economic roles. Land use in urban districts was committed to industrial development while rural counties focused on providing the city with industrial raw materials and agricultural products. A household registration system was introduced to categorise Chinese people as either urban or non-urban. Most people who held urban household registration concentrated in the urban districts, while the agricultural population was bound to rural counties. Rural–urban migration was strictly controlled. Central plans rather than market forces determined rural–urban relations, with the government purchasing almost all agricultural products while distributing industrial products to rural inhabitants.

In 1959 and 1960, ten rural counties from Jiangsu and Zhejiang provinces were put under the jurisdiction of Shanghai, increasing the municipality from 590 km² to 6,340 km². During most of the period between 1960 and 1980, Shanghai's metropolitan area was organised into ten urban districts with an area of approximate 280 km², and ten suburban rural counties covering the rest of the metropolitan area. These rural counties are referred to as the peri-urban area (PUA) in this chapter (see Fig. 1).

Fig. 1. The administrative structure of Shanghai

Source: Jiaping Wu

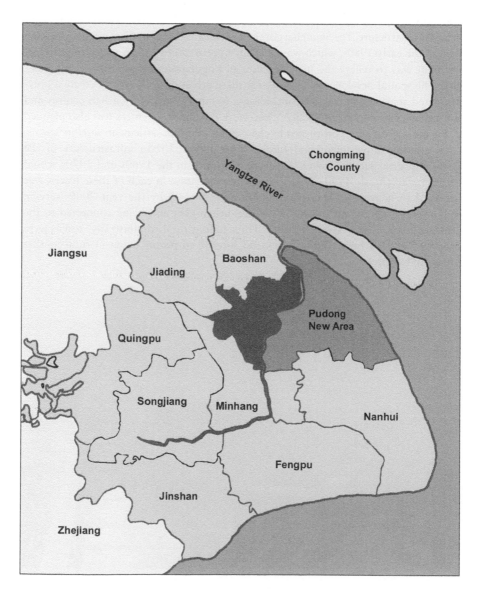

The peri-urban area was a significant environmental resource in support of the city's self-sufficiency in vegetables, edible oils, milk and freshwater fish, as well as providing a great proportion of grain needed by the city (see Fig. 2). It was organised into 230 people's communes, the basic units formed to carry out agricultural production plans and to bond rural labour to their localities in order to fulfil those plans and avoid overpopulation in urban districts. Up to 1978, there were 5.5 million people living in the rural counties, accounting for half of the city's total. Some 76% of rural labourers were engaged in agricultural activities. The boundaries between rural counties and urban districts constrained the central- led 'urban' expansion.

Strategies of global integration

In the 30 years of socialist planning, the concentration of urban economic activities and population in the urban districts left the peri-urban area without significant urban infrastructure. The Shanghai government created a comprehensive metropolitan plan in the early 1980s, which was centrally approved in 1986. A major objective of the plan was to relieve the high densities of population and industry in the inner city. The spatial development strategy for these urban districts was to create a poly-nucleated structure for the city, including a number of ancillary urban centres and sub-urban centres in the inner city. A network of suburban centres was also planned in the peri-urban area, constituted by three types of urban settlement: satellite towns, rural administrative towns and rural market towns. Urban infrastructure of the seven industrial satellite towns that were developed in the 1960s and 1970s would be improved (Atash, 1990; Yan, 1985). The population in each of these towns was expected to increase to between 100,000 and 300,000 by the year 2000, serving as comprehensive urban centres in the peri-urban area and being connected to the central city by an expressway network. Three to four rural administrative towns were proposed in each rural county, to extend services to people living in rural market towns and villages.

Fig. 2.
Agricultural landscape in peri-urban Shanghai
Source: Jiaping Wu

Table 1. Planning divisions and core business in Pudong's themed zones

Source: Jiaping Wu

	Lujiazui–Huamo division	Jinqiao–Qinningshi division	Waigaoqiao–Gaoqiao division	Zhangjiang–Baicai division	Zhoujiadu–luli division
Planned Area (km²)	30	33	62	27	35
Expected Population	500,000	450,000	300,000	220,000	550,000
Themed zones	Financial and Trade Zone	Export Processing Zone	Free Trade Zone	High-Tech Park	

The proposal was challenged soon by both the formulation of FDI-oriented planning policies and ways of deciding on FDI locations. The three Economic and Technological Development Zones (ETDZs) were first established in Minhang, Hongqiao and Caohejing areas in the peri-urban area in the middle of 1980s. A number of targeted development zones with different labels and industrial specialisations, and designated by different levels of government, have been established in the peri-urban area. The opening of the Pudong New Area in 1990 as a Special Economic Zone (SEZ) was the most significant thrust; it is a huge triangular area of 520 km². The planning area was divided into five divisions. The development of each division was centred on a themed zone, designed to respond to a particular type of FDI. These zones were separated by greenbelts, with each independently developing its residential, commercial, recreational and educational facilities. Four themed zones were centrally approved: these are the Waigaoqiao Free Trade Zone, the Jinqiao Export-Processing Zone, the Lujiazui Financial and Trade Zone and the Zhangjiang High-Tech Park (see Fig. 3 and Table 1).

Fig. 3. The planning of Pudong New Area and location of centrally approved themed zones

Source: Jiaping Wu

Key

a- Waigaoqiao Free Trade Zone

b- Jingqiao Export Processing Zone

c- Lujiazui Finance and Trade Zone

d- Zhangjiang High-Tech Park

e- Zhoujiadu-luli division

f- Pudong International Airport

Fig. 4. Emerging
industrial
landscape in
peri-urban
Shanghai
– the Songjiang
Industrial Park
Source: Jiaping
Wu

The municipal government designated its first Industrial Park (IPK), with an area of 20 km², at Songjiang Town, thirty kilometres from the downtown on the south-west of the peri-urban area in 1994 (see Fig. 4). Eight other industrial parks in Jiading, Jinshan, Fengpu, Xinghuo, Zinzhuang and Chongming, Baoshan and Qingpu have been established successively. Each is on a similar scale and has a similar role in attracting foreign capital and rearranging the city's industrial spatial system. By 2006, the total area of ETDZs, and the themed zones in Pudong and IPKs reached 320 km², larger than the area of the inner city. In addition to this,

Name	Location	Planned area in 2005 (km²)
Hongqiao ETDZ	Hongqiao area in Changning district	0.65
Minhang ETDZ	Minhang district	3.5
Caohejing ETDZ	Caohejing area in Xuhui district	14
Songjiang IPK	Songjiang county seat	20.56
Jinshan IPK	Shanyang town in Jingshan district	58
Kangqiao IPK	Zhoupu town in Pudong	26.8
Xinzhang IPK	Xinzhang town in Minhang district	18
Fengpu IPK	Xidu town in Fengxian county	20.8
Jiading IPK	Jiading county seat	24.8
Qingpu IPK	Qingpu county seat	56
Baoshan IPK	Xulian town in Baoshan district	5
Xinhuo IPK	Nanhui county	20

Table 2. ETDZs
and Municipal
industrial parks
in Shanghai
Source: Jiaping
Wu

Note: The area of the ETDZ and IPK has been extended from their original plans. The data is updated from the official website of Shanghai government (www.sh.gov.cn) in 2007.

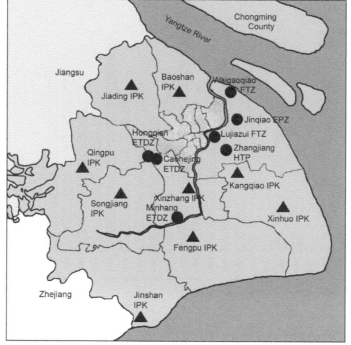

Fig. 5. The location of ETDZs, Pudong themed zones and municipal-designated IPKs in Shanghai Source: Jiaping Wu

Key

● Central-designated ETDZs and themed zones in Pudong

▲ Municipal designated Industrial Park

lower level governments established their own development zones within their jurisdictions to compete with the higher governments. By 2001, there were thirty-six development zones established by urban districts and counties, and one hundred and ninety six established by townships, in addition to the numerous areas designated by many other institutions and private firms. Shanghai's peri-urban area has become a patchwork of areas; some are oriented to foreign investment and others that are not, while the governance of the area as a whole has multiplied and fragmented (see Fig. 5 and Table 2).

The metropolitan area has been divided into 'central designated places', 'municipal-designated places', 'district-designated places', 'town-designated places' and general urban places. An administration commission is usually established to manage urban development of the designated areas. The administration commissions for industrial parks, designated by the municipal government, have the same authority as the urban district/county governments, while the authority of ETDZ administration commissions, including the themed zones in Pudong, are equal to that of the municipal government in terms of approving FDI. For example, an Administration Commission of Pudong New Area (ACPNA) was established in 1993 to take over responsibility for Pudong's urban development after it was designated as a SEZ. The ACPNA, headed by a vice mayor of Shanghai, functioned administratively as an urban district but not as a legislative government body until the urban district government of Pudong New Area was officially established in 2000. Development Corporations were established below the ACPNA in each themed zone to manage and facilitate land development of the zones in an entirely entrepreneurial manner. The management of the rest of the area in Pudong has

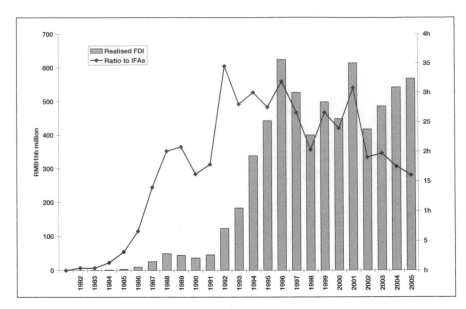

Fig. 6. The growth of FDI in Shanghai and its ratio to the city's total investment in fixed assets (IFAs), 1982–2005

Source: Shanghai Statistical Yearbook (2001; 2005; 2006)

remained with rural town administrations. This dual management has led to sharp contrasts between the FDI-oriented themed zones and the rest of the area. The four themed zones have been the growth engines of Pudong, which together contributed 87% of its total industrial output by value in 2000 – the remaining area stayed rural in character.

The growth of FDI and restructuring the space of production

The past two decades have witnessed a huge amount of FDI flowing into the city, in particular since the early 1990s (see Fig. 6). By the end of 2005, the city's cumulative FDI totalled US$98.8 billion in over 36,000 projects. It has attracted nearly one tenth of the national total FDI annually since the mid 1990s, albeit with fluctuations in some years. Some 51.5% of FDI in the city concentrated on manufacturing (Shanghai Statistical Bureau, 2006). Shanghai has also become a locus of FDI in the global context, attracting 1% of the world's FDI and over 2% of FDI directed to developing countries in the 1990s (UNCTAD, 2000; Shanghai Statistical Bureau, 2001). The utilisation of foreign capital has become the significant driving force of urban transformation in terms of both capital formation and location, increasing from almost none in early 1980s to 30% of Shanghai's total investment in fixed assets in the 1990s. The concentration of foreign capital is unlikely to follow the central allocation, which has gradually transferred the direction of industrial production and distribution processes to match the preferences of foreign investors.

The growth and distribution of FDI in the peri-urban area has been very important in the establishment of development zones, in particular those designated by the central government (Wu and Radbone, 2005). The first three ETDZs, in Hongqiao, Minhang and Caohejing have been the foci of foreign investment in the city. Pudong has played a predominant role in attracting FDI since the 1990s. In 2004, the city successfully attracted 4,321 foreign invested projects with an investment of US$11.69 billion; of these, 39% of the projects and 55.5% of investment went to Pudong (Shanghai Statistical Bureau, 2005).

FDI in other urban districts and rural counties in the peri-urban area has increased since 1992 due to the decentralisation of FDI project approvals and the restriction of pollution-intensive FDI projects in the inner city. The municipal government published guidance for FDI utilisation that restricted some polluting industries from locating in the inner city. In 2004, the area attracted 37.7% of the city's projects and 27.6% of its total investment. FDI in this area is dominated by small size projects and manufacturing activities. Over four fifths of total FDI in the area went to manufacturing activities, 16% to tertiary industries and only 2% to agriculture. By contrast, 80% of FDI that went to the inner city concentrated on tertiary activities over the same period.

The spatial change of production was reinforced by land use reform and the relocation of domestic manufacturing. The introduction of a land market in 1988 resulted in high land prices in the inner city areas, which made many manufacturers relocate their operations to the peri-urban area. In 1990, 58% of Shanghai's industrial output was produced in the ten inner urban districts. A planning policy was proposed in the mid 1990s that permitted only the third of these manufacturing activities that were compatible with inner city's functions to stay, while another one third might be closed down due to poor performance, and the remaining third would be moved out of the inner city (McGee *et al.*, 2007).

The past two decades have witnessed a rapid growth of the peri-urban area's share of total industrial output and a decline in the share of that coming from the inner city. The share of industrial output produced by the inner urban districts decreased from 71.6% in 1985 to 11.4% in 2005, while the peri-urban area's share increased by 60% during the same period (see Table 3). This spatial shift of industrial location is accompanied by the change in industrial ownership. In 1978, the state monopolised 92% of the city's industry by output value, with the remaining 8% produced by collective industries (Shanghai Statistical Bureau, 2000). Sino-foreign joint and foreign-solely owned manufacturing enterprises controlled over 70% of the city's total industry by output value in 2005 (ibid., 2006).

		Inner city		PUA	
Years	Total Output	Output	% of total	Output	% of total
1985	83.0	59.4	71.6	23.6	28.4
1990	105.9	61.5	58.1	44.4	41.9
1995	466.3	162.8	34.9	303.5	65.0
2000	660.4	145.0	22.0	514.2	77.9
2005	1576.8	179.8	11.4	1338.9	88.6

Table 3. Changing spatial shares of industrial output by value (RMB billion)

Source: Shanghai Statistical Yearbooks (1986; 1991; 1996; 2001; 2006)

Note: Shanghai experienced many administration changes. The area of inner urban districts was 261.4 km² in 1985, then extended to 280 km² in 1995 and 289 km² in 2000

Population movement and growth patterns

The redistribution of production space has resulted in changes in the geography of the population in Shanghai. In 2005, foreign firms directly employed 1.4 million of people, accounting for 53.1% of Shanghai's labour force. Shanghai's long-term inhabitants increased from 13.01 million in 1995 to 17.78 million in 2005. 96.2% of the newly increased population lived in the peri-urban area in particular urban districts of Pudong, Minhang, Baoshan, Jiading and Songjiang. The ten-year period saw over 1.3 million more people in Pudong, increasing its population from 1.49 million in 1995 to 2.79 million in 2005. 1.16 million people added to the Minhang district while population in Baoshan and Songjiang districts increased by 0.61 million and 0.39 million respectively over the decade.

The peri-urban area has become the favourite destination of a 'floating population', who change their living places for jobs, business and social services purposes for a period of six months or less, but without changing their household registration. Since 1983, when restrictions on rural-to-urban migration were lifted, the increase of the 'floating population' has been an important phenomenon in China's urbanisation (Solinger, 1999). The 'floating population' in Shanghai increased from 1.06 million in 1988 to 3.87 million in 2000, mainly coming from the neighbouring provinces of Jiangsu (35.8%), Zhejiang (12.8%) and Anhui (27.8%). Two thirds of them were found in the peri-urban area and 78% of them chose to live in farmer's rental housing or temporary quarters (Shanghai Statistical Bureau, 2001). The concentration of floating populations in Minhang urban district for example was greater than that of permanent local residents. Common features of their housing conditions are overcrowding and lack of facilities. Per capita housing is less than eight square meters and fewer than 20% have their own independent kitchen and bathroom. Housing ownership among these migrants was only 0.9% (Wu, 1999b).

The competition between development zones in attracting FDI, along with the movement of the population, has determined the growth patterns in the peri-urban area. The emergence of specialisation centres as a result of centrally-designated development zones has characterised the growth of the inner peri-urban area. The Hongqiao area, located 6.5km from the city centre, has been transformed into a major FDI-oriented office centre with a huge amount of apartment space (see Fig. 7). Infrastructure improvement and the concentration of manufacturing activities have also resulted in rapid development of Caohejing and Minhang ETDZs and themed zones in Pudong. The emergence of Manhattan-style high-rises in the Lujiazui Finance and Trade Zone has formed a new part of the central business district of the city.

Development zones designated by the government at municipal level and below have been growth engines for the far flung peri-urban area. The nine IPKs designated by the municipal government produced RMB 245 billion or 18.3% of total industrial output of the peri-urban area in 2005. However, many district- and town-designated development zones were suddenly 'transferred' from rice fields to 'urban areas' overnight. Some of these zones were developed with paved roads, buildings and other facilities, and attracted a few foreign or domestic manufacturing projects, while others had little development and were fenced around only. The intent of simultaneous development of all towns has prevented the growth of the central towns marked for expansion under the municipal urban plan. There were 161 towns

Fig. 7. Emerging specialised centres in peri-urban Shanghai – the Hongqiao Office Centre Source: Jiaping Wu

and townships in Shanghai's peri-urban area in 2001. Only three of them had a population over 100,000 with a further six towns having a population between 40,000 and 100,000. Urban development, centred on the satellite towns proposed by the municipal government in the 1986 metropolitan comprehensive plan, has not been seen. For example, Songjiang town, expected in the plan to hold 300,000 people, had only 100,000 people by 2001 and was rescheduled to reach the target by 2010. Growth instead has been shared by numerous smaller towns in the area.

Challenges to urban sustainable development

There has been a great deal of research into the environmental problems of Shanghai (e.g. Lam and Tao, 1996, Shi *et al.*, 2001; Shi *et al.*, 2004). While a number of measures have been taken, the environmental quality of the city has become progressively worse. The causes of the environmental degradation of the city are complex. This section does not assess the economic performance and environmental change of the city. Instead, it focuses on the way in which the FDI-driven urban development in the peri-urban area has created the environmental problems.

FDI and environmental pollution

The overall environmental quality of Shanghai was revealed by the national comprehensive urban environmental quality and management assessment for 37 cities. This has been undertaken annually since 1979. Shanghai has increased its scores in terms of the measures taken and infrastructure developed to mitigate environmental problems, but has maintained a very low absolute score in terms of environmental quality. Environmental costs accounted for 3.1% of the city's GDP in 2005, increasing by a factor of almost thirty since 1991 (Shanghai Statistical Bureau, 2006). The per unit area emission of total particles in suspension (TPS) and SO_2 of the city is still one of highest in the nation. Industry has been the major source of pollutant discharge into the environment, accounting for over 93.3% of the gaseous emissions in 2004 (ibid., 2005).

The connection between environmental deterioration and FDI is complicated. Many attempts have been made to test the so-called 'pollution haven' hypothesis which assumes that pollution abatement costs at home are large enough to significantly affect the location and magnitude of foreign investment abroad. Some studies have found a significant impact of environmental factors on industrial transformation, so supporting the hypothesis (e.g. Mani and Wheeler, 1999). Lax environmental regulations and low standards do significantly raise the probability that 'dirty' projects will be concentrated. But some argue that the pollution control cost differential does not provide OECD firms with a strong enough incentive to move offshore (Jaffe *et al.*, 1995).

Notwithstanding, by the end of 2004, there was a cumulative total of over 20,000 manufacturing FDI projects in Shanghai. A large proportion of these FDI projects has been located in the peri-urban area, and has concentrated on the higher pollution intensity groups (high levels of abatement expenditure per unit of output) including iron and steel, non-ferrous metals, chemicals, pulp and paper, leathers, plastics and transportation equipment.

An environmental impact assessment is required for every significant new construction and reconstruction project in Shanghai, but it does not determine whether a project is accepted or not. Environmental impact assessments for FDI projects have been particularly sloppy in the peri-urban area. During the 1990s, many small factories in the peri-urban area dumped directly into the environment with little treatment. Most of these waste sites are an environmental hazard to nearby residents and agricultural activities. Huge volumes of wastewater have been discharged directly into urban water bodies without any treatment. In 2001, it was reported that 3 million m^3 sewage discharge directly to the East China Sea per day. The black water belt can be seen and smelt over a dozen kilometres from sewage outlets (Shi *et al.*, 2001). Some are even close to water intake points; for example, to the Dianshan Lake at the upper reaches of the Huangpu River. Recent research shows that the environmental quality of Shanghai dramatically decreased after 1992 (Shi *et al.*, 2004). The decreasing trend of Environmental and Resource Sustainability Index in this research coincides with the rapid growth of manufacturing FDI intakes in the city.

Urban structure and sustainability

The connection between urban structure and sustainable development of cities is currently a crucial component in debate. It is argued that compact cities contribute significantly to sustainable development in terms of efficiency of land use, energy consumption and gaseous emissions and less automobile dependence (e.g. Burgess, 2000; Thinh *et al.*, 2002; Camagni *et al.*, 2002). However, the development of a city which is too dense and lacks quality public spaces, makes it unpleasant to inhabit.

Shanghai, as well as the whole of China, has long been known for its widespread use of bicycles in urban transportation. The dispersal of urban development in the peri-urban area as a result of decentralisation and competition for FDI has changed the structure of the city and the transportation modes used. The scattered nature of productive sites has increased demand for motorised transport in Shanghai; the distances are too large for the bicycle to be useful. The past decade has seen a rapid increase in the number of private cars and passenger vehicles, which have increased from 190,000 in 1996 to 1,296,000 in 2005. The number of motorcycles has increased from 93,000 to 1,204,000 during the same period (Shanghai Statistical Bureau, 2006). Another factor of diffuse development is the high consumption of land for road infrastructure. In Songjing district alone there were twenty-six development zones and industrial parks scattered in nineteen townships, involving a total of nearly seventy km² of formerly arable land in 2001. The need to provide new roads to take cars and trucks is staggering. The development of rural community infrastructure has become one of the most important reasons for land consumption. The city as a whole lost 783 km² of arable land from 1990 to 2005.

Urban health in the peri-urban area

Rapid and unplanned urban growth is the cause for many of the environmental hazards faced by the peri-urban area. The changing ecological system, that is intensive use of land but absence of adequate pollution management and health amenities in the peri-urban area, has been at the forefront of newly emerging infectious diseases (WHO, 1992). The most recent examples include the outbreak of the Severe Acute Respiratory Syndrome (SARS) in 2002–03 and the subsequent resurgence of highly pathogenic Avian Influenza.

Peri-urban areas in China's cities often share problems with low-income urban inhabitants, in particular the 'floating population'. These migrants are highly mobile and particularly vulnerable to infection, and are excluded from the healthcare system of the cities. The SARS outbreak demonstrated how the mobility of this population could amplify the potential impact of emerging infectious diseases on social and economic development. The illness was rapidly spread to surrounding cities such as Heyuan and Zhongshan after the first case was recorded in Foshan, the peri-urban area of Guangzhou, on 16 November, 2002. And it spread to more than two dozen countries all over the world in a couple of months. A total of 8,400 people worldwide were infected and over 810 died (Zhong *et al.*, 2003). During the SARS period, China faced a national crisis, with thousands fleeing cities like Guangzhou, Shanghai and Beijing, possibly carrying infection with them across many SARS-free areas. Normal life in many parts of the nation ceased. Although the cause of these diseases needs more investigation by health professionals, the environmental deterioration of the cities' peri-urban areas and the inability to provide adequate healthcare to the urban poor perpetuate such health crises. This can be detrimental to growth not only of the area or the city, but also of development worldwide.

Conclusion

With twenty years of global integration, Shanghai has re-emerged in the global economy. One important consequence is the rapid urbanisation and industrialisation of the peri-urban area of the city. All suburban rural counties, except Chongming Island, have been granted urban district status, enlarging the urban area of the city more than two dozen times.

Urban development of the peri-urban area in Shanghai is characterized by the importance of development zones. Various specialisation centres, based on the development zones designated by the central and municipal government, have emerged; these have not only changed the production space of Shanghai but also turned the urban form from a compact city to a dispersed city. The growth in number of development zones has fragmented the whole peri-urban area of Shanghai into a patchwork of 'central designated places', 'municipal-designated places', 'district-designated places', town-designated places' and general rural towns. These are in contrast to the application of FDI-oriented policy and by manipulations of land development. Administration commissions for the development zones have been created to manage and facilitate FDI attraction and land development of the designated areas in an entrepreneurial manner, while development of non-designated areas remained with rural town administrations.

The location of development zones has been widely distributed. Urban growth has been scattered throughout the peri-urban area instead of the concentration of manufacturing activities and population movements being encouraged into planned areas as intended under municipal urban plans. The original planning intentions of the 1986 Shanghai Comprehensive Metropolitan Plan that attempted to create a much more controlled use of urban space have been overridden. The resulting spread of productive sites has brought about increases of transport demand and the costs of providing services. The traditional transportation mode of the bicycle became ineffective for people to use for their daily activities. Transportation in the city has increasingly depended on the automobile, seeing a rapid increase of car and passenger vehicle numbers over the past few years.

The growth of manufacturing activities, often characterised by higher pollution intensities, flowing into the area has had a seriously deleterious effect on the environment of the peri-urban area of Shanghai. Its agricultural supply role and environmental value has been supplanted by manufacturing production sites and waste dumps. The scattered and fragmented nature of production sites has made environmental management more difficult. This, combined with the concentration of the urban poor and 'temporary migrants', has turned the peri-urban area into a potential crisis point that threatens the social and economic development of the city.

The creation of development zones to accommodate the increase in manufacturing activities for the global market and for the inner city's restructuring in the peri-urban Shanghai is similar to that in other megacities in China and other Asia countries. The experience of Shanghai's peri-urban area has much to offer to environmental management in these emerging metropolises which share similar growth patterns of rapid peri-urbanisation.

Notes

1. The peri-urban area (formerly under municipal jurisdiction but administered by rural communities), has become the most politically, socially, economically and environmentally dynamic area, moving from its rural base through the implantation of a mix of industrial estates and urban residential sprawl amongst the ongoing rural activities.
2. Jabotabek is the metropolitan region around Jakarta with a population of nearly 24 million. It comprises Jakarta and the three regencies of Bekasi, Bogor and Tangerang.

References

Atash, F, W. (1990) Satellite Town Development in Shanghai, China: an Overview. *The Journal of Architectural and Planning Research*, 7(3): 245–257.

Burgess, R. (2000) The compact city debate: a global perspective, in *Compact cities: sustainable urban forms for developing countries* (eds M. Jenks, M and R. Burgess), Spon Press, London.

Camagni, R., Gibelli, M. *et al.* (2002) Urban mobility and urban form: the social and environmental costs of different patterns of urban expansion. *Ecological Economics*, 40: 199–216.

Douglass, M. (2000) Mega-urban regions and world city formation: globalisation, the economic crisis and urban policy issues in Pacific Asia. *Urban Studies*, 37(12): 2315–2335.

Fuchs, R. J. and Pernia, E. M. (1989) The influence of foreign direct investment on spatial concentration, in *Urbanisation in Asia, spatial dimension and policy issues* (eds J. Costa and A. Dutt *et al.*), University of Hawaii Press, Honolulu.

Gale, R. P. and Cordray, S. M. (1994) Making sense of sustainability: nine answers to what should be sustained? *Rural Sociology*, 59: 311–332.

Jaffe, A. B., Peterson, S. R., Portney, P. N. and Stavins, R. N. (1995) Environmental regulation and the competitiveness of U.S. manufacturing: what does the evidence tell us? *Journal of Economic Literature*, **XXXIII**: 132–163.

Lam, K. and Tao, S. (1996) Environment quality and pollution control, in *Shanghai* (eds Y. Yeung and Y. Sung), the Chinese University Press, Hong Kong.

Lin, G. C. S. (2001) Metropolitan development in a transitional socialist economy: spatial restructuring in the Pearl River Delta, China. *Urban Studies*, 38(3): 383–406.

Mani, M. and Wheeler, D. (1999) In search of pollution havens? Dirty industries in the world economy, 1965–1995, in *Proceedings of the OECD conference on foreign investment and the environment*, The Hague, Netherlands.

Marcotullio, P. (2001) Asian urban sustainability in the era of globalisation. *Habitat International*, 25: 577–598.

McGee, T. G. (1991) The emergence of desakota regions in Asia: expanding a hypothesis, in *The extended metropolis: settlement transition in Asia* (eds N. Ginsburg and T. McGee, *et al.*), University of Hawaii Press, Honolulu.

McGee, T. G. and Robinson, I. (ed.) (1995) The mega-urban regions of southeast Asia: policy challenges and response, University of British Columbia Press, Vancouver.

McGee, T., Lin G. C. S. *et al.* (2007) *China's Urban Space: Development Under Market Socialism*, Routledge, London.

Ng, Mee Kam and Hills, P. (2003) World cities or great cities: a comparative study of five Asian metropolises. *Cities*, 20(3): 151–165.

Satterthwaite, D. (1997) Sustainable cities or cities that contribute to sustainable development? *Urban Studies*, 10: 1667–1691.

Shanghai Statistical Bureau (2000) *Shanghai Statistical Yearbook (SSY)*, China Statistical Press, Beijijng.

Shanghai Statistical Bureau (2001) *Shanghai Statistical Yearbook (SSY)*, China Statistical Press, Beijijng.

Shanghai Statistical Bureau (2005) *Shanghai Statistical Yearbook (SSY)*, China Statistical Press, Beijijng.

Shanghai Statistical Bureau (2006) *Shanghai Statistical Yearbook (SSY)*, China Statistical Press, Beijijng.

Shi, C. and Hutchinsom, S. M. *et al.* (2001) Towards a sustainable coast: an integrated coastal zone management framework for Shanghai, China. *Ocean & Coastal Management*, 44: 411–427

Shi, C., Hutchinson, S. M. and Xu, S. (2004) Evaluation of coastal zone sustainability: an integrated approach applied in Shanghai municipality and Chong Ming Island. *Journal of Environmental Management*, 71: 335–344.

Shui, B. and Harriss, C. (2006) The role of CO_2 embodiment in US–China trade. *Energy Policy*, 34(18): 4063–4068.

Solinger, D. J. (1999) *Contesting citizenship in urban China: peasant migrants, the state, and the logic of the market*, University of California Press, Berkeley.

Thinh, N. X., Arlt, G., Heber, B. *et al.* (2002) Evaluation of urban land-use structure with a view to sustainable development. *Environment Impact Assessment Review*, 22: 475–492.

UNCTAD (2000) *World investment report*, UN publication, New York and Geneva.

Webster, D. (2002) *On the edge: shaping the future of peri-urban east Asia*, Asia/Pacific Research Centre, Stanford University. Retrieved from: http://APARC.stanford.edu.

World Health Organisation (1992) *Our Planet, Our Health*, Geneva.

Wu, F. (2000) The global and local dimensions of place-making: remaking Shanghai as a world city. *Urban Studies*, 37(8), 1359–1377.

Wu, J. and Radbone, I. (2005) Global integration and the intra-urban determinants of foreign direct investment in Shanghai. *Cities*, 22(4): 275–286.

Wu, W. (1999a) Shanghai. *Cities*, 16(3): 207–216.

Wu, W. (1999b) Temporary migrants in China's urban settings: housing and settlement patterns, a paper presented at international conference on the future of Chinese cities, Shanghai.

Yan, Z. (1985) Shanghai: the Growth and Shifting Emphasis of China's Largest City, in *Chinese Cities* (eds V. Sit), Oxford University Press.

Yueng, Y. (2001) Coastal mega-cities in Asia: transformation, sustainability and management, *Ocean & Coastal Management*, 44: 319–333.

Zhong, N. S., Zheng, B. *et al.* (2003) Epidemiology and cause of severe acute respiratory syndrome (SARS) in Guangdong, People's Republic of China. *Lancet*, 362(9393): 1353–1358.

Zhou, Y. and Shi, Y. (1993) Toward establishing the concept of physical urban area in China. *ACTA Geographical Sinica*, 50(4): 289–301 (in Chinese).

12
Shih-wei Lo

Taichung the Waiting Metropolis and its Campaign Towards a 'World Class' City:

A case of glo*collision*, glo*coalition*, or glocalisation?

Introduction: a displaced conception of urban imagineering

Rutheiser describes Atlanta as paradigmatic of ageographic and generic urbanism, a phantasmagorical landscape characterized by fragmentation, near-instantaneous communication, privatization of public spaces, highly stylized simulations, and the subordination of locality to the demands of a globalizing market culture (Rutheiser, 1996, p. 4). It seems that Taichung has been confronted with a similar trend and, even intentionally, boosted to be such by the public and private sectors over the past two decades. The very dramatic operation of marketing the city during 1995–2007 through international competitions, inviting and sifting out the global élite architects to design the new civic buildings, was a particularly unprecedented attempt to project the city to be amongst those of the 'world class'.

Rutheiser's notion of 'urban imagineering', which was addressed specifically to the 1996 Olympic preparations in Atlanta, seems to be a helpful concept for explaining those moves of international city 'boosterism' in Taichung. The concept implies the intangible dimension of self-consciously promoting and marketing a city by its citizens, politicians and investors (ibid., pp. 4, 9–15). This also conforms to what Soja contends as one of his trialectics of city space: the real, the imagined and the lived, reflected from Lefebvre's theory on the production of space (Soja, 1996); whereas, for Soja, the imagined space pertains more to conceived perspectives, like that of urbanism.

In this chapter, the idea of imagineering denotes the collective imagination guided through organized public events and mass media by urban agents, particularly by the public sector. While confronted with Taichung's often delayed and substituted modes of development, this idea seems to be able to encompass the anxious but conservative, and the fictitious but practical process of city marketing. Building a 'world class' city through mobilizing global brand architects has become the key strategy for imagineering Taichung over the past four terms of its mayoralty. The main focus is on three international architectural competitions and invitations. These were held for the New Civic Centre (1995), the Guggenheim Museum (2003) and the Opera House (2006). The Taichung local government persisted with such global city marketing strategies and claimed itself to be an internationalized city. This seems to be a common practice for those in a developing country context, yearning for 'world class' status.

However, imagineering a city in the sphere of a global cultural economy should be treated with caution, especially for a fringe city like Taichung. To elaborate:

stories about the city's self-marketing maybe tentatively summed up as a 'city glocal movement'; for example, Appadurai's idea of trans-locality and Soja's real-and-imagined spatiality can be taken as theoretical starting points and references. Between the dynamics of the local and global sectors, there are subtly embedded oppositions and conspiracies. The campaign to become world class always means a city evolving from a naïve mode to the more sophisticated one of city governance, and this process may also become a must, judged upon the omnipresent global/local interaction.

The main argument of this chapter is that opening the door to the world is unavoidable, but global élite architects appointed to take charge of local practice do not guarantee a sustainable way of urban development. Such a mode of imagineering the city often turns out to be capitalist-biased rather than environment-friendly, and local government, like that of Taichung, tends to weaken its critical role in balancing private speculation and public welfare, becoming an unwitting catalyst of the commodification of public spaces. Such observations may not only be helpful to Taichung in the aftermath of its marketing activities, but also be applicable to those non-Western contexts which often deem 'world class' status as a measure of success.

The substitutive urbanity

Taichung is currently the third largest city in Taiwan, with a population of about 1.5 million. However, if including the adjacent townships which already form together a metropolitan region centred on Taichung, the total population is more than 3 million.

The city was originally appointed as the provincial capital by the Ching government in 1887, taking into account its advantageous location in the middle of Taiwan Island. A walled city was then built for the capital, but eventually only partly completed. Due to threats from European and Japanese imperialism in the north of Taiwan, the focus of construction concentrated on Taipei.

During the colonial occupation (1895–1945), Taichung was almost rebuilt as a brand new city by the Japanese, while the other important cities built upon the old Chinese walled urban entities. Taichung was dubbed 'Little Kyoto' during the Japanese rule. However, in the late 1930s, under the Southbound Move Policy, driven by the militarist dream of the Greater East Asian Co-prosperity Sphere, the Japanese constructed Kaohsiung, the southern port-city in Taiwan, as the logistical invasion base towards Southeast Asia. The urban development of Taichung was thus put aside again by the ruling power.

Therefore, historically, Taichung's advantage had been superseded first by Taipei, and then Kaohsiung. It has been the third biggest city in the island, with Taipei as the capital and Kaohsiung the heavily industrialized centre. While Taipei was the international city in politics and business, and Kaohsiung the international port, Taichung was a relatively local city in the middle of Taiwan. Over the last two decades, the distinctive feature of regional development in Taiwan has been the transformation from a polarized pattern of Taipei (in the north) and Kaohsiung (in the south) to a tri-centric structure, in which Taichung played the role of parvenu. It boosted itself as the city of consumer culture. In academic circles it was viewed as a speculative city, with twelve phases of land consolidation projects from 1965 onwards, with rapid implementation during 1987–93.

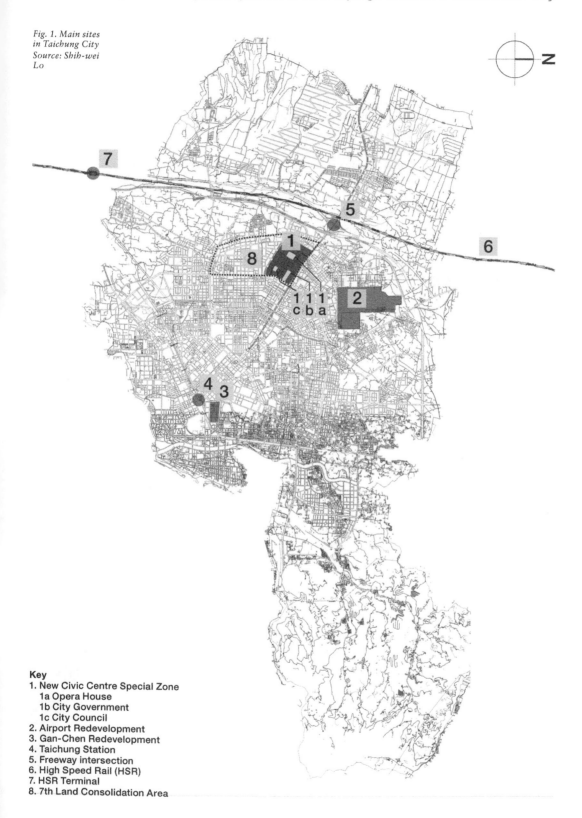

Fig. 1. Main sites in Taichung City Source: Shih-wei Lo

Key
1. New Civic Centre Special Zone
   1a Opera House
   1b City Government
   1c City Council
2. Airport Redevelopment
3. Gan-Chen Redevelopment
4. Taichung Station
5. Freeway intersection
6. High Speed Rail (HSR)
7. HSR Terminal
8. 7th Land Consolidation Area

In the 1980s, while the real estate business bloomed, Taichung turned from its long lasting fame as a city of culture, to be a city of the erotic. The seventh land consolidation area (335 hectares in total) was the first district to be subjected to floor area ratio regulations (eroding long-term profit margins) and subsequently it became a waiting area, exploited mostly as short-lived and instant-profit businesses, especially for night-time entertainment and catering. The area was fantastic at night, with omnipresent neon lights and twinkling signs, and was often associated with impressions of Las Vegas.

In the early 1990s, Kenzo Tange's team from Japan won the Gan-cheng Redevelopment Project on the site of some two hundred hectares previously settled by military dependents in row houses. This project was to attract intensive commercial, business and residential investment. But the economic depression in Taiwan in the late 1990s meant the project was suspended.

Taichung was a waiting metropolis, with successive lost opportunities for reasons outside its own control. Nevertheless, since 1995, the city government has kept organizing significant international competitions for large-scale urban redevelopment in Taichung (Fig. 1). No matter how alien the result might be, these imagineering works seem to justify the emerging significance of Taichung. It reflects the need for the city to be internationalized to promote itself as the new nucleus in the west of Taiwan, to form, together with Taipei and Kaohsiung, all now linked

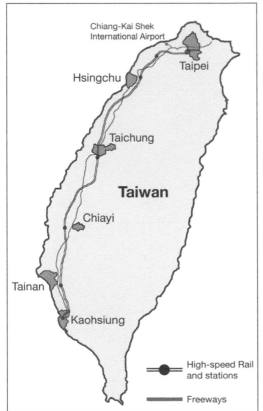

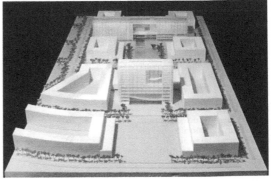

Fig. 2. (left) High speed train line, and corridor metropolis
Source: Shih-wei Lo

Fig. 3. Site Model of Taichung New Civic Centre Project
Source: Taichung City Government (1995)

together by a high-speed train system, a corridor metropolis in East Asia (Fig. 2). Such a polycentric structure of mutually complementary functions can thus facilitate Taiwan becoming competitive enough to cope with the growing situation in the East Asian region, especially in terms of economic growth. Taichung as a waiting metropolis may finally turn out to be an acting metropolis!

New civic centre project

In order to move the city government and council out of the old city centre, where streets and blocks were not wide enough for motor vehicle movement, the New Civic Centre Project was initiated with an international competition organized in 1995 during Mayor Lin's tenure. The new site was located about 3km to the north-west of the old city centre. This project was claimed to be the first international architectural competition in Taiwan and it showed off the ambition of Taichung to be a 'world class' city.

The site for the project, including two buildings and the parkway in between, was 6.6 hectares in area (Fig. 3). The area for the Music Hall (later to become the National Opera House) was 5.77 hectares, linked by a spacious parkway to the municipal buildings (including the municipal government office and the municipal council). This golden trio formed the very core of the Seventh Land Consolidation area.

The New Civic Centre Project drew much attention from abroad. The Swiss team Weber + Hofer AG Architects won the competition with their controversial scheme – two clean and prism-like glass boxes – which was criticized by a local juror as not fit for the tropical climate of the city, and could result in a waste of energy (Figs 4 and 5). It was only in October 1997, at the last moment of Mayor Lin's term, that both sides signed the contract for planning, design and supervision of the project. However, over the next two terms of mayoralty, the project was almost shelved.

The main reason behind the reluctance of the authority to implement the project was concern about the decline of the old city centre when the existing city government should move out to the new project once it was completed. Another reason was that it was financially not feasible to support a project costing 180 million US dollars. During the first term of Mayor Hu (2001–05) the New Civic Centre project stagnated again, due largely to his campaign promise to promote

Fig. 4. (below–left) Model of Taichung City Government Building
Source: Taichung City Government (1995)

Fig. 5. (below–right) Model of Taichung City Council Building
Source: Taichung City Government (1995)

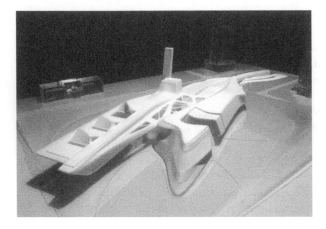

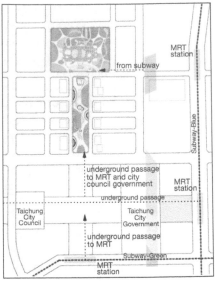

the Guggenheim Museum project. It was not until the end of 2005 when Mayor Hu ran his second electoral campaign for mayoralty that the project was resumed and construction begun with a scheduled completion in 2009.

Clearly, it was the real estate sectors which would profit from the City Civic Centre project. They would be expected to boom along the parkway in the area assigned to the project. The halt to implementing the project after the mayoral election in 2001 somewhat unexpectedly yielded to the new programme of marketing the city – the Guggenheim Museum Project – a project which was never imagined before then.

Fig. 6. (above-left) Model of Taichung Guggenheim Museum Source: A+U (2004)

Fig. 7. (above-right) Site Plan for Taichung Civic and Cultural Center Source: Taichung City Government (2006)

Guggenheim Museum Project

The Guggenheim Museum Project was so far the most sensational international event in the city's international campaign. This was also part of Guggenheim Foundation's global manoeuvring within the cultural economy.

In 2001, Mr. Hu Chi-chiang was elected mayor of Taichung, and one of his campaign policies was to build a Guggenheim Museum. He signed a feasibility assessment agreement in June 2002 with Thomas Krens, chief executive of the Solomon Guggenheim Foundation, in Bilbao, Spain. In July 2002, Krens, Jean Nouvel and the Curator of the Guggenheim Museum in Bilbao visited Taichung for a site survey, and the site that had been designated for building the New Municipal Centre was assessed as an ideal choice for the Museum.

Soon after, the Guggenheim Foundation acted as a go-between to suport Frank Gehry, Jean Nouvel and Zaha Hadid to visit Taichung in December 2002. The Guggenheim Project proposal was to be extended to include an Opera House (by Nouvel), a New Municipal Centre (by Gehry), and the Museum (by Hadid). A project of 12.4 billion US dollars was proposed by the 'Two Kings and one Queen' – a dream for Taichung to link to the world.

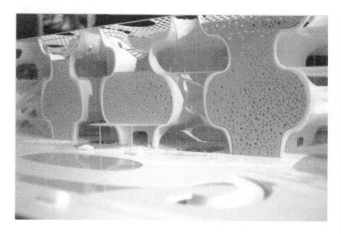

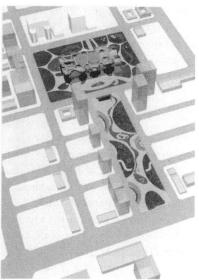

Fig. 8. (above-left) Model of Taichung Opera House
Source: Taichung City Government (2006)

Fig. 9. (above-right) Perspective of Taichung Opera House
Source: Taichung City Government (2006)

In July 2003, Hadid came to Taichung with her design model for the Guggenheim Museum, published in the January 2004 issue of A+U magazine (in both English and Japanese)(Fig. 6). Hadid's design included a 50m deep cantilever feature marking the entrance of the museum, and the streamlined free form of the museum was certainly a brand new landmark for the city.

Mayor Hu briefed President Chen, who was running his second national election campaign, about the Guggenheim Taichung Museum Project in August 2003. In September 2003, the Executive Yuan approved a 170 million US dollar subsidy for the Guggenheim Museum Taichung Project. Because of the presidential election to be held at the end of the year, no one in the contesting parties dared to stop the Guggenheim Project. This was the most optimistic moment for the project's realization. However, soon after the presidential election, in January 2004, the budget subsidy was cut in parliament under the 'Five Years Fifty Billion NT (about 1.7 billion US dollars) Special Budget'. Although the Executive Yuan earmarked some 10 million US dollars for down payment to the Guggenheim Foundation, Taichung City Council decided to turn down the Guggenheim Project by the end of 2004. The reasons were concern about the operational and maintenance costs, especially under the conditions dictated by the Guggenheim Foundation.

The Bilbao model was far removed from the circumstances of Taichung, where the MRT system and other urban infrastructure were neither planned or funded at the time. The Guggenheim Project was an isolated and instant 'vision-eering' exercise, plagued by malfunctioning politics; finally it was no wonder it was aborted.

Metropolitan Opera House project

In the early spring of 2006, Toyo Ito won the competition for the Metropolitan Opera House in Taichung. Other super-stars who entered the competition were Zaha Hadid, de Portzamparc, Richard Rogers, and Hans Hollein. The design contract was signed in August 2006 and the project was scheduled for completion in 2009. It seems that this time the project can be expected to progress more smoothly than in previous cases. After the controversial precedents of the New Civic Centre and Guggenheim Museum, this time the city government has matured enough both to host such an international competition and its subsequent implementation.

This is a project of a 100 million US dollars budget (initial allocation), located at the end of the parkway perpendicular to the axis of the New Civic Centre (Fig. 7). Ito's work was a cheese-like solid-void composition in a rigid profile of a rectangular glass box, revealing an inside-out and outside-in transparency. It was conceptually the same as the Ghent Music Forum, for which Ito won second prize in the competition of 2004. This practice simply exemplified the global game played by élite architects to peddle their 'fantastic' ideas around the world, despite local differences and particularities.

The key concept of the Opera House is the 'Sound Cave', a horizontally and vertically continuous network, connecting seamlessly the interior with the outside (Wen, 2006, p. 12)(Figs 8 and 9). It is really a work of epochal talent, but it can be put anywhere, anytime in the world. Also, seemingly absurdly enough, just because this is a universal scheme worked out by a global brand architect, it is a scheme with which the city expects to step out into the spotlight of worldwide attention.

The edge-cutting spatial experience offered by Ito can indeed bring forth an excellent arena for debate in the global cultural economy. The configuration of space flowing in-and-out incorporates technological innovation and may result in worldwide fame. Nevertheless, does this guarantee the quality of a public space? The inside-out or outside-in space flow is only conceptual and a visual effect. In reality, the seemingly non-existing glass forms a barrier for movement between the interior and exterior. There are no intermediate spaces (such as an arcade, a popular urban space element in Taiwan) for the public standing or lingering about without being exposed directly to the sub-tropical sunshine and rainfall.

At the beginning of February 2007, the Premier Su promised to increase the subsidy for this project. The overtone was very clear: that he was preparing his own way to the presidential election the next year and starting his warm-up campaign by awarding favours. Again, thanks to an election, the Opera House hopefully may be implemented.

Conclusion: a city of glo*collision* or the city glocal movement?

Finally there will be three sharp rectangular glass boxes standing at the ends of parkways in Taichung. They may not be the last cases of globally marketing the city. Taichung airport was moved in 2004 and left 247.22 hectares area for redevelopment. 'The Former Taichung Airport Site Redevelopment International Planning Contest' was organized in January 2007. This may result in another run of global architectural competitions in the coming years. Now should be the right time to reflect on the consequences of the past competitions, to provide lessons for the coming decade.

This chapter considered the Guggenheim Project and The New Civic and Cultural Project. The latter was a real project of three public buildings and was implemented legally by the city government, while the former was virtually a speculative venture for both Taichung City Government and the Guggenheim

Foundation. For Taichung, the failure of the Guggenheim Project was justifiable. It was a scheme deprived of local opinion, which would have almost no way to steer the future museum management, but bear the whole of the costs and risks. The price for marketing the city globally as such with international brand architects was too high a price to afford

As for the real projects – the Civic Buildings and the Opera House – some differences could be found between the two within a little more than ten years. The local jurors for the competitions increased from two to three, somewhat strengthening local opinion. In addition, the jurors of the later competition were a generation younger than for the previous one. However, the argument of the local conditions against international style, like the issue of tropical climate versus energy consumption, did not seem to appear among the jury members for the Opera House competition.

The developers profited from the aggressive promotion of these projects by the public sector. Along this New Civic and Cultural Centre Zone, elegant high-rise apartment buildings emerged within the last decade, and now, very oddly, they stand by and wait for the completion of the three alien, brand new box-objects (Fig. 10). The area, at the very heart of the new city vision, is to be fed with global fashion fodder (see 'knowledge fodder' in Appadurai, 2002, p. 32–47), not to mention the tremendous cost of maintaining its actual operation.

Taichung, like most cities in Taiwan, had been built by 'others' over different régimes. These alien buildings might at first seem out-of-place, but eventually look in place some time afterwards. In fact, beyond narrow nationalism or immediate

Fig. 10. View along Parkway Source: Shih-wei Lo

autochthonous concerns, there is a subtle clue implying Taichung's own agenda, set for the near future. To think on a bigger or extra-urban scale, the global brand operation for marketing Taichung may not necessarily be wrong for local identity. It may be a fair strategy, taking into account its role in a trans-urban and even a trans-regional development across the Taiwan Strait. The new high speed train may make West Taiwan a corridor-metropolis, of which Taichung becomes the middle city as a new centre, dealing with cross Strait mobility, together with Amoy, and Fu-chou in coastal Fu-chien, China, where both sides share the same dialect. The polycentric corridor-metropolitan Taiwan restructured with Taichung as a new hub, backed up by over ten million educated people, should join in the thriving momentum in southeast China, and hopefully build up a cross-Taiwan Strait co-prosperity circle. The alien buildings of global fashion may therein provide a cosmopolitan capacity for a more vital city to come.

Indeed, they bring revolutionary visual impacts with prism-like neat *zeilenbau*, streamlined amorphous forms, and interior/exterior transparent continua. All these bring a new sense of capacity. Through them people can expect a metropolitan mode of urbanity and way of life. However, with such a homogeneous global look, these objects of nowhere cultivate less sense of here and now concerning the place-ness identifiable for a particular city.

Today, what does the 'world class' city mean? In contrast to the global production of locality, as contended by Appadurai (2002), Taichung reveals a particular process of 'the local production of globality'. The strategy of imagineering the city with global brand architects has not yet proved to be a win-win. It has been so far more like a global-local-collision (glo*collision*) than a global-local-isation (glocalisation). The opposition can be found not only between isotropic boxes designed by global elites and snobbish high-rise towers by local developers, but also between the city government (led by the former minister of foreign affairs) and the city council (represented by fraction-minded powers), and further between the real city and the imagined city.

Here, as it is more concerned with 'the restructuring-generated crisis', I would argue that Soja's contention of the real-and-imagined city is mainly addressed to 'spatiality' (Soja, 1996). Whereas Taichung could evolve from a waiting metropolis to a developing metropolis, and its 'becoming-generated possibilities' seem instead to be the key concern. The point is that Taichung has always been in its own transitional state, ever evolving with interventions from outside its normal capacity. 'Time' should be taken into account, and time-space, or temporal-spatiality – different from Giedion's definition which implied the *zeitgeist* idea – should be the theme to be addressed.

Temporality here is concerned with problems about: how good will a new civic building be if administrative efficiency or even corruption is not improved within the city government and council? And it is also about: how will it fare if there is still no good local team to regularly host appealing performances in the new opera house? And, furthermore: how will it be if there is still no convenient mass transportation for local residents to visit it? In general: how to make people feel they are having a good 'time' during the developing process of those projects, and then enjoying a good 'time' to visit, to work, to handle business, and to enjoy cultural activities? When the city has world class hardware eventually: how will it be if it still lacks the world class software, which always takes a much longer 'time' to put into place?

A city like Taichung does not seem to lie in any fixedly defined category, but in its ever-changing mode of urbanity, always subjected to alien and global influences. For Appadurai, the global horizon can become a material thing, through media, imagination and migration. He suggests that the global is a sort of expansion of the local horizon (Appadurai, 2002, pp. 32–47). Here and now Taichung is striving more for middle-class affluence. This is fully demonstrated in the quality high-rise apartments emerging in the New Civic and Culture Centre zone. The three glass boxes[1] designed by the global brand architects look quite in accordance with middle-class aesthetics. They are clear-cut boxes without much shading or overhanging to provide shelter from the strong sunshine and rainfall to allow the masses to stay in comfort; they show little welcoming gesture to the unprivileged underclass. In fact, they are public only in a selective sense.

In terms of middle-class publicity, those three boxes of civic and cultural architecture are in a conspiracy with local developers. They both engage in a glo*coalition* – a global-local-coalition. In this sense, the civic buildings and the Opera House become commodities in themselves, and they join to increase the exchange value of the surrounding land without giving enough care for, or feedback to, the true public.

Today, the idea of the local is not an inert backdrop, nor is it a given condition. The local is a process and a project, just like other things (Appadurai, 2002, p. 32). Therefore, the 'world class' city discussion should not be limited just to an intra-urban sphere. Networking in global or a regional sphere is becoming irresistible. A globally élite-driven mode of marketing the 'real-and-imagined' city, borrowed from Soja's terminology, must be criticized in its trans-local and/or trans-regional context.

Just here, William Lim's lament based on his radical postmodern perspective is so genuine to the situation happened in Taichung: 'The essentiality of local creative energy and dynamic interaction to anchor architectural discourses and practices has yet to be fully understood or developed' (Lim, 2005, p. 33). In the case of Taichung, it has proved a great pity that the local public sector was not self-conscious enough to envision potential opportunities released from the city glocal movement, or was not alert enough to the highly capitalist-driven nature of the glocalisation, and then missed making maximum profit for the general public through negotiating the dynamics of glo*collision* and glo*coalition* in the process of glocalisation.

Note

1. The most updated source suggests that the glass curtain-wall may probably be changed to a stone slate wainscot, due to considerations of energy consumption. The debate was still ongoing when this chapter was written.

References

Appadurai, A. (2002) The Right to Participate in the Work of the Imagination in *Trans-Urbanism* (eds J. Brouwer, P. Brookman and A. Mulder), NAI, Rotterdam, pp. 32–47.
Cheng, Ming-yu and Wang, Chun-hsiung (1995) Report on the International Competition of Taichung New Civic Center. *Chinese Architect*, July: 59-133
Hadid, Z. (2004) Zaha Hadid. *A+U*, January.
Lim, W. S. W. (2005) *Asian Ethical Urbanism: A Radical Postmodern Perspective.* World Scientific, Singapore.
Rutheiser, C. (1996) *Imagineering Atlanta: The Politics of Place in the City of Dreams.* Verso, London.

Soja, E. (1996) *Third Space: Journeys to Los Angeles and other Real-and-Imagined Places.* Blackwell, Oxford.

Tseng, H. (2006) *Viewing at the Urban Governance of Taichung with Guggeheim Museum Project as a Mirror,* Master Thesis, Graduate Institute of Planning and Building, Taiwan University.

Wen, M. (ed.) (1995) *The New Taichung City Civic Center International Competition Entries.* Leader of The Garden AD, Taichung.

Wen, M. (ed.) (2006) *Taichung Metropolitan Opera House International Competition Award-winning and Participant Projects.* Taichung City Government, Taichung.

'World Class' Vancouver:
A terminal city re-imagined

Introduction

A dense cluster of towers on a peninsula rises against a spectacular backdrop of unspoiled forests, mountains and ocean – this is the iconic image of Vancouver, Canada's third largest city (City of Vancouver, 2006a; Fig. 1). It contains over two million inhabitants and consists of twenty-one separate municipalities and one electoral area, forming the Greater Vancouver Regional District (GVRD). The City of Vancouver, referred to in this chapter as 'Vancouver' or 'the city', is the central municipality of the GVRD, with 600,000 residents, or approximately 30% of the total metropolitan population.

In the city's current housing boom, claims of Vancouver's 'world class' status abound. Advertisements for new condominium developments elevate Vancouver to a city that 'the world is watching' because of its high-quality urban lifestyle, an assertion borne out by Vancouver's consistently high rankings on international quality-of-life indices (Westbank and Peterson Group, 2006; Simon, 1995; Fong, 2005; Mercer, 2006).[1]

Fig. 1.
Vancouver
Downtown
Peninsula
from Central
Broadway, 2001
Source: May So

Greater Vancouver has been striving to offer a high quality of urban life for over three decades. With the exception of a few brief periods, the region has consistently experienced healthy population growth due to natural increases, inter-provincial migration and international migration, which peaked at 3% in the late 1980s and early 1990s as the pending hand-over of Hong Kong to China loomed. From less than 1 million people in the 1970s, the region grew to 2.2 million in 2006, and projections suggest that it will be 2.7 million by 2021 (Kane, 2005).

To combat the low-density sprawl and the increasing automobile-reliance that accompanies growth, in 1972 Greater Vancouver began developing a growth management strategy centred on the vision of a 'liveable city' (Seymoar and Timmer, 2005). As currently defined in Greater Vancouver's 1996 Liveable Region Strategic Plan (Fig. 2), created from four years of public and intergovernmental consultation, 'liveability' is the 'quality of life' experienced by a city or region's residents (Greater Vancouver, 1996).[2] In a 'liveable city,' residents have equitable access to infrastructure and amenities that enhance their quality of life: transportation, communication, water and sanitation, food, clean air, affordable housing, meaningful employment, and green space. Additionally, they have access to participation in decision-making (Seymoar and Timmer, 2005).

To achieve liveability, Greater Vancouver's first liveable regional plan recommended directing new growth towards the development of a compact polycentric system of comprehensive communities connected by transportation (Greater Vancouver, 1980).[3] The 1996 plan proposed four fundamental strategies to

Fig. 2. Liveable Region Strategic Plan

Source: Adapted from Vancouver Regional District Planning Department (1996)

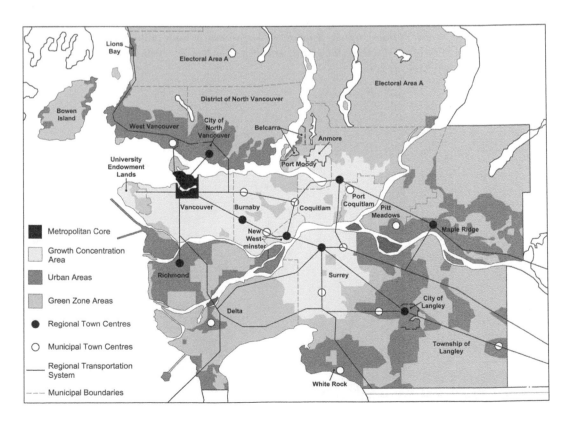

manage growth: protect green space and natural resources such as watersheds and agricultural land in a 'Green Zone'; create complete town centres at the metropolitan, regional and municipal levels; focus residential growth in a 'growth concentration area' to achieve a compact metropolitan region; and increase transportation choice through a transit-supportive and automobile-restrained transportation system (Greater Vancouver, 1996).

Achieving regional liveability relied heavily on voluntary coordination of growth between municipalities after regional planning ceased to be a statutory function of the British Columbia municipalities in 1983 (Seymoar and Timmer, 2005). The impact on municipal planning was to distribute and concentrate Greater Vancouver's jobs, new housing, culture, and recreation in higher densities in the region's town centres. Using the uncommon powers possessed by the City of Vancouver's council[4] to direct development and public benefits through policy, discretionary approvals, and a responsive design-led planning system, Vancouver absorbed growth in a dense, regenerated, 'liveable downtown,' often referred to as the 'Vancouver achievement'.

This chapter discusses Vancouver's successes in densifying downtown and the limits it has encountered in attempting to densify neighbourhoods outside the downtown area. The latter has led to the recent shift towards mandating sustainable practices for all new developments as the mechanism for achieving a more widespread 'liveable city'. The chapter concludes with implications for other cities striving to gain world recognition as 'liveable cities'.

Policy for a liveable Vancouver

Vancouver's planning achievements have been delivered through the public regulation of private development, that is, 'urban design as public policy'. First coined by Jonathan Barnett in 1974 to describe his work in New York City, the phrase refers to the safeguarding of a city's interests through systems of design review, special district zoning, comprehensive design of the public realm and participatory planning in private developments, as opposed to direct public initiatives (Punter, 2003). Despite criticism of its 'aestheticizing' and exclusionary tendencies, its relative weakness to influence strong economic and social processes and its vulnerability to political manipulation (ibid.), urban design as public policy became widespread in North America after the withdrawal of federal funding in the 1960s for large-scale urban renewal projects and the subsequent loss of ability by cities to directly reshape their built environment.

1970–1985: The liveable downtown and urban regeneration

While Vancouver is one of Canada's youngest cities, established in the 1870s as a sawmill settlement and incorporated in 1886 as the terminus of the Canadian Pacific Railway, it is the core of Canada's third largest metropolitan area. Recognizing the particular challenges Vancouver faced as a fast-growing urban area, the provincial government of British Columbia granted the Vancouver Charter in 1953, giving the city council ownership of public areas within its boundaries and control over development and public benefits through wide ranging policy formulation and authority to approve plans, re-zonings and design guidelines (Province of British Columbia, 1953).[5] The Charter gave the council authority over municipal planning, allowing it to effect urban reform in response to public opposition to large-scale, unsympathetic development of Vancouver's downtown in the 1960s. The triple-tower Bentall Centre on Burrard Street, begun in 1965, had greatly exceeded the zoning provisions and created minimal public space (Punter, 2003). At the same time, large-scale clearance and redevelopment were planned for the low-income communities of Strathcona and Chinatown, along with a proposed highway that would bring commuters downtown. In the style of Jane Jacobs' opposition to the Lower Manhattan Expressway, pressure from the public and community organizations successfully halted these redevelopment projects.

The pro-development city council was defeated in 1972 and the new council reorganized planning practices, focusing initially on downtown urban renewal and improving the control and design of its major developments. A downtown study team led consultations with the public and special interest groups; from this general planning values emerged and were articulated in the 1975 Downtown Official Development Plan as principles for a more 'liveable downtown' (Fig. 3). The plan stated a commitment to repopulating the central area as 'an attractive place to live, work, shop, and visit', and encouraged 'more people to live within the Downtown District' (City of Vancouver, 1975). The plan supported the objectives of Greater Vancouver's plan for enhancing regional liveability by decentralizing some office employment from downtown Vancouver to regional centres in other parts of Greater Vancouver (Greater Vancouver, 1975). Learning from the uncontrolled densification of the adjoining residential West End in the 1950s, the plan used five basic discretionary controls: land use, density, building height, parking, and amenities, with no outright exceptions on any site (City of Vancouver, 1975; Fig. 4). Retail use continuity, mixed-use developments and walking were encouraged,

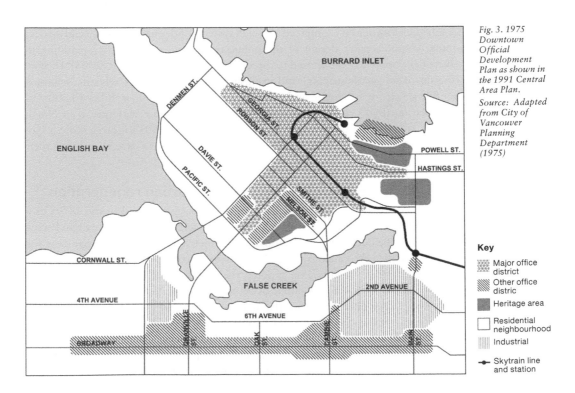

Fig. 3. 1975 Downtown Official Development Plan as shown in the 1991 Central Area Plan.

Source: Adapted from City of Vancouver Planning Department (1975)

and the physical diversity of distinct areas of downtown was maintained through character area guidelines.

The council also undertook reforms of the planning system. An overlapping system of discretionary controls was established to improve the design quality of the public realm, exercised through a Development Permit Board, an advisory panel of public representatives and developers, and an Urban Design Panel of architects and planners. A series of reforms between 1980 and 2000 created a permit system with a single point of inquiry. It identified and solved potential problems at the pre-application stage before plans proceeded to one of four streams in the approval process, dependent on the complexity of the application (Punter, 2003; Fig. 5).

1986–1991: 'Living first', downtown and globalization

By the mid-1980s, the main design principles for creating liveable, high-density residential neighbourhoods on the margins of downtown had been refined to include safe and lively streets, diverse and denser building forms, and residential complexes with private and public amenities. Pressure for residential re-zoning was rising as office development declined, inducing the city to commission a sequence of urban design studies to study appropriate densities and urban forms, amenities, affordable housing, park provision, liveable streets and heritage conservation. These issues were only partly resolved when Vancouver was launched onto the global stage with Expo '86, an international transportation fair for Vancouver's centennial celebrations held on False Creek North, former industrial lands on downtown's south shore. The fair is credited with boosting Vancouver's economy and international profile.

Fig. 4. 1975 Height Limit on Downtown Buildings

Source: Adapted from City of Vancouver Planning Department (1975) Punter (2003)

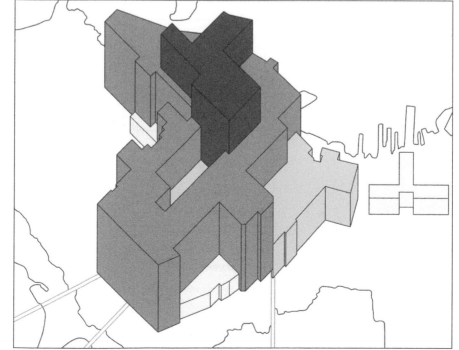

Key

■ 450 feet (135 metres)

▨ 300 feet (90 metres)

▨ 150 feet (45 metres)

□ 70 feet (21 metres)

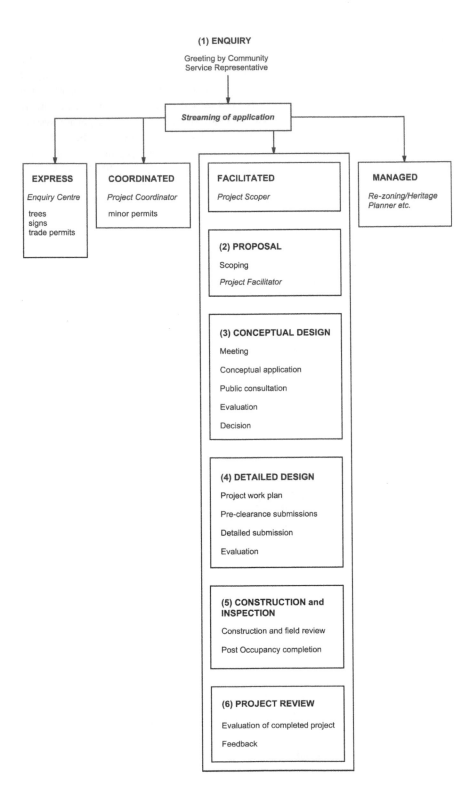

(1) ENQUIRY

Greeting by Community
Service Representative

Streaming of application

EXPRESS

Enquiry Centre

trees
signs
trade permits

COORDINATED

Project Coordinator

minor permits

FACILITATED

Project Scoper

MANAGED

*Re-zoning/Heritage
Planner etc.*

(2) PROPOSAL

Scoping

Project Facilitator

(3) CONCEPTUAL DESIGN

Meeting

Conceptual application

Public consultation

Evaluation

Decision

(4) DETAILED DESIGN

Project work plan

Pre-clearance submissions

Detailed submission

Evaluation

**(5) CONSTRUCTION and
INSPECTION**

Construction and field review

Post Occupancy completion

(6) PROJECT REVIEW

Evaluation of completed project

Feedback

*Fig. 5. Proposed
permitting
process under
the new
Development
and Building
Regulation
Review, 2000
Source: Adapted
from Punter,
2003, p. 304*

Compared to New York, London and Tokyo, the triad of global capitalism's command centres named by Sassen (1991) as 'global cities', Vancouver plays a peripheral role in the global economy, attracting executives primarily in forestry, mining, high-tech and biotechnology (Fong, 2005). While initial theorizing of the connections between globalization and cities focused on the concentration of command functions of the global economy in a few predefined 'world cities' (Beaverstock *et al.*, 1999), recent research on globalization and cities proposes that all cities are potentially affected by globalization and can act as transmitters, or 'gateways', of global and local flows in intercity networks (Beaverstock *et al.*, 2000; Short *et al.*, 2000; Smith, 2005).

With a recent federal commitment to expand Vancouver's port and highway system over the next twenty years, Vancouver's function as a gateway in the global economy will intensify (Transport Canada, 2005; Port Vancouver, 2006). Asian immigration to Vancouver had been steady since the early 1980s, and the city council used the growth to solidify Vancouver's role as a Pacific Rim 'gateway' in 1989 with the sale of the Expo site to Victor Li, a Hong Kong tycoon looking to expand his international property investment portfolio (Olds, 1996). Vancouver had to suddenly contend with a large influx of money, people, commodities, ideas, information and culture, accompanied by new relationships to places, intensified connections of daily activities to far-away events, and concentrated diversity (Allen *et al.*, 1999). Resisting the homogenization that often occurs when cities imitate the functions and practices of successful 'world cities' (Taylor *et al.*, 2002), Vancouver exhibited its unusual attitude to development during the planning of the Expo site, summarized by then co-director of planning, Larry Beasley, as 'If you don't measure up, we're not afraid to say *No* in this city. We want quality of life first' (Sandercock, 2005).

Concord Pacific's original master plan for the Expo site development, named Pacific Place, required excavating much of the shoreline to create two large islands, extending into False Creek, of residential tower blocks set in parkland. While increasing the amount of valuable waterfront property and, for Asians, the amount of *ch'i* (positive energy), the scheme conflicted with many of False Creek North's key planning principles, including integration of the project in the existing city pattern and maintenance of the 1987 base shoreline. Efficient resolution of planning issues in Concord Pacific's scheme was achieved by the settlement of large-scale conceptual issues before proceeding to specifics and through cooperation between public and private sectors to develop the scheme.

Up to this point, density increases had been negotiated on a project-by-project basis with the affected community. With the entire Expo parcel being planned at once, however, establishing its overall density at the outset yielded a compact neighbourhood. A dedicated inter-departmental city staff team collaborated with Concord Pacific and their designers on site plans, floor-space allocations and design guidelines, a deviation from the typical practice of planners to merely critique developers' proposals. The city council adopted a public involvement programme that gathered three different public groups – neighbouring communities and property owners, special interest groups, and the general public – to advise on the project and build support for False Creek North's plan through fulfilling community needs (Punter, 2003). A second key factor in the development's final approval and public acceptance was early resolution of the community facilities and social housing to be provided by the developer. In 1990, the council began imposing development-

cost levies based on increased land value from rezoning and community-amenity contributions to fund parks, day-care, replacement housing and soft infrastructure. These amenities comprised the Major Project Public Amenity Requirements subsequently imposed on all mega-projects. In the successful fusion of civic and private ambitions, the development community learned that certain 'quality of life amenities,' while costing more to build, resulted in increased market demand and local appreciation of high-density, high-rise residential urban environments. At Pacific Place, this 'humane density' is expressed in widely spaced residential point towers set back on two-to-three storey podiums containing retail, family housing and community amenities (Sandercock, 2005; Figs 6 and 7).

Building on the successes of Pacific Place, the 1991 Central Area Plan created a variety of mixed-use and high-density residential areas in the rest of downtown by converting eight million square feet of excess commercial office capacity to allow residential development through a major stroke of rezoning (Beasley, 2000; Fig. 8). The plan defined a metropolitan core, proposed a more compact central business district for offices downtown and uptown on Broadway, and supported the development of surrounding dense liveable residential neighbourhoods in a 'living first' downtown strategy (Greater Vancouver Regional District, 1996; Fig. 9).

So successful is 'living first' that downtown doubled to 100,000 residents or 20% of Vancouver's total population over two decades, and exacerbated housing affordability problems. With return on investment for condominiums five times as high as the return on office space and few commercial sites available, new commercial development was locating to the suburbs, leaving Vancouver in danger of becoming a 'resort city' of living and recreation (Boddy, 2006). Given that

Fig. 7.
Marinaside
Crescent
Neighbourhood,
Pacific Place
from Cambie
Street Bridge
Source:
Courtesy City of
Vancouver

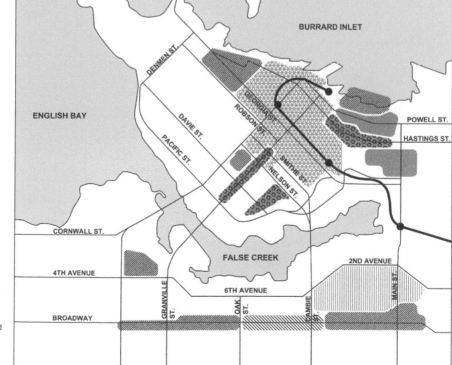

Fig. 8. 1991
Central Area
Plan
Source:
Adapted
from City of
Vancouver
Planning
Department
(1991)

Key

▒ Central
   Business
   district

▨ Uptown office
   district

■ Heritage area

▨ Heritage
   character area

▨ Choice of use/
   mixed use

□ Residential
   neighbourhood

▥ Light industry

●— Skytrain line
   and station

BURRARD INLET

ENGLISH BAY

DENMAN ST.

DAVIE ST.

PACIFIC ST.

GEORGIA ST.

ROBSON ST.

SMITHE ST.

NELSON ST.

POWELL ST.

HASTINGS ST.

CORNWALL ST.

FALSE CREEK

2ND AVENUE

4TH AVENUE

6TH AVENUE

GRANVILLE ST.

OAK ST.

CAMBIE ST.

MAIN ST.

BROADWAY

commercial properties account for approximately 40% of Vancouver's tax base and projections that the downtown core could run out of space for jobs within five to 25 years under existing zoning policies, the city placed a temporary moratorium on office conversions to study the problem. (Boddy, 2005; Bula, 2006a; Ehrenhalt, 2006). Having successfully achieved downtown's urban regeneration with high-density neighbourhoods on derelict industrial lands around False Creek, Burrard Inlet and the margins of the downtown core, developers began to increase pressure on existing residential areas to densify.

1992–Present: Ecodensity and the sustainable city

Outside downtown, Vancouver was struggling with issues of development capacity, housing affordability, rising poverty, amenity provision, protection of neighbourhood character and environmental concerns through the 1980s and 1990s (Merill Cooper, 2006).[6] Most of Vancouver's land base is zoned for single-family housing and the city identified densification of existing residential neighbourhoods as the single-most important strategy for addressing its issues. The city proceeded with legalization of secondary suites,[7] through re-zoning but most west-side neighbourhoods opposed being re-zoned through a vote. Discussions of densification only raised

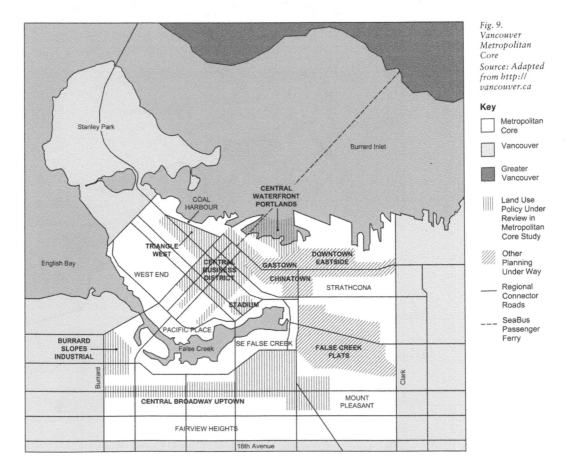

*Fig. 9.
Vancouver
Metropolitan
Core
Source: Adapted
from http://
vancouver.ca*

neighbourhood demands for more restrictive zoning and design guidelines, sparking criticism from the development community and furthering the crisis of housing affordability (Punter, 2003). Without further densification of Vancouver, its vision for a 'liveable' city was stalled.

The 1990 Clouds of Change report (City of Vancouver, 1990) on atmospheric quality provided the impetus for a citywide growth management strategy. The report found that a quarter of all air pollution in the Fraser delta comes from Vancouver through the prevailing winds, and recommended that Vancouver take leadership on atmospheric change in the region by creating energy-efficient land-use policies. The City of Vancouver responded in 1992 with City Plan (City of Vancouver, 1992), a process which brought together citizens representing different interests and developers in round-table discussions covering key issues such as the preferred level and pattern of urban intensification and the consequent location and form of new housing.

Participants were exposed to the inevitable conflicts and trade-offs which the city normally faced and then asked to make hard choices (Punter, 2003). The discussions produced four alternative futures: a traditional city which limited housing to existing and planned, a central city which put housing on industrial lands, a city of mixed residential and main streets, and a city of neighbourhood centres, which ultimately garnered 80% of public support. A second phase of public and government consultation generated eighteen specific neighbourhood visions for implementation by neighbourhood committees (ibid.).

The neighbourhood visions addressed movement, new public places, community services, neighbourhood centres, housing variety and cost, neighbourhood character and financial accountability. Revisions to commercial re-zoning along the main arterial roads in 1990 created mixed commercial–residential strips. Medium-term planning priorities were identified, such as the design of neighbourhood centres

Fig. 10.
Craftsman single
family home
in Grandview
Woodland with
townhouses
behind and
downtown in the
distance
Source: May So

for significant retail, community services and housing, the review of zoning and building code with regard to secondary suites, intensification of public transit along main streets, and the accommodation of conversions, laneway housing, additional units and infill on larger residential sites (ibid.). Consensus was reached on modest intensification to absorb another 140,000 residents in 80,000 housing units across the city (Bula, 2006b). In the last decade, Vancouver has added approximately 47,000 people citywide, but lost its share of growth in the region, growing only 8% compared with Greater Vancouver's 13% (Berelowitz, 2006).

Densifying existing single-family neighbourhoods presents many challenges including residents' fears of losing neighbourhood character (Fig. 10), the requirement of long-term incremental development to achieve density and amenities, and the difficulty of consolidating several contiguous sites for denser building forms. The city realized that any attempt to gain public acceptance of densification as official city policy would have to address residents' legitimate fears while providing a compelling vision of why densification is not only in their immediate interest of liveability, but in their long-term interest. In June 2006, the city launched the EcoDensity Initiative, which frames high-quality densification and housing affordability in the environmental language of 'single-planet living' (City of Vancouver, 2006b). The initiative strives to accelerate the implementation of existing policy to densify neighbourhoods sustainably,[8] develop new tools through a public process to guide future decisions on planning and development, and create evaluation criteria against which EcoDensity proposals can be measured (Fig. 11).

In specific terms, EcoDensity proposes to achieve sustainable urban growth by increasing accessibility to transport, making better use of existing infrastructure and implementing community energy, water and waste systems. This is being developed in the model for a sustainable Southeast False Creek community (Grdadolnik, 2006). Southeast False Creek's official development plan aims to minimize automobile use by providing bicycle lanes, pedestrian paths, bus routes and a nearby light rail transit

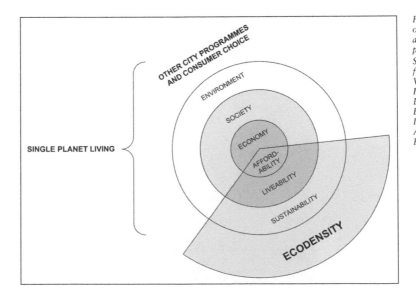

Fig. 11. The role of EcoDensity in achieving single-planet living Source: Adapted from City of Vancouver Planning Department, Ecodensity Initiative: Adminstrative Report, 4–5

stop. Community amenities will include a grocery store, three day-care centres, a mix of restaurants and shops on the waterfront, sports fields, a seawall and re-naturalized areas. The initial progressive housing mix of one-third each low, middle and market housing was amended to a minimum of 20% low-income housing across the site, but with a goal of 33%. The most innovative aspect of the policy is the requirement for community-level energy generation, water management, waste management and urban agriculture strategies to be monitored for stewardship issues, and then applied to other communities in Vancouver (City of Vancouver, 2005).

Through the EcoDensity initiative, liveability will continue to be realized through complete compact neighbourhoods, and promotion of neighbourhood character and public safety through design. To attain housing affordability, EcoDensity proposes to increase the housing supply, introduce new housing types, reduce the cost of construction and operation through green building technology, and investigate new financial tools (City of Vancouver, 2006b).

Will EcoDensity be a more effective policy than liveability for increasing density in Vancouver's existing neighbourhoods? Only time will tell, but a strength of the initiative is that it explicitly links density to sustainability, a cause that is gaining global attention. EcoDensity builds on the environmental activism already evidenced in Vancouver, such as public advocacy for cycling to be integral in Vancouver's transportation culture and for protection of the Agricultural Land Reserve.[9] EcoDensity has afforded many opportunities through EcoDensity fairs, lectures, and online discussions, for the public to be educated and involved in living more sustainably.[10]

Mayor Sam Sullivan launched the EcoDensity Initiative with the insight that:

> ... in order to become a truly sustainable city, we must do more to increase the densification of our neighbourhoods, and today I am inviting all Vancouverites to join the discussion on how we can meet this challenge. High quality densification is one of the most critical steps to reducing our ecological footprint as a city, and it will also enable us to expand housing choices and improve affordability for all residents, including young families and seniors. The EcoDensity Initiative will also provide for new investment in public amenities such as parks, green space, cultural facilities and community centres – all of which will contribute to greater quality of life (City of Vancouver, 2006c).

Here at last is an ambitious new vision, and not a claim, for a 'liveable' Vancouver.

Conclusion

Vancouver has been embroiled in debate and experimentation around liveability and its relation to densification for more than thirty years, beginning with urban reform to combat the negative effects of uncontrolled growth. Greater Vancouver's regional plan first linked quality-of-life and densification by concentrating growth in multiple town centres in a compact metropolitan area to improve residents' access to infrastructure and amenities. Vancouver first applied the principles of liveability supported by densification to downtown as the core strategy of urban regeneration. The 1975 Downtown Official Development Plan introduced the basic discretionary controls of land use, density, building height, parking and amenities that would be used to direct private development into the creation of liveable neighbourhoods.

The refinement of these controls during the 1980s facilitated a fusion of private and civic ambitions whereby the increased land value of a high density allowance through re-zoning provided the developers with incentive to build and allowed the city to extract a high level of public amenities. Occurring on former industrial lands and rail yards, the new downtown neighbourhoods encountered minimum public opposition. On the contrary, demand for downtown condominiums rose to the point of exacerbating problems of housing affordability, diminishing downtown's office capacity, and increasing development pressures to densify existing single-family neighbourhoods. Density had successfully been made liveable in downtown and Vancouver became globally recognized as a 'liveable' city.

By the early 1990s, however, the limit of 'liveability' as the benchmark for approval of new developments became apparent in increasing air pollution, undeterred sprawl outside of the growth concentration area, inflated housing prices and neighbourhood resistance to densification. The goals of Greater Vancouver's liveable regional plan had eluded Vancouver despite its successes in creating dense liveable downtown neighbourhoods. To address the complexity of environmental, social and economic issues facing Vancouver, EcoDensity provides a compelling long-term citywide vision for accommodating urban growth. EcoDensity attempts to move Vancouver beyond a preoccupation with its own quality-of-life to embrace larger concerns for the global environment through adoption of sustainable building practices and urban forms. While EcoDensity may position Vancouver competitively on the global stage, this time as a 'sustainable city', it will, more importantly, enhance the awareness Vancouverites have of the planet on which they reside, stir their responsible participation in the world's political, social and environmental realities, and in the process transform Vancouver into a truly 'world' city.

Acknowledgement

Phoebe Wong for assisting with illustrations.

Notes

1. In 2006, Greater Vancouver was rated the most liveable city by the Economist Intelligence Unit, which assessed living conditions of 127 cities using 40 indicators grouped in 5 categories: stability, healthcare, culture and environment, education and infrastructure (City Mayors, 2005). The same year, Greater Vancouver tied second with Vienna on Mercer Human Resource Consulting (2006) Quality of Living index, which measured 39 tangible quality-of-living factors grouped under political, social, economic and environmental factors; personal safety and health; education; transport and other public services; recreation; consumer goods; housing and the natural environment.
2. While the Liveable Region Strategic Plan (LRSP) builds upon the Official Regional Plan (1966), the Liveable Region 1976/1986 (1975), and the Plan for the Lower Mainland of British Columbia (1980), Creating Our Futures (1990, 1993, and 1996) in particular provided the necessary vision for promoting the region's liveability by addressing environmental health, growth management, transportation issues, and social and economic challenges.
3. The Liveable Region Program (LRP) was concerned with voluntary coordination of growth and change in jobs, population, housing and transportation in the Greater Vancouver metropolitan area. The LRP augmented the 1973 Official Regional Plan.
4. Referred to in this chapter as 'city council' or 'the council'.
5. See also City of Toronto Chief Administrative Officer, Canada's Cities Unleash our Potential: The Relationship of Five Canadian Cities and their Provinces 5 September 2000: 21 November 2006 <http://www.canadascities.ca/caoreport_092000.htm> for comparisons with charters for other Canadian cities.
6. Social and economic trends show that Vancouver is moving away from social inclusion and social sustainability. Although average incomes have been rising, Vancouver's rate working

poor is one of the highest of any major city in Canada because most of the growth has benefited high-income households.

7. Secondary suites are self-contained rental units frequently located in the basement or ground floor level of single-family dwellings. While the development of secondary suites had been encouraged in the 1940s to relieve wartime housing shortages, protection of neighbourhood character became the city council's predominant concern by 1956 and surveys by the planning department in 1974 showed that two thirds of homeowners did not want secondary suites in their area. Illegal suites proliferated, however, to meet demands for modest rental housing, so in the late 1980s, a new residential zoning category was created for secondary suites and introduced in Vancouver's affordable east-side without resistance.

8. The City of Vancouver already pursues a mandatory regulated green building strategy for most new development in the city. Rather than providing incentives for building green, the approach normalizes a high standard of sustainability for new development in the entire city. Green development standards will eventually be established within existing building by¬laws and codes to ensure a new baseline for all non-combustible buildings.

9. See Vancouver Area Cycling Coalition's website, <http://www.vacc.bc.ca/> and Smart Growth BC's website on the Agricultural Land Reserve, <http://66.51 .172.1 16/AboutUs/Issues/AgriculturalLandReserveALR/tabid/1 1 1/Default.aspx>

10. See <http://www.vancouver-ecodensity.ca/> for online consultation of the Vancouver EcoDensity Planning Initiative.

References

Allen, J., Massey, D. and Pryke, M. (1999) Introduction, in *Unsettling Cities* (eds J. Allen, D. Massey and M. Pryke), Routledge, London.

Beasley, L. (2000) 'Living First' in Downtown Vancouver. *American Planning Association, Zoning News*, April 2000.

Beaverstock, J.V, Smith, R.G. and Taylor, P.J. (1999) A Roster of World Cities. *Cities*, 16(6): 445–458.

Beaverstock, J.V, Smith, R.G. and Taylor, P.J. (2000) World City Network: A New Metageography? *Annals of the Association of American Geographers*, 90(1): 123–134.

Berelowitz, L. (2006) The Myth of Dense Vancouver. *The Tyee*, 7 March 2006. Retrieved in November 2006 from: http://thetyee.ca/Views/2006/06/2 1/DenseVancouver.

Boddy, T. (2005) Downtown Vancouver's Last Resort: How Did 'Living First' Become 'Condos Only?'. *ArchNews*, Now 11 August 2005. Retrieved in November 2006 from: http://www.archnewsnow.com/features/Feature174.htm.

Boddy, T. (2006) Downtown's Last Resort: A Critique of the Last 20 Years of Vancouver's Approach to Downtown Living Asks Some Difficult Questions About Where the Future Lies for Canada's Ocean Playground. *Canadian Architect*, August 2006: 20–22.

Bula, F. (2006a) Downtown Is Running Out of Working Space: Planners Try to Rebalance After Explosion of Residential Development. *Vancouver Sun*, 5 December 2006, final edn: B4.

Bula, F. (2006b) 'Ecodensity' Mayor Sam's Newest Plan: Sullivan Wants the Quality Development Going on Downtown to Encompass Other Vancouver Neighbourhoods. *Vancouver Sun*, 17 June 2006, final edn: B1.

City Mayors (2005) Vancouver, Melbourne and Vienna Named World's Most Liveable Cities: A Report by the Economist Intelligence Unit. *City Mayors*, 5 October 2005. Retrieved in October 2006 from: http://www.citymayors.com/environment/eiu_bestcities.html.

City of Toronto (2000) *Canada's Cities Unleash our Potential: The Relationship of Five Canadian Cities and their Provinces*, City of Toronto Chief Administrative Officer, September 2000. Retrieved in November 2006 from: http://www.canadascities.ca/caoreport_092000.htm.

City of Vancouver (1990) *Clouds of Change: Final Report of the City of Vancouver Task Force on Atmospheric Change*. Vols 1 and 2, Vancouver.

City of Vancouver (2006a) *Visitor Information*, City of Vancouver. Retrieved in October 2006 from: http://vancouver.ca/visitors.htm.

City of Vancouver Planning Department (1975) *Downtown Official Development Plan*, City Planning Department, Vancouver.

City of Vancouver Planning Department (1991) *Central Area Plan: Goals and Land Use Policy*, City Planning Department, Vancouver.

City of Vancouver Planning Department (1992) *CityPlan Toolkit*, City Planning Department, Vancouver.

City of Vancouver Planning Department (2005) *Southeast False Creek Official Development Plan*, City Planning Department, Vancouver.

City of Vancouver Planning Department (2006b) *EcoDensity Initiative: Administrative Report*, City Planning Department, Vancouver.

City of Vancouver Mayor's Office (2006c) *Mayor Sullivan Launches Vancouver EcoDensity Initiative*, Press Release, 16 June 2006. Retrieved in October 2006 from: http://www. vancouver.ca/ctyclerk/councillors/mayor/announcements/200 6/06 1 606.htm.

Ehrenhalt, A. (2006) Extreme Makeover: After Transforming Its Downtown into a Residential Mecca, Vancouver is Trying to Find the Right Balance Between Condos and Commerce. *Governing.com*, June 2006. Retrieved in October 2006 from: http://www.governing. com/articles/7down.htm.

Fong, P. (2005) Vancouver 'Most Livable' Again. *Globe and Mail*, 4 October 2005: A8.

Greater Vancouver Regional District Planning Department (1975) *The Livable Region 1976/1986: Proposals to Manage the Growth of Greater Vancouver*, Greater Vancouver Regional District, Planning Department, Vancouver.

Greater Vancouver Regional District Planning Department (1980) *The Livable Region from the 1970s to the 1980s*, Greater Vancouver Regional District, Planning Department, Vancouver.

Greater Vancouver Regional District Planning Department (1996) *Livable Region Strategic Plan*, Greater Vancouver Regional District, Planning Department, Vancouver.

Grdadolnik, H. (2006) Our World Class Olympic Village? Southeast False Creek Can Showcase a Better Future. *The Tyee*, 7 March 2006. Retrieved in October 2006 from: http://thetyee.ca/Views/2006/03/07/WorldClassVillage/.

Kane, M. (2005) Population Growth Predicted to Double. *Vancouver Sun*, 14 April 2005, final edn: D1.

Mercer Human Resource Consulting (2006) *Highlights from the 2006, Quality of Living Survey*, Press Release, 10 April 2006. Retrieved in October 2006 from: www.mercerhr. com/summary.jhtml;jsessionid=DECA2YSDXFMZ2CT GOUGCII.

Merrill Cooper (2006) *Social Sustainability in Vancouver*, Canadian Policy Research Networks, Ottawa.

Olds, K. (1996) *Developing the Trans-Pacific Property Market: Tales from Vancouver via Hong Kong*, Research on Immigration and Integration in the Metropolis Working Paper Series, Vancouver Centre of Excellence, Vancouver.

Port Vancouver (2006) *Vancouver Port Authority Applauds Immediate Allocation of Federal Funding for Transportation Infrastructure Development*, Press Release. Retrieved in October 2006 from: http://www.portvancouver.com/media/news_20061011 -2.html.

Province of British Columbia (1953) *Vancouver Charter [SBC 1953]*, Queen's Printer, Victoria, British Columbia.

Punter, J. (2003) *The Vancouver Achievement*, University of British Columbia Press, Vancouver.

Sandercock, L. (2005) An Anatomy of Civic Ambition in Vancouver: Towards Humane Density. *Harvard Design Magazine*, **22**: 36-43.

Sassen, S. (1991) *The Global City*, Princeton University Press, Princeton, NJ.

Seymoar, N.-K. and Timmer, V. (2005) *The Livable City*, Vancouver Working Group Discussion Paper for the World Urban Forum 2006, International Centre for Sustainable Cities, Vancouver.

Short, J.R., Breitbach, C. Buckman, S. and Essex, J. (2000) From World Cities to Gateway Cities. *City*, **4**(3): 317-340.

Simon, D. (1995) The World City Hypothesis: Reflections from the Periphery, in *World Cities in a World System* (eds P. L. Knox and P. J. Taylor), Cambridge University Press, Cambridge, pp. 132–155.

Smith, R.G. (2005) Networking the City. *Geography* 90(2): 172–176.

Taylor, P.J., Catalano, G., Hoyler, M. and Walker, D.R.F. (2002) Diversity and Power in the World City Network. *Cities*, **19**(4): 23 1–241.

Transport Canada (2005) Government of Canada Announces Pacific Gateway Strategy, News Release, GC No. 013/05. Retrieved in October 2006 from: http://www.tc.gc.ca/ mediaroom/releases/nat/2005/05-gc0 1 3e.htm.

Westbank and Peterson Group (2006) Woodward's. Retrieved in October 2006 from: http:// www.woodwardsdistrict.com.

Planning a 'World Class' City without Zoning:

The experience of Houston

Introduction

Houston is a city that has assumed considerable status globally. It is ranked as one of the third tier Gamma World Cities in the Globalization and World Cities Network (GaWC) inventory of world cities (Beaverstock *et al.*, 1999), in which cities are ordered according to their 'world city-ness' – generally meaning their relative strength as global service centres. Friedmann's (1986) hierarchy identified Houston as one of the core secondary world cities. Houston is also defined as a specialist city because of its status as the world's energy capital. The city is ranked 10th of all the US cities in terms of global network connectivities (Taylor and Lang, 2005).

Despite its world city status, Houston is the only major city in North America without an overall plan or land use zoning.[1] The growth of Houston illustrates a traditional free market philosophy in which land use zoning is seen as a violation of private property and personal liberty. Although attempts were made in 1929, 1948, 1962 and 1993 to introduce land use zoning ordinances, all failed to be enacted. In such a *laissez-faire* city, public-sector-initiated urban planning policies are discouraged or limited to a patchwork of local land use regulations and codes on aspects of development such as lot sizes and parking. Urban development policies and plans are made by investors, developers, builders, realtors, homeowners, architects, and planners in the private sector and by business associations organized by city enterprise élites. Except for the limited daily urban needs that are the responsibilities of the public sector such as water, sewage, health and education – planning, especially that which effects economic growth, is initiated, developed, and monitored by leading voices in the private sector (Fisher, 1989).

How then does local land use policy and urban planning practice work in this unique political and economic setting? This chapter explores how local land use policy made by both the local government and non-governmental sectors impacts on urban development in Houston which was born out of several anti-zoning battles. On the one hand, despite the city's lack of zoning, local land use regulatory policies made by the municipality such as minimum lot sizes, minimum parking requirements and setbacks, street width and large freeway mileage have had significant influence on urban physical development. On the other hand, civic and private organizations such as super neighbourhoods (explained later below) and homeowner associations effectively fill the gaps left by the lack of land use zoning. This study examines how these aspects contribute to the city's planning, and whether this kind of planning matters in a 'unzoned' city like Houston.

Houston's political environment of both pro-economic growth and minimal government intervention has fascinated scholars. In spite of the city's unique status of 'non-zoning', relatively little has been written about land use in Houston. Real estate lawyer Bernard Siegan's *Land Use Without Zoning* (1972) remains the definitive document on Houston's 'non-zoning'. According to Siegan, the market-place provides economic incentives for segregation of uses and produces patterns of development similar to what is found under zoning. As he put it, 'economic forces tend to make for a separation of uses even without zoning' (Siegan, 1972, p. 75). Siegan also puts forward the argument for Houston's unplanned, unregulated development in a set of articles defending the city's refusal to enact a zoning code. He asserts that land use regulation in Houston is extremely modest compared to most zoning ordinances because Houston has no ordinance placing specific restrictions on the uses that may be established on any property.

More recent studies address Houston from different perspectives, such as urban geography (Kirby and Lynch, 1987; Vojnovic 2003), political science (Gainsborough, 2001), public policy (Fisher, 1989) and legal and economics (Berry, 2001), but few from land use planning. For instance, the most recent research explores the political and social forces that have shaped Houston's local governance (Vojnovic, 2003). It employs two theoretical interpretations – public choice and political economy, and concludes that the new directions in Houston's policy are a reflection of a different growth strategy reflecting changing demographics and a diversifying economy. Gainsborough (2001) explores the politics of regional cooperation in Houston, focusing on the role of the state in facilitating or inhibiting metropolitan-wide approaches to urban problems, and argues that while generous annexation rules have facilitated regionalism, these rules are themselves only as powerful as the political consensus to use and maintain them. Berry (2001) explores the relationship between land use regulation and residential segregation by comparing an unzoned Houston and a zoned Dallas. By examining indices such as race, tenure, and housing type, the study found no significant differences in residential segregation between the two cities. In addition, urban sociologists often portray Houston as an archetype, a free enterprise, capitalist, or *laissez-faire* city (Feagin, 1998; Lamare, 1998; Lin, 1995).

All those previous studies provide a solid background and politico-economic context for this research on Houston's land use planning approaches in its lack of zoning regulation. This chapter first depicts the growth of Houston. It then analyzes the political culture behind the city's urban development. Critical investigations of both land use regulatory policies from public sectors and neighbourhood land use control efforts from private sectors demonstrate the uniqueness of land use planning in Houston without zoning. The study concludes by highlighting the key contributors to Houston's urban development, the impact of public land use intervention, the diversity of various deed restrictions on land use controls, and their implications for Houston as a 'world class' city.

The growth of Houston

Founded in 1836, Houston was originally located inland with no natural outlet to the sea. Prior to 1900 Houston was a mercantile city, a connecting point between an agricultural hinterland and regional, domestic, and international markets. After the Spindletop oil discovery ninety miles east of the city in 1901, the city became the leading oil refining centre in the country. But Houston remained a small city until its rapid growth in the Second World War era (Shelton *et al.*, 1989). During the Second World War, it became the world's largest petrochemical manufacturing site. Houston ranked sixth in federal wartime plant investments (Mollenkopf, 1983). These investments were also a stimulus for private investment in advanced technology, defence, oil and natural gas, tourism, and property during and after the war. The driving force behind the 1970s post-industrial economy was the influx of managerial and technical employees into the oil and gas industries in the city. The local economy in other fields also benefited – the Texas Medical Center (Fig. 1) became a world-renowned research and treatment centre and NASA located its manned space control centre in the city. The energy industry was still the primary economic contributor as it accounted for over 80% of local employment in the early 1980s. The oil industry did not develop without local government intervention in that public regulation involved numerous benefits to oil entrepreneurs, including tax breaks, price fixing and subsidies directed to petroleum infrastructure. However, in the early and mid-1980s the oil industry slumped with the sharp oil price drop: the Houston area lost about 150,000 high-paying jobs in 1982 and 1983, and suffered more losses in 1986. Houston gradually recovered in the early 1990s. Besides the energy sector, there was significant growth in other sectors such as high technology industries, professional services, medical research and oil related technology industries. While in 1990 energy accounted for 60% of Houston's economic base, it accounted for only 49% in 1999 (Smith, 2000). Houston-based Compaq Computers became one of the largest PC manufacturers in the world. Continental Airlines also chose Houston as its headquarters in the 1980s. The city enjoyed an improving economic climate in the late 1990s as new commercial and residential buildings and a sharp increase in property and sales taxes generated more public revenue. The city also built one of the nation's leading ports. In 1998, the port ranked first in the country in foreign tonnage and second in total tonnage (Port of Houston Authority, 1999).

The city may be characterized as a futurist emblem of the metropolis, a decentred profusion of 'suburbs in search of a city' (Shelton *et al.*, 1989). Houston's identity is mirrored in the profusion of adjectives used to describe the metropolis, such as

Fig. 1. Texas Medical Center skyline from the Women's Hospital of Texas, with apartments and condominiums in front of the Medical Center Source: Zhu Qian

magnolia city, freeway city, strip city, mobility city, high-tech city, space city, oiltown. Between 1990 and 2000, Houston had the third largest population growth in the US, remaining the fourth largest city in terms of population (US Census Bureau, 2000). Houston is far less densely populated than most other cities of a similar size, with only 3,372 people per square mile (1,297 per km²), less than half the density of any of the three cities larger than Houston (Table 1 and Fig. 2). Los Angeles has 7,877 residents per square mile (3,030 per km²), while Chicago and New York have over 10,000 residents per square mile (3,850 per km²) (Lewyn, 2005).

Table 1. Population and Density: Major US Cities, 1990/2000 Source: US Census Bureau (1990; 2000)

City	Total Population			Square Miles		Persons Per Square Miles	
	1990	2000	Change %	1990	2000	1990	2000
New York	7,322,564	8,008,278	9.4	308	469	23,775	17,080
Los Angeles	3,485,398	3,694,820	6.0	469	498	7,430	7,415
Chicago	2,783,726	2,896,016	4.0	227	234	12,252	12,376
Houston	*1,631,766*	*1,953,631*	*19.7*	*581*	*618*	*2,807*	*3,161*
Philadelphia	1,585,577	1,517,550	–4.3	135	140	11,736	10,877
Phoenix	983,403	1,321,045	34.3	420	475	2,342	2,781
San Diego	1,110,549	1,223,400	10.2	324	372	3,428	3,288
Dallas	1,006,877	1,188,580	18.0	342	377	2,941	3,156
San Antonio	935,933	1,144,646	22.3	333	412	2,811	2,778
Detroit	1,027,974	951,270	–7.5	139	143	7,411	6,655

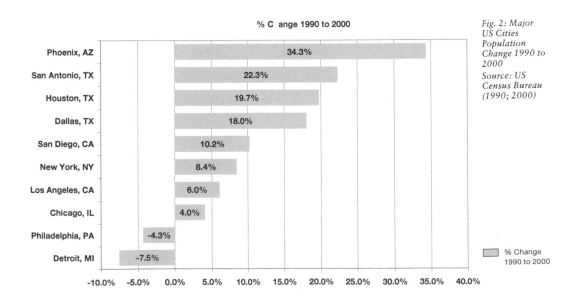

% C ange 1990 to 2000

City	% Change
Phoenix, AZ	34.3%
San Antonio, TX	22.3%
Houston, TX	19.7%
Dallas, TX	18.0%
San Diego, CA	10.2%
New York, NY	8.4%
Los Angeles, CA	6.0%
Chicago, IL	4.0%
Philadelphia, PA	-4.3%
Detroit, MI	-7.5%

Fig. 2: Major US Cities Population Change 1990 to 2000 Source: US Census Bureau (1990; 2000)

□ % Change 1990 to 2000

With vacant land accounting for 24% of the city's total land use, it presents opportunities for the city to guide future growth and leverage infrastructure investments. Clearly these areas with high vacancy rates and land available for infill development can help accommodate future population growth. Between 1990 and 2000, Houston experienced approximately a 20% change in population. Overall, the spatial growth and change trends for the last decade have been: 1) significant redevelopment within Loop 610, west of Downtown; 2) new development on the fringes of the city, outside Loop 610 in the west, southwest, and southeast; 3) and east of downtown, near the ship channel and Highway 90, large undeveloped parcels have been developed since 1990 (Fig. 3).

Political culture

Houston's dominant political culture for the governance of planning and development is to assist private economic needs. Growth and development have long been goals among Houston's élites, and the city supports programmes that enhance private economic expansion with only minimal supportive programmes for public services in matters like public transport, health care and welfare. Nevertheless, the political élite as defined by the economic marketplace, and successful business leaders, are expected to exercise great influence in local decision-making. For instance, the business organization Greater Houston Partnership provided support for the Metropolitan Transit Authority's plans to build a 7.5-mile light rail system linking two of the city's major employment centres with its major league sports facilities, art centres, educational institutions and local neighbourhoods. The links between political economy and cultural change demonstrate that the political economy of Houston is increasingly decentralized and polyglot (Lin, 1995).

The implications for 'privatized politics' in urban policy-making in Houston are clear: with planning removed from public debate and discussion, most people in the

Fig. 3. City of Houston map: three super neighbourhoods Source: Harris County Appraisal District (HCAD), City of Houston Planning and Development Department (2000)

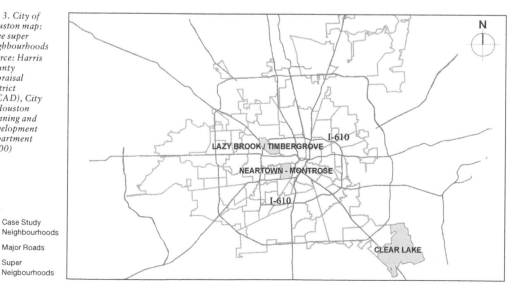

Key

Case Study Neighbourhoods

—— Major Roads

Super Neigbourhoods

city, especially the marginalized, have little access to planning and policy decisions affecting them (Fisher, 1989). The *laissez-faire* capitalism and élite political culture results in poor social services. A number of costs are typically passed on to certain groups of residents, or external forms of government. For instance, Kirby and Lynch (1987) blamed the city's lack of zoning restrictions for the 300-plus toxic waste sites identified by the Environmental Protection Agency.

Houston's rhetoric is for limited government intervention, low taxes and low expenditure on public welfare, and a disinterest in social service and income redistribution programmes. Such an urban policy philosophy is supported by a strong belief in self-reliance and individualism (Lin, 1995). The government has adopted the entrepreneurial spirit of the private sector in the public realm in which certain issues such as government performance at low cost, increased flexibility in agencies and personnel, decentralization, and privatization are highly regarded (Osborne and Gaebler, 1992). The belief governing Houston is that the lack of interest in social service programmes would be compensated for by the support of pro-growth urban policies. Only a few groups such as Blueprint Houston have challenged Houston's pro-growth agenda. As Ross *et al.* commented, the penetration of government by local business élites has been so considerable that the borderline between business and government is no longer clearly discernable (1991, p. 56).

Government intervention in land use

Despite the *laissez-faire* rhetoric, government intervention in Houston's growth has had a significant impact on urban development. The City of Houston draws up park and library master plans, neighbourhood plans, major thoroughfare plans, and various infrastructure plans. Houston Galveston Association Council (HGAC) makes regional transportation plans. Harris County Flood Control District drew up the Brays Bayou[3] Improvement Plan. Houston's municipal government exercises some regulation of land use in a variety of ways, including: minimum lot sizes, minimum parking requirements and setbacks, street widths and block sizes. Moreover, building lines regulation helps preserve the residential character of existing blocks in inner city neighbourhoods by requiring new development to comply with the most frequently constructed building line along the block. Similarly, prevailing lot size regulation requires new development to comply with the most frequently occurring lot size. In fact, Houston regulates land use almost as intricately as cities with zoning by mandating suburban-style low densities, arranging large parking areas in front of commercial buildings, and facilitating land use segregation.

Minimum lot sizes

Until 1998, Houston's municipal land use code set the minimum lot size for detached single family residences at 5,000 square feet (465m^2). In addition, the municipality made it impossible for developers to build large numbers of non-detached single family homes, such as townhouses, because it required townhouses to sit on at least 2,250 square feet (209m^2) of land (Siegan, 1972). The city's townhouse regulations were significantly more restrictive than those of other American cities (Allbee, 1998; Skrzycki, 1983). The townhouse regulations and minimum lot size requirements meant that almost all single family development was low density (i.e. 8.7 houses per acre (21.5 per hectare)), and it was not uncommon to find only two residences per acre (Williams, 2003). These requirements may have unsustainable impacts on urban development. For instance, such low density: makes improved public transit impractical (Nichols, 1992; Hanson, 1999); increases the cost of providing infrastructure and public utilities such as water and sewerage services (Speir and Stephenson, 2002): and, to certain extent, encourages population growth to shift away from the city centre.

In 1998, the municipality amended the minimum lot size requirement regulations so that the 5,000 square foot minimum now applies to areas outside Interstate Highway 610 (I-610), which is about five miles from downtown Houston. Within the I-610 ring, the minimum lot size has been decreased to 3,500 square feet (325m²), and the minimum lot size for townhouses has been changed to 1,400 square feet (130m²). However, given that only 1.4% of city dwellings were built in 1998 (US Census Bureau, 2004) and that about 25% of Houston residents live inside the I-610 highway (Roth, 1991), the impact of the amended regulation on urban development has been limited from the beginning.

The reform of Houston's land use policies was a response to the real estate industry in the city. In the late 1990s, homebuilders urged the city, when it was rewriting its subdivision ordinance, to allow more compact development through reducing lot sizes (Schwartz, 1998). Developers and homebuilders did make it happen. Those changes have recently resulted in some positive development, particularly in the downtown area: townhouses are showing up throughout Houston's inside I-610 area, inner city population is starting to grow, and urban land value is rising significantly. The change featured a mini-boom of loft conversions and townhouses that has transformed a derelict central city area into one of Texas's fastest growing residential neighbourhoods. The affluent and middle classes have begun to return to inner city neighbourhoods.

Minimum parking requirements and setbacks

Houston also requires an ample supply of parking for apartment buildings, single family homes, office buildings, hospitals and supermarkets. For example, apartment buildings must provide 1.33 parking spaces for each 1-bedroom apartment, meaning that property owners must in effect supply more than one parking space for every apartment even though 17% of Houston renters do not own even one car (US Census Bureau 2000). Houston bars must accommodate drivers by providing 10 parking spaces for each 1,000 square feet gross floor area. Furthermore, the city also requires that structures abutting major thoroughfares be at least 25 feet from the street.[4] The combination of mandatory setbacks and minimum parking requirements limits the population density in the case of residenial use, and the employment density in the case of business and commercial use. Landlords pass at least some of the cost of parking spaces on to society through higher prices for goods and services.

Minimum right-of-way widths and miles of freeways

Houston's land use code requires that major thoroughfares must have a 100 feet unobstructed right-of-way for traffic, and all other streets must generally have a 50 to 60 feet right-of-way (Feldstein, 1998). With the addition of the usual 4 feet wide sidewalk on both sides of streets, Houston's wide streets contrast with most American streets which are around 35 feet wide or even narrower (Colby, 2000; Coden, 2003; Swift, 2003). Wide streets reduce the amount of land available for housing and commerce, and reduce residential and employment density.

Houston has more overall freeway mileage than other American regions of comparable size. With only about 10% more population than the Boston urbanized area, Houston has almost twice as many lane-miles of freeway (Houston's 2,460; Boston's 1,310), and about the same lane-miles as Chicago with less than half its population (Chicago's 2,655) (Texas Transportation Institute, 2003). The large mileage of freeway nevertheless does not help the transportation situation as Houston's roads are more congested than those in either Boston or Chicago. Houston is known as one of the most congested cities in the US. The low density development, little by way of compact urban form, the shift of development from city centre to urban fringe and suburbs, and the lack of mass transport have resulted in private vehicle-dependent lifestyles. The 2004 average household in Houston spends $9,566 per year on transportation, which is over $3,000 per year more than residents of Boston metropolitan area (Surface Transportation Policy Project, 2004).

Private sector controls on land use

Given anti-government, anti-public planning, anti-regulation ideology, much of the planning in Houston has been carried out by either the private sector, or the public sector at the request of, and under the guidance of, private sector leadership (Fisher, 1989). Many American cities use single use zoning to prohibit the creation of business or shops in residential zones, or residences in commercial zones, thus separating residential areas from employment areas. Houston has no such government zoning code. The city council turned down zoning in 1929 and 1938. Then Houstonians rejected zoning in each of the three referenda proposing zoning in 1948, 1962 and 1993. The mass media, supported by the anti-zoning groups, described zoning as 'socialism' and even 'communism' in earlier days (Feagin, 1988).

Instead, Houston residents separate residences from businesses through deed restrictions (covenants) that specify the appropriate use for each lot in a subdivision, and enable owners to sue in the event of a violation. Since covenants are created by private contract rather than by government, it could be argued that 'zoning' is achieved by the private sector. Nevertheless, covenants are heavily facilitated by government involvement as the city regulations allow the city attorney to sue to enforce restrictive covenants (Berry, 2001). Houston requires its taxpayers to subsidize the enforcement of covenants, and by subsidizing enforcement of use restrictions, the municipality subsidizes segregation of land use. However, the city cannot determine with absolute certainty whether each individual lot of a neighbourhood is covered by a covenant or not, or the status of the covenants (whether they have expired, changed, or been reinstated), due to a lack of data availability. However, at the larger neighbourhood scale, it is generally known whether or not most of the land parcels are covered by covenants. In fact on average it is estimated that only 50%–60% of residential land use is covered by covenants in Houston. Nevertheless, despite some uncertainties, Houston has created a 'single use zoning' disguised as covenant enforcement.

Deed restrictions are a primary method of private land use control and consist of an agreement between parties to limit the use and operation of one or more parcels of real estate. Those restrictions are binding on all future purchasers and creditors as covenants running with land ownership (Jacobus, 2005). It is apparent that deed restrictions go far beyond the requirements of zoning or of building permits.

Covenants are created by subdivision developers or homebuilders and attached as terms to the properties when they are sold to the first homebuyers. During the early stages of development, the developer normally acts as the primary watchdog for deed restriction violations. Once all the properties are sold, covenants are monitored by neighbourhood residents without further intervention from the developers or homebuilders. Subdivision residents then organize homeowner associations and contribute funds for enforcement. In most cities, deed restrictions are used so that land use controls can be more restrictive than those specified in the zoning ordinance. However, Houston's deed restrictions operate without any zoning and have been the single most important approach in land use control. Since covenants are private contracts, they vary greatly in terms of land use conditions. A covenant for residential property might regulate the type of use, the number of structures, lot size, height, setback, maintenance, number of occupants, green area size, tree height, minimum square footages, roofing and building materials. Once all property owners in an area sign the same covenant, normally entered into voluntarily, they can effectively achieve land use control in that area (Fischel, 1985). Moreover, as Fisher (1989) argues, in a context where the local government does not direct or regulate land use, the primary function of civic clubs has been not the delivery of services, but rather the protection of the neighbourhood against changes which would decrease property values.

Houston's reliance on deed restrictions has led to several legal institutions that facilitate those private contracts (Shelton *et al.*, 1989). At the state level, Texas has the Restrictive Covenants Enforcement Acts, under which city enforcement is limited to violations of restrictions on use, setback requirements, size of lots, and size, type, and number of buildings. Many violations of covenants are not within the purview of city enforcement. Instead, homeowner associations remain the primary enforcers of covenants, monitoring and enforcing the many restrictions not covered by the government. Berry's (2001) study on land use regulation and residential segregation argues that deed covenants achieve in Houston what zoning achieves elsewhere and that zoning simply does not matter, at least for residential segregation.

Non-government organizations like homeowner associations and super neighbourhoods seek to restrain commercial development in and near their neighbourhoods even when there are no covenants prohibiting such development. A development may be stopped if it can be shown that it will produce traffic congestion or water runoff or contamination problems. In this sense, neighbourhood organizations serve a land use function similar to zoning.

Super neighbourhoods and homeowner associations

As defined by the City of Houston's Planning and Development Department, 'a Super Neighbourhood is a geographically designated area where residents, civic organizations, institutions, and businesses work together to identify, plan, and set priorities to address the needs and concerns of their community' (City of Houston, 2006). By getting residents of individual communities to focus their attention on areas that do not affect only their immediate neighbourhood or subdivision, they are encouraged to broaden their communities by identifying, prioritizing, and addressing the needs and concerns of the wider neighbourhood (City of Houston, 1999). These communities are grouped together, according to shared common physical characteristics, identity or infrastructure that connects one area to the next

(City of Houston, 2006). Currently there are 88 super neighbourhoods in Houston, encouraged by the city to form Super Neighbourhood Councils. These councils serve as a forum where a representative group of residents and stakeholders can discuss issues impacting on their communities, reach a consensus on projects, and develop a Super Neighbourhood Action Plan for community improvements.

Homeowner associations are private organisations, who create rules and regulations (e.g. deed restrictions) that are enacted to protect the property values, amenities, and homeowners' quality of living in a neighbourhood. A homeowner association is most commonly created by a developer or house builder even before a community is built (Berry, 2001; Jacobs, 2005). They play an important role in curbing the effects of the lack of zoning (Berry, 2001). Deed restriction, landscape ordinances, and regulations formed by these private organisations shift the burden from the public to the private sector. The city government of Houston likes homeowners' associations because they reduce costs, and take on some of local government's responsibilities.

Those neighbourhood organisations may effectively work to form different land use patterns. Houston is unlike many other urban areas in which the most affluent neighbourhoods are outside the central city. This city has a number of up market neighbourhoods within the incorporated city, including the most affluent – River Oaks. Not surprisingly, the poorest neighbourhoods are also in the city, as well as a number of middle-class neighbourhoods, but with relatively few Anglo[5] blue collar communities.

The rest of this section illustrates three different land use patterns at neighbourhood level that reflect the operation of private covenant controls over land use. The three examples range from land with strictly enforced covenants, and are therefore effectively zoned, to neighbourhoods that have few covenants or weak control over them where land use has moved from segregated use to a mixture of land uses. The examples are Clear Lake with covenant controlled land use, the Lazy Brook/Timbergrove Neighbourhood with both mixed land use and covenant-controlled land use, and Montrose which is primarily mixed land use. There is no zoning control in any of these three cases.

Clear Lake Neighbourhood

Clear Lake lies in the south-eastern most portion of the city of Houston (Fig. 4). Before NASA's Manned Spacecraft Center was built in the early 1960s, it was largely coastal prairie used for ranching. Today, the area includes the deed-restricted privately master planned community of Clear Lake City, the adjacent communities of Pipers Meadow and Sterling Knoll, and a large shopping mall. Clear Lake Neighbourhood provides a definitive example of deed restrictions separating uses, which is in effect like zoning. The neighbourhood is home to numerous aeronautics contractors attracted by NASA. New house construction continues in the northern part; however, the northern and western edges of the area are undeveloped because of major roads and a nearby oil field. The neighbourhood has seen a great deal of single family residential development despite the fact that Clear Lake had higher median house prices ($115,063) than the city at-large ($79,300). Clear Lake's population has third highest median income in the city (Houston Planning and Development Department, 2000). The population in the community increased from 45,875 in 1990 to 57,117 in 2000 – a 24.5% increase.

Fig. 4. Plan of Clear Lake
Source: Houston Planning and Development Department

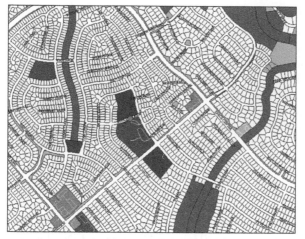

Note: The differences in tone on the built-up areas indicate the extent of variation between single and mixed uses.

Lazy Brook/Timbergrove Neighbourhood

Lazy Brook and Timbergrove are deed-restricted subdivisions located along the wooded banks of White Oak Bayou (Fig. 5). They have both controlled, and therefore segregated land uses (south/west), and mixed uses (north/east). This community offers ranch style brick homes in attractive settings built in the 1950s and 1960s, and are easily accessible to I-610. Its north-western edge includes a business park, a large shopping mall and Houston Independent School District's Delmar Stadium complex. The population in the community increased from 10,869 in 1990 to 11,655 in 2000 – a 7.2% increase (Houston Planning and Development Department, 2000).

Fig. 5. Plan of Lazy Brook/ Timbergrove
Source: Houston Planning and Development Department

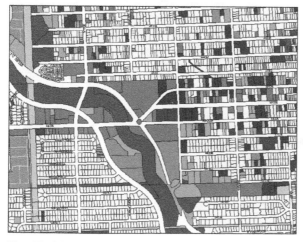

Note: The differences in tone on the built-up areas indicate the extent of variation between single and mixed uses.

Montrose Neighbourhood

Montrose is an eclectic neighbourhood where cottage housing exists side by side with burgeoning town-house developments, large luxury apartment complexes and older duplexes (Fig. 6). In the 1980s, Montrose, which contained many of Houston's historic mansions, was the first neighbourhood in Houston to redevelop. It has less land covered by valid covenants than the other two examples above, and expiring deed restrictions and development pressures of the fast-growing city led to the conversion of much of the area's finest homes into businesses. The area is populated with restaurants, bars and unique retail shops. The neighbourhood illustrates a desirable mixed use urban area, hard to achieve in zoned city (Figs 7 and 8). The neighbourhood has an unusually high proportion of working age and young (0–17 years) population. Montrose has seen significant development activity, particularly of high income rental units. Housing values in this area exceed the median for Houston ($79,300). The population of the community increased from 26,733 in 1990 to 28,015 in 2000 – a 4.8% increase (Houston Planning and Development Department, 2000).

Figures 4–6 clearly indicate the diversity of land use of the three neighbourhoods. When comparing the three neighbourhoods ranging from controlled use, mixed use together with controlled use, to mixed use, a few socioeconomic differences can be observed (Table 2). Household income appears to be one of the most important factors that relates to the level of deed restriction implementation. 85.9% of Clear Lake's residents have a household income of $25,000 or more, higher than the other two neighbourhoods. Deed restriction implementation is also associated with homeownership. Clear Lake has the most owner-occupied housing (63.5%), over twice the proportion of Montrose (29.7%).

This suggests a relationship between the socio-economic and perhaps even the educational level of inhabitants and their ability, or willingness, to enforce private restrictive covenants in their own interests. In areas with fewer covenants, or where

Fig. 6. Plan of Montrose
Source: Houston Planning and Development Department

Note: The differences in tone on the built-up areas indicate the extent of variation between single and mixed uses.

Fig. 7. A 'site specific' sculpture on Montrose Blvd. Montrose Neighbourhood in Houston, Texas by Dean Ruck and Dan Havel. The Art League of Houston owns this building. It was demolished in June 2005 Source: Zhu Qian

covenants expire, especially in less socio-economically advantaged areas, residents may not have the power, resources or education to enforce them themselves. Without covenant protection, residential land use may transform to high density residential use, commercial use or even industrial use. That might explain the highly mixed use and somewhat 'kaleidoscope' style land use pattern in Montrose. Overall, the comparison of three neighbourhood plans shows that the socio-economic characteristics of different neighbourhoods can make significant differences in the operation of deed restrictions and thus greatly influence the land use controls and land use patterns.

Fig. 8. High rise residential building in Montrose Super Neighbourhood Source: Zhu Qian

	Clear Lake	Lazy Brook/ Timbergrove	Montrose
Total Population	57,117	11,655	28,015
White	70.7%	54.1%	67.6%
Black	4.9%	6.2%	3.6%
Hispanic	10.2%	36.4%	23.2%
Asian	11.9%	2.0%	3.8%
Other	2.2%	1.3%	1.7%
Educational Attainment			
Persons 25 years and over	37,259	8,190	21,976
No High School Diploma	5.8%	25.6%	13.1%
High School Diploma and higher	94.2%	74.4%	86.9%
Household Income			
Total Households	22,619	5,240	16,300
Below $25,000	14.1%	29.5%	28.0%
Above $25,000	85.9%	70.5%	72.0%
Labour Force			
Persons 16 years and over	31,328	6,188	20,321
Employed	96.1%	94.5%	96.4%
Unemployed	3.9%	5.5%	3.6%
Housing			
Total Occupied Units	22,442	5,418	16,239
Owner Occupied	63.5%	46.4%	29.7%
Renter Occupied	36.5%	53.5%	70.3%

Table 2. Comparison of three neighbourhoods Source: Houston Planning and Development Department (2000)

Discussion and Conclusions

Houston's land use was, and is, determined by transportation infrastructure, mega and major projects, private land use controls, and a few land use regulatory policies. Large federally funded infrastructure such as Port of Houston, Houston Ship Channel, Big Inch and Little Inch oil pipeline, interstate highways, airports, and NASA Space Center, partially defined the city's historic landscape. More recently, large projects still dominate urban form, especially in downtown, including: Enron Field Stadium, Reliant Stadium, 7.5 mile Main Street light rail, Hobby Performing Arts Center, and the Texas Medical Center expansion.

The civil societies in Houston play important roles in land use due to the relatively low level of leadership by and participation of the public sector. They work to fill the public sector vacuum. Deed restrictions have been principal reasons why Houston's physical appearance and land use patterns are not greatly different, at least in a general sense, from those in other major cities: nevertheless there is more commercial strip development along major arterial streets than in zoned cities, a larger than average number of oddly mixed land uses (Feagin, 1988), and one of the tallest buildings in the world outside of a downtown area (Fig. 9).

Some potential land use compatibility problems arise from deed restrictions. As subdivisions have grown older and as deed restrictions expire, other uses have encroached upon them, resulting in a change in land use character. For instance, heavy commercial and industrial uses exist alongside single family residences; small bungalows are adjacent to commercial, industrial and vacant land (City of Houston, 1992) (Fig. 10). In many cases residential uses are directly adjacent to heavy industries, toxic sites, and landfills. Such land development strategies might satisfy pro-growth ideology, but they fail to alleviate the living circumstances of the poor, and transfer the costs generated from growth to them (Vojnovic, 2003).

A key problem with the deed restrictions is their variation from area to area in content and enforcement. In particular, deed restrictions in minority and/or lower income communities might be simply ignored by landowners and developers. This is a critical problem given that over 500,000 citizens (24% of 2.1 million Houston residents as of 2006) are living at or below the poverty line, ranking Houston worst of major Texas cities (Brown, 2006).

If conventional planning means significant governmental intervention in the public interest for land use and related socio-economic problems of development and growth, then Houston is indeed 'unplanned'. Planning in Houston has been privatized, largely due to the dictates of the investment market and economic growth. Houston's lack of zoning and generally weak planning laws have been used as a defence of the viability of planning with only limited public intervention. The co-existence of private planning and public intervention was supported by the belief that economic growth would result in the correction of dysfunctional conditions on an *ad hoc* basis (Fox, 2003). Such planning is clearly not without social and environmental costs.

The world city thesis has come under sharp criticism in recent years (e.g., White 1998; Hill and Kim, 2002). Some of the criticisms are mostly of macro-economic explanations of urban development patterns, and that researchers tend to overlook

Fig. 9. (below–left) The Williams Tower (formerly the Transco Tower) is Philip Johnson's 1983 masterpiece. The 64-storey skyscraper claims to be the world's tallest building outside of a downtown area
Source: Zhu Qian

Fig. 10. (below–right) One of the views of Downtown Houston skyline. Note the white townhouse at the left corner
Source: Zhu Qian

local variations and are indifferent to indigenous political cultures that affect urban form. Houston's planning and development indicate that local politics are a primary factor in its development. Local political history and culture determine distinct urban development patterns. Local élites devise strategic plans and make political choices that shape the city's development.

The privatization of land use planning, with weak and partial regulations and controls and no formal zoning, has lead to a unique form of development in Houston. It is also one that is changing. It favours the wealthier residents who have the power, education and resources to maintain a high quality for their own environment through enforcing private covenants and controlling land use in their neighbourhoods. However, where covenants expire, or communities are too disadvantaged or too poor to wield sufficient influence, powerful private interests can either ignore private covenants or overcome them. The result is interesting. In the richer areas, private covenants result in environments that are effectively strictly zoned, and effectively become single use mono-cultures. In the poorer areas, where covenants are less likely to be enforced, mixed use areas begin to emerge, with many of the much claimed urban design benefits of viability and vitality, but with social and economic disbenefits of inappropriate mixtures of uses that may over-ride community amenity.

Up to today, Houston is still a free enterprise city. Houston exemplifies a unique 'world class' city without land use zoning. Transportation plans and large projects define the main framework of its urban form. Deed restrictions fill the gap of non-zoning, but not without problems. Private land use controls are weak in addressing social and environmental concerns in lower income communities. Downtown redevelopment brings people back to the inner city but creates gentrification. Hence as Houston develops further towards a quintessential world city, if public sector planning and maybe zoning is still impossible for the city to achieve, a new type of planning in a free market context needs to be developed to address current problems.

Notes

1. Houston is the only major city, and only world city in the US without zoning, but there are also some smaller cities (e.g. in Texas) without land use zoning.
2. 1 square mile equals 2.6km².
3. A bayou is a small, slow-moving stream or creek. Bayous are usually located in low-lying areas, for example in the Mississippi River delta region of the southern United States. Houston is known as the 'Bayou City', primarily because of the massive, muddy, miles-long Buffalo Bayou that twists and turns its way through the city. Other major bayous in Houston include Brays Bayou, Sims Bayou, White Oak Bayou, and Greens Bayou.
4. The city has allowed setbacks of less than 25 feet under certain defined conditions since 1998
5. In the southwest United States, Anglo, short for Anglo-American, refers to non-Hispanic European Americans, most of whom speak the English language but are not necessarily of English descent.

References

Allbee, L. P. (1998) Building Neighborhood from Scratch. *Dallas Morning News*, August 10: 1C.

Beaverstock J.V., Smith R.G. and Taylor P.J. (1999) A roster of world cities. *Cities*, 16(6): 445–458.

Berry, C. (2001) Land Use Regulation and Residential Segregation: Does Zoning Matter. *American Law and Economics Review*, 3: 251–274.

Brown, P. (2006) *Build A Better Houston*, Peter Brown's Councilman Newsletter of June 2006.

City of Houston. (1992) *Demographic and land use profile for Houston, Texas*, Planning and Development Department, Houston.

City of Houston, (1999) *Guidelines for Organizing Super Neighborhoods and forming Super Neighborhood Councils*, Planning and Development Department, Houston.

City of Houston (2006) Retrieved from www.houstontx.gov.

Coden, A. B. (2003) *Narrow Streets Database*. Retrieved from: http://www.sonic.net/abcaia/narrow.htm.

Colby, R. (2000) How Narrow a Street is Safe, Officials Ask. *Portland Oregonian*, 21 August.

Feagin, J, (1988) *Free Enterprise City: Houston in Political-economic Perspective*, Rutgers University Press, New Brunswick.

Feagin, J. (1998) *The New Urban Paradigm: Critical Perspectives on the City*, Rowman and Littlefield Publishers, Lanham, MD.

Feldstein, D. (1998) High-Style, Wide and Handsome. *Houston Chronicle*, 12 June.

Fischel, W. (1985) *The Economics of Zoning Laws*, The Johns Hopkins University Press, Baltimore.

Fisher, R. (1989) Urban Policy in Houston, Texas. *Urban Studies*, 26: 144–154.

Fox, S. (2003) Planning in Houston: A Historic Overview, in *Ephemeral City: Cite Looks at Houston* (eds B. Scardino *et al.*), University of Texas Press, Austin, TX, pp. 34–40.

Friedmann, J. (1986) The World City Hypothesis. *Development and Changes*, 4: 12–50.

Gainsborough, J. F. (2001) Bridging the City-Suburb Divide: Stages and the Politics of Regional Cooperation. *Journal of Urban Affairs*, 23: 497–512.

Hanson, E. (1999) Voter's Guide: City Council At-Large Races. *Houston Chronicle*, 24 October.

Hill, R. C. and Kim, J. W. (2000) Global Cities and Developmental States: New York, Tokyo and Seoul. *Urban Studies*, 37(12): 2167–2195.

Houston Planning and Development Department. (2000) *Houston Land Use and Demographic Profile 2000*.

Jacobus, C. J. (2005) *Texas Real Estate Law*, 9th Edition, Thomson South-Western Publisher, Mason, OH.

Kirby, A. and A. K. Lynch. (1987) A Ghost in the Growth Machine: the Aftermath of Rapid Population Growth in Houston. *Urban Studies*, 24: 587–596.

Lamare, J. (1998) *Texas Politics: Economics, Power, and Policy*, West/Wadsworth, New York.

Lewyn, M. (2005) How Overregulation Creates Sprawl (Even in a City without Zoning). *Wayne Law Review*, Winter: 1171–1208.

Lin, J. (1995) Ethnic Places, Postmodernism, and Urban Change in Houston. *The Sociological Quarterly*, 36: 629–647.

Nichols, B. (1992) Houston Rail Plan Apparently Heading Nowhere. *Dallas Morning News*, 26 July, 41A.

Osborne, D. and Gaebler, T. (1992) *Reinventing Government: How the Entrepreneurial Spirit is Transforming the Public Sector*, Addison-Wesley, Reading, MA.

Port of Houston Authority. (1999) *The Port Report*, Port of Houston Authority, Houston.

Ross, B. Levine, M., and Stedman, M. (1991) *Urban Politics: Power in Metropolitan America*, F. E. Peacock Publishers, Ithaca, IL.

Roth, B. (1991) Urban and Suburban Houston: A Tale of Two Cities. *Houston Chronicle*, 7 July.

Schwartz, M. (1998) Revised Subdivision Ordinance Sent to Panel. *Houston Chronicle*, 8 September.

Shelton, B.A. *et al.* (1989) *Houston: growth and decline in a sunbelt boomtown*, Temple University Press, Philadelphia.

Siegan, B. (1972) *Land Use without Zoning*, Lexington, Lexington, MA.

Skrzycki, C. (1983) If You're Looking for a House You Can Afford. *US News & World Report*, 5 December.

Smith, B. (2000) *The past decade/the next decade: Differences and Similarities*, University of Houston, Center for Public Policy, Houston.

Speir, C. and K. Stephenson. (2002) Does Sprawl Cost Us All? Isolating the Effects of Housing Patterns on Public Water and Sewer Costs. *APA Journal*, 68(1): 56–70.

Surface Transportation Policy Project (2004) *Transportation Costs and the American Dream-Spending Table*. Retrieved from: http://www.transact.org/report.asp?id=225.

Swift, P. (2003) *Residential Street Typology and Injury Accident Frequency*. Retrieved from: http://www.sierraclub.org/sprawl/articles/narrow.asp

Taylor, P. J. and R. E. Lang (2005) US Cities in the 'World City Network'. *Metropolitan Policy Program Survey Series*, The Brookings Institution, February 2005.

Texas Transportation Institute (2003) *Urban Mobility Study*. Retrieved from: http://mobility.tamu.edu/ums/report

US Census Bureau (2000) *Statistical Abstract of the United States*, Government Printing Office, Washington, DC.

US Census Bureau (2004) *American Fact Finder, Quick Tables, Table QT-H7*, Houston, Texas. Retrieved from: http://www.factfinder.census.gov.

Vojnovic, I. (2003) Governance in Houston: Growth Theories and Urban Pressures. *Journal of Urban Affairs*, 25: 589–624.

White, J. W. (1998) Old Wine, Craked Bottle? Tokyo, Paris, and the Global City Hypothesis. *Urban Affairs Review*, 33(4): 451–477.

Williams, J. (2003) Mayoral Campaign Revs Up In Garage. *Houston Chronicle*, 13 January, A15.

Section Three

Aspects of Fragmentation and Polycentrism

15

Daniel Kozak

Assessing Urban Fragmentation:
The emergence of new typologies in central Buenos Aires

Introduction: world cities and recent urban transformations

Many recent studies have identified similar processes of urban transformation within those cities currently classified as world cities (e.g. GaWC, 2006). Although there is a significant level of coincidence amongst the descriptions of these transformations, the causes and outcomes of them remain contentious. While some studies explain them as a direct consequence of the impact of globalization, others are more inclined to look to local explanations. Nevertheless the ubiquitous nature of these processes appears to be generally acknowledged. One of the commonalities found throughout these studies is the emergence of new typologies that challenge traditional understanding of the urban centre and periphery. Concepts such as 'grid erosion' (Pope, 1996) and 'rebundled city' (Dick and Rimmer, 1998; Graham and Marvin, 2001) are used to grasp these urban transformations and form part of a new theoretical framework centred around the concept of *urban fragmentation*. Not surprisingly, 'fragmentation' is both one of the most quoted terms in current urban texts and also one of the most polysemous. The question of the meaning, roots and consequences of urban fragmentation is a very compelling topic in the contemporary urban debate.

This chapter critically examines existing concepts associated with urban fragmentation and attempts to further develop the theoretical framework for understanding the phenomenon through empirical research in Buenos Aires, and more specifically in Abasto, a central area of the city. Following a trend common to many contemporary metropolises, the area of Abasto has been recently transformed with the conversion of a fruit and vegetable central market into a shopping centre and the development of a gated tower complex and a hypermarket in its centre. Abasto is analysed and related to the concept of urban fragmentation.

Globalization and urban fragmentation

Fig. (opposite page). Poverty and wealth divided by a wall; a graphic example of urban fragmentation in the border of the Paraisópolis Favela in São Paulo
Source: Tuca Vieira

Explanations of urban fragmentation vary, and here two contrasting points of view from studies by Welch Guerra (2005) and Carmona (2000) are used to illustrate the differences. In a recent work about urban transformations in Buenos Aires, Welch Guerra (2005) argues that, at least since the 1990s, similar urban phenomena have emerged in most large cities and urban regions in the world. These phenomena include:

a strong increase in the dispersion of urbanisation, even in cities which have traditionally tended to compaction but currently experience a *fragmented* territorial extension; a new type of socio-spatial segregation, due to the relative changes in the distribution of jobs and incomes and also to the adoption of new lifestyles; and the emergence of spatial relations within the cities and in what was previously considered the hinterland that cannot be described and explained in terms of the traditional categories of centre and periphery (Welch Guerra, 2005, p. 10, author's translation, emphasis added).

'Although these phenomena have already emerged several decades ago in some cities, especially in North America', Welch Guerra says, 'over the last fifteen years they have concurrently become more strongly apparent' in the vast majority of large cities and urban regions throughout the world (ibid., pp. 10–11). That is for him, as well as for many other authors, an exceptional and remarkable feature. Having stressed the seemingly ubiquitous nature of these urban transformations, Welch Guerra challenges the dominant explanation for these phenomena in Latin America: that these processes are a direct outcome of globalization. This contrasts with Carmona's (2000) view that the 'spatial structure of most Latin American metropolitan areas has changed considerably in the last two decades as a result of the impact of globalization' (ibid., p. 53). According to Carmona,

metropolitan regions have reacted more quickly to the shifts in the world economy than have smaller agglomerations … [T]hey have benefited from greater foreign and national investment, and investment in infrastructure; they have a pool of skilled labour, and access to advanced technology and markets … [Furthermore] they express the contradictions of globalization, inasmuch as these megacities are socially and spatially *fragmented*, with high levels of unemployment and a wide range of informal activities with significant social, economic and spatial side-effects (Carmona, 2000, p. 57, emphasis added).

The reason why Welch Guerra is inclined to look at local rather than global explanations for these phenomena is because he aims to find factors and causes that are possible to modify. However, he does not refute the 'globalization argument'. He simply asserts that he finds this explanation 'unproductive, as it seeks the causes of current urban and territorial transformations in processes that are seemingly impossible to influence from a local or national position' (Welch Guerra, 2005, p. 11, author's translation). This methodological justification does not seem entirely convincing; rather than look for an explanation where there are more indications that it could be found, the author seeks out one where he would like it to be found because it could be influenced. In this case, the recognition that certain urban phenomena may be influenced by global forces does not imply a denial of the local agency and its ability to endorse or counterbalance this influence. Nevertheless, Welch Guerra and Carmona agree to a large extent in the diagnosis and interpretation of what they both name 'recent urban transformations' in Latin American cities, and which resonate with the worldwide literature (e.g. Fainstein *et al.*, 1992; Graham and Marvin, 2001; Marcuse and van Kempen, 2000, 2002; Mollenkopf and Castells, 1991; Soja, [1997] 2002).

A key concept connecting this literature is that of fragmentation. Specifically, these works, as well as many other recent urban studies, relate to a particular reading of contemporary metropolises which stress a current dynamic that can be labelled as the *integration–fragmentation dialectic*. Nevertheless, the term fragmentation is also related to several other conceptualisations, and its broader meaning is reviewed below.

Fragmentation in the urban discourse

Over the last three decades the concept of fragmentation has appeared with increasing frequency in the urban discourse and it has been directly related to two broad frameworks highly influential in urban studies: postmodernism and globalization. The meaning and connotation of fragmentation in both frameworks varies significantly. During the 1980s it acquired a firmly optimistic connotation, closely associated with theories about the 'end of grand narratives', crystallized in Jean-François Lyotard's (1979) *La condition postmoderne*. After the 1990s, discussions about postmodernism gave way to debates about globalization, and in the same way as in the 1980s the *urban fragment* became a recurrent idea – especially in architectural circles – a less optimistic vision of contemporary cities depicting them as *socially* and *spatially fragmented* has gained credence in the urban discourse.

Despite their differences both frameworks are part of a conceptual shift in urban studies in which *the city* ceased to be conceptualised as a whole. Cities are now primarily understood as either a collection of 'urban fragments' that enhance the 'diverse' and 'fragmentary urban experience', or as 'interrelated nodes' that 'by-pass' 'unconnected fragmented areas'. Whereas postmodernism overtly celebrates the 'city of fragments', globalization alternates between lamenting the 'fragmented city' and accepting it as a natural outcome of globalization processes. In both cases, the idea of fragmentation plays a central role (Kozak, 2004, 2005).

Urban fragmentation

Generally, 'fragmentation' is defined in two main ways: as a *process*, 'the action of breaking or separating into fragments', or 'the separation into parts which form new individuals or units' (Brown, 1993, p. 1018), or as a *state*, 'the state of being fragmented' (ibid.). Likewise, fragmentation in the urban discourse is also used in two ways: 1) as a *generating process* or *a way of operating in the city*, and 2) as a *spatial phenomenon* or *state*. Although most references to fragmentation within urban literature fall into these two main categories, there is a third worth considering: 3) fragmentation as an urban experience; as *a way of experiencing* or *perceiving the city*. The common point of reference for these three categories, and central to subsequent analysis, is urban fragmentation as a spatial state. The processes that produce this type of space – either through the breaking up of, or from disjunctions within an existing integrated urban area, or by producing fragmentation *ex novo* – are those which are perceived and experienced. These three categories can be used to group the different concepts of urban fragmentation immersed in the frameworks of postmodernism and globalization.[1]

It can be argued that within postmodernism the starting point is recognition and celebration of a 'fragmentary living' in contemporary metropolises (de Certeau, [1980] 1984); an idea that reclaimed some concepts previously developed by the *Situationists* in the late 1950s (Debord, [1957] 1981) and also with echoes of the figure of the *flâneur* as characterised by Walter Benjamin ([1939] 1999). This was followed by both a particular reading of the city and a prescription for operating in it. For example: Rowe and Koetter's (1978) conceptualisation of the 'city as collage' and their initiation of the 'architect-bricoleur', Reyner Banham's (1959) conceptualisation of the 'city as scrambled egg', Aldo Rossi's ([1966] 1982) notion of the 'city of parts', and Léon Krier's (1984) 'cities within the city' (see Fig. 1).

By contrast globalization's starting point seems to begin with a process of acting 'globally and locally' (Castells, [1996] 2000), erasing old lines of division but drawing new boundaries which are increasingly hard – the integration–fragmentation dialectic. These processes tend to be dominated by accounts relying heavily on technological explanations (e.g. Castells, [1996] 2000; Graham and Marvin, 2001). But there also exist political-economic accounts that analyse these processes by linking them to the process of neoliberalisation begining in the late 1970s (e.g. Brenner and Theodore, 2002; Burgess *et al.*, 1997; Harvey, 2005). Following these macro explanations, several works describe and analyse the socio-spatial outcomes. These concentrate on increasing socio-spatial fragmentation in contemporary cities, and stressing different aspects of this phenomenon. They conceptualise it with various terms, such as 'partitioned city' (Marcuse and van Kempen, 2002), 'quartered city' (Marcuse, 1989, 1993), 'metropolarities' (Soja, 1989), 'archipelago city' (Davis, 1998; Soja, 2000), 'dual city' (Mollenkopf and Castells, 1991), and 'divided cities' (Fainstein *et al.*, 1992; Scholar, 2006) (see Fig. 1).

Fig. 1. Different understandings of urban fragmentation within the frameworks of the debates on postmodernism and globalization Source: Daniel Kozak

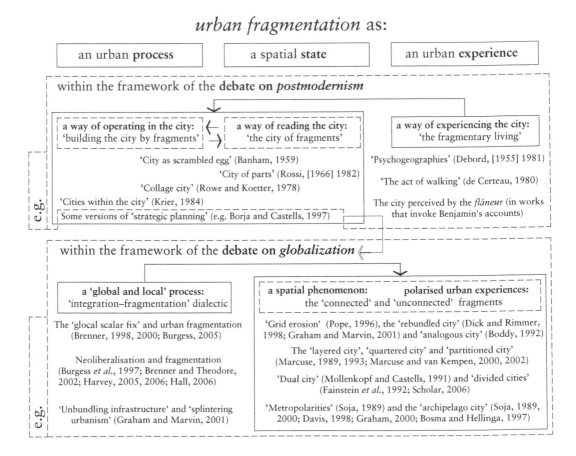

urban fragmentation as:

| an urban **process** | a spatial **state** | an urban **experience** |

within the framework of the debate on *postmodernism*

a way of operating in the city: 'building the city by fragments' ← → a way of reading the city: 'the city of fragments'

a way of experiencing the city: 'the fragmentary living'

'City as scrambled egg' (Banham, 1959)
'City of parts' (Rossi, [1966] 1982)
'Collage city' (Rowe and Koetter, 1978)
'Cities within the city' (Krier, 1984)
Some versions of 'strategic planning' (e.g. Borja and Castells, 1997)

'Psychogeographies' (Debord, [1955] 1981)
'The act of walking' (de Certeau, 1980)
The city perceived by the *flâneur* (in works that invoke Benjamin's accounts)

e.g.

within the framework of the debate on *globalization* ←

a 'global and local' process: 'integration–fragmentation' dialectic

a spatial phenomenon: polarised urban experiences: the 'connected' and 'unconnected' fragments

The 'glocal scalar fix' and urban fragmentation (Brenner, 1998, 2000; Burgess, 2005)

Neoliberalisation and fragmentation (Burgess *et al.*, 1997; Brenner and Theodore, 2002; Harvey, 2005, 2006; Hall, 2006)

'Unbundling infrastructure' and 'splintering urbanism' (Graham and Marvin, 2001)

'Grid erosion' (Pope, 1996), the 'rebundled city' (Dick and Rimmer, 1998; Graham and Marvin, 2001) and 'analogous city' (Boddy, 1992)

The 'layered city', 'quartered city' and 'partitioned city' (Marcuse, 1989, 1993; Marcuse and van Kempen, 2000, 2002)

'Dual city' (Mollenkopf and Castells, 1991) and 'divided cities' (Fainstein *et al.*, 1992; Scholar, 2006)

'Metropolarities' (Soja, 1989) and the 'archipelago city' (Soja, 1989, 2000; Davis, 1998; Graham, 2000; Bosma and Hellinga, 1997)

e.g.

Of course, there are continuities and dialogue between the postmodern and globalization debates. 'Strategic planning' is an example: while originally a consequence of the crisis of the modernist model – *masterplanning* – it can be considered postmodern, as well as attached to globalization theories such as that of 'local and global' (Borja and Castells, 1997). It stresses the role of 'large-scale urban projects' as the main producers of the contemporary metropolis, thus is related to postmodern concepts such as Rossi's 'city of parts'; but at the same time it bases much of its argument on the 'opportunities and challenges' brought by globalization.

Having introduced the general context in which the term fragmentation has been used in the urban discourse in the past, the following analyses a specific and more current use of this term: the integration–fragmentation dialectic.

The *integration–fragmentation* dialectic in globalization studies

Two general conceptualisations underlie most globalization studies concerned with cities: these are the dialectical pairs of *concentration–dispersion* and *integration–fragmentation*. The first pair, concentration-dispersion, is linked to the argument which states that 'the more globalized [i.e. dispersed] the economy becomes the higher the agglomeration [or concentration] of central functions in a relatively few sites, that is, the global cities' (Sassen, [1991] 2001, p. 5). The urban-spatial counterpart of this argument explains the apparently contradictory trends of spatial concentration at certain points of the city combined with dispersion across the metropolis. This explanation dismissed one of the first assumptions about the impact of the new information technology and globalization in cities which acquired a considerable momentum in urban studies in the early 1980s. As Sassen put it:

> The extremely high densities evident in the business districts of [global] cities are one spatial expression of this logic. The widely accepted notion that density and agglomeration will become obsolete because global telecommunications advances allow for maximum population and resource dispersal is poorly conceived (Sassen, [1991] 2001, p. 5).

The second pair, integration–fragmentation, can be also exemplified by another quote from Sassen (1998, p. xxvi): 'cities that are strategic sites [i.e. the most integrated] in the global economy tend, in part, to become disconnected [or fragmented] from their region and even nation'. This suggests why cities highly integrated within the 'global network' are barely related to their immediate geographical context. At the urban scale this is linked to the idea that contemporary metropolises have splintered in uneven fragments, divided between those connected to the urban network and those that become increasingly marginalised. In Graham and Marvin's (2001, p. 9) words: 'contemporary urban change seems to involve trends towards uneven global connection combined with an apparently paradoxical trend towards the reinforcement of local boundaries'. Both explanations are potent theses in the globalization debate and could be analysed extensively, but for the purposes of this chapter the focus is on integration and fragmentation.

'Integration–fragmentation' and the 'glocal scalar fix'

The term fragmentation in globalization studies is characterised as the opposite to integration, this being the concept that lies at the core of the globalization argument.

However, fragmentation is not used as a counter-argument or as a sign of the limits of the globalization processes. Rather, it is used to explain the dynamics of these processes. The more connected and consolidated the global network becomes, that is, the more integrated the 'globalized world', the more fragmented and disconnected becomes the 'non-globalized world' – all those countries and cities that are not part of the global network.

The concept of scale is fundamental for this argument. Given that what is under discussion is the integration – or lack of integration – of certain parts of the world to a global system, the unit of scale considered becomes crucial. It is by now virtually unchallenged that it is not entire countries that are part of a global network but cities or urban regions. Many authors take this argument further by arguing that 'global cities' are not entirely integrated into the global system but only certain areas of these cities, while other areas remain completely excluded.

> Global cities are a result of transactions that fragment space, such that we can no longer talk about cities as whole cities – instead, ... we have ... bits that are highly globalized – and bits juxtaposed that are completely cut out ... In this sense, some parts of cities can have more in common with parts of other global cities or cities in the same region than with the part of the city [that are most closely] juxtaposed (Bridge and Watson, [2000] 2003, p. 255).

Brenner (2000, pp. 375-376) argues that the question of scale 'has become increasingly central to critical urban theory in the contemporary period of global restructuring'. Thus, 'multiscalar methodologies are now absolutely essential for grasping the fundamental role of cities as preconditions, arenas and outcomes of the current round of global capitalist restructuring' (ibid., p. 375). In this context Brenner 'elaborate[s] the notion of a "scalar fix" to theorize the multiscalar configurations of territorial organization within, upon, and through which each round of capital circulation is successively territorialized, deterritorialized, and reterritorialized' (Brenner, 1998, p. 459). The notion of 'fix' exploits the double meaning of the word: 1) fix as opposed to loose or mobile; and 2) as a solution provider.

This argument rests on an understanding of 'geographic scale ... being socially constructed rather than ontologically pre-given' (Delaney and Leitner, 1997, p. 93), or that 'geographical scales are both the realm and the outcome of the struggle for control over social space' (Swyngedouw 1992, p. 60). According to Brenner (1998, p. 459), 'the current round of globalization can be interpreted as a multidimensional process of re-scaling in which both cities and states are being reterritorialized in the conflictual search for "glocal" scalar fixes'. The idea of 'glocal' is central to the integration–fragmentation dialectic. As Burgess explains:

> The emergence of a global network society (i.e. a society that is increasingly being configured as a global network) unlocks a dialectic of 'local spatial fragmentation' and 'global spatial integration' ... [T]he primacy of establishing global/local connections over intra-urban or national level connections is asserted in line with the concept of cities as 'staging-posts' for the organisation of global flows, exchanges and mobility. At the local level the process involves the rupturing of the spatial integrity of the bounded city and the selective reconfiguration of its spaces and networks (Burgess, 2005, p. 9).

The conceptualisation of this increasing tension between global integration and local fragmentation was anticipated by Lefebvre in the 1970s, who ' ... describes globalization as an intensely contradictory integration, fragmentation, polarization and redifferentiation of superimposed social spaces' (Brenner, 2000, p.361). In The Production of Space Lefebvre asks:

> How and why is it that the advent of a world market, implying a degree of unity at the level of the planet, gives rise to a fractioning of space – to proliferating nation states, to regional differentiation and self-determination, as well as to multinational states and transnational corporations which, although they stem from this strange tendency towards fission, also exploit it in order to reinforce their own autonomy? Towards what space and time will such interwoven contradictions lead us? (Lefebvre, [1974] 1991, p. 351)

These last sections reviewed some of the most discussed concepts in the globalization debate, which in turn permeated current conceptualisations of urban transformations in contemporary metropolises. These are particularly about the arguments of scale, integration and fragmentation. The next sections attempt to relate these theoretical constructs with a case of a recent urban transformation in Buenos Aires.

Spatial fragmentation in Buenos Aires

Arguably there are three dimensions of inquiry to an in-depth study of urban fragmentation. First a *discursive dimension* to address the question: *how is urban fragmentation conceptualised?* Its analysis is based on data collected from urban-discourse producers (e.g. architects, planners, social scientists) and urban texts. The first part of this chapter has briefly addressed this question. The second is centred on the *socio-spatial dimension*, and the main question: *how is urban fragmentation perceived and experienced?* – data would come from urban users. The third is the *physical-spatial*; and the question simply: *what is physical-spatial fragmentation?* – its data obtained from the built environment. The second part of this chapter is mostly concerned with the third question, but also refers to the second.

Having seen how many and diverse meanings the term fragmentation has in urban discourse, it is pertinent to begin by asking what urban fragmentation means in the context of a specific case; namely in Buenos Aires. The table in Figure 2 was drawn up after a series of interviews with architects, planners and social scientists in Buenos Aires,[2] all asked to provide local physical-spatial examples of urban fragmentation.

The first four rows of the table show cases of fragmentation existing in Buenos Aires well before the mid-1980s. The last three rows show cases appearing only in the last two decades. These later examples also illustrate a change in the production and planning of the city in which the private sector became much more influential.

Urban fragmentation in the urban periphery (the cases on the right of the table) has been widely discussed and is more easily understood. The most explicit example in the Metropolitan Area of Buenos Aires are the gated communities side by side with squatter settlements, giving a clear illustration of what urban fragmentation is about. Separated by a wall or a fence, life in those two microcosms could not be more different. But what is urban fragmentation in the centre in an open-grid context? Is it coterminous with urban fragmentation in the periphery?

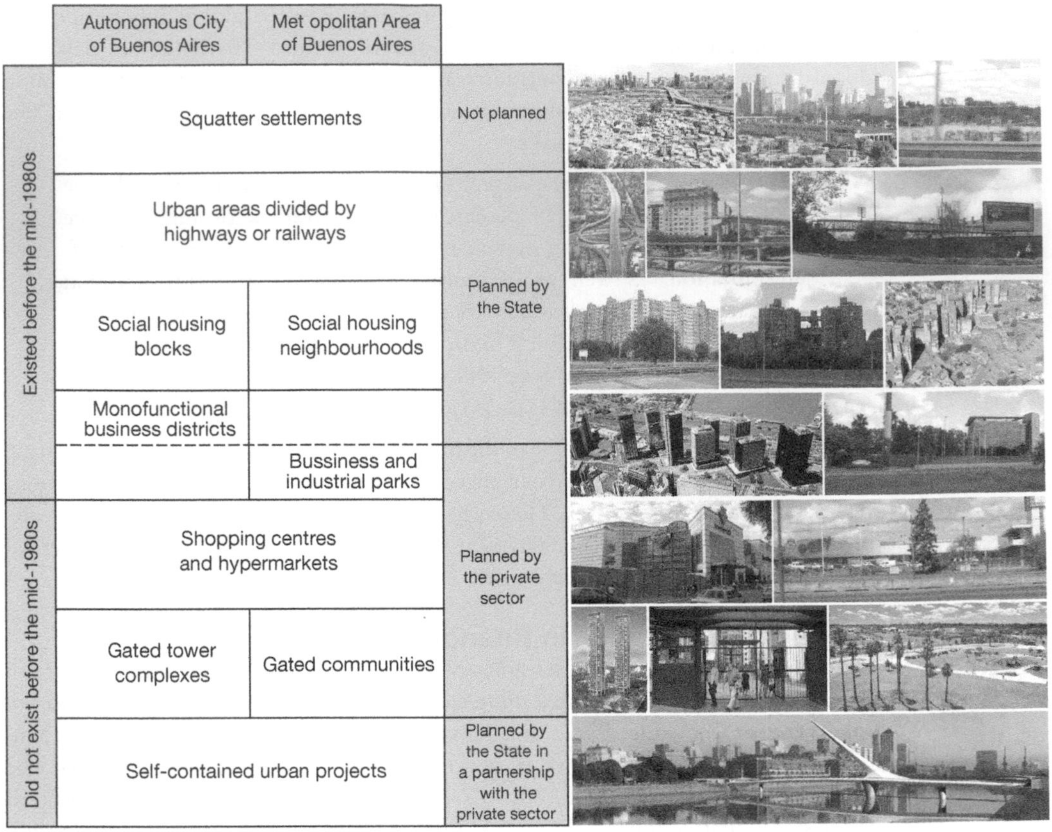

	Autonomous City of Buenos Aires	Met opolitan Area of Buenos Aires	
Existed before the mid-1980s	Squatter settlements		Not planned
	Urban areas divided by highways or railways		Planned by the State
	Social housing blocks	Social housing neighbourhoods	
	Monofunctional business districts		
		Bussiness and industrial parks	
Did not exist before the mid-1980s	Shopping centres and hypermarkets		Planned by the private sector
	Gated tower complexes	Gated communities	
	Self-contained urban projects		Planned by the State in a partnership with the private sector

Fig. 2. Physical-spatial examples of urban fragmentation in Buenos Aires
Source: Daniel Kozak

The next section examines a case in the centre of Buenos Aires, repeatedly referred to as urban fragmentation during the interviews.[3] This case also demonstrates the appearance of new typologies, usually associated with the periphery, in the centre.

Abasto in central Buenos Aires

The area of Abasto owes its name to the former central fruit and vegetable market of the city, the *Mercado de Abasto*. From its foundation in 1889, the market of Abasto became the heart and engine of this part of Buenos Aires (Berjman and Fiszelew, [1984] 1999). The prime years of Abasto coincided with a booming Argentine economy. In the 1930s, when Argentina ranked among the wealthiest countries in the world, the private society that owned the market built a new and expanded building – thereafter known as the 'new market' – which promptly became an icon of Buenos Aires (Figs 3 and 4). When in 1984 the market of Abasto was closed as a result of the opening of a modernised central market in the outskirts of the city, a heated debate took place about what to do with the exceptional but redundant building. The closure of the market was followed by a rapid decline of the area which became one of the most marginalised in Buenos Aires. After a series of failed initiatives meant to transform the market into a number of different purposes, ranging from a cultural centre to a home for the National General Archives, the former market of Abasto reopened in 1998, turned into a shopping centre (Kozak, Forthcoming).

The buildings that formerly housed the market, along with several properties and plots in the area, were bought by a local investment group associated with the global financier George Soros. The new shopping centre, branded *Abasto de Buenos Aires*, was part of a larger investment that included a gated tower complex and a hypermarket. Ostensibly, the transformation of Abasto had most of the components that, as seen above, are associated with the concept of fragmentation in Buenos Aires (i.e. shopping centre, hypermarket, gated tower complex). It is also an example related to the discussion of a merger between the concepts of 'centre' and 'periphery' in contemporary cities. The large fruit and vegetable markets were functions typical of the centre that have since moved out to the peripheries. The hypermarkets and shopping centres, when they first appeared, were typical of the peripheries but are now also common in centres. The gated tower complex is a variation of a typology originally existing only outside the cities: the suburban gated community. Furthermore, the case of Abasto seems also to be an archetypical example of what is commonly understood by the impact of globalization in cities. It is a direct result of

Fig. 3. (below–upper) 'Old market' of Abasto ca. 1900
Source: Former SAMAP Archive, currently at the American Art Institute Archive (IAA-FADU-UBA)

Fig. 4. (below–lower) 'New market' of Abasto at its inauguration in 1934
Source: Berjman and Fiszelew (1999)

foreign direct investment (FDI) and its production involved various 'global actors'. The architecture office in charge of the main project, for example, is an international firm based in Boston, *BTA Architects Inc.* (www.bta-architects.com), with expertise in 'global projects' (according to its website) ranging from Europe and the US to the Pacific Rim and Latin America.

Having briefly introduced the context of this case, the next sections analyse the urban transformation of Abasto to assess the extent to which it represents a new form of urban fragmentation.

The 'peripherisation' of the centre

Abasto comprises 88 blocks, distributed over an area of 132 hectares, and has a population of 43,451 inhabitants.[4] Thus, with a density of 32,825 persons per km^2, it is one of the densest areas in Buenos Aires City.[5] Abasto is located almost in the exact geographical centre of the city, slightly displaced to the east, and thus closer to the real existing financial, commercial and administrative centre standing next to the *de la Plata River* (Fig. 5).

Figure 6 shows the transformation of Abasto from 1940 (six years after the inauguration of the 'new market') to 2004 when, except for its façade, the 'old

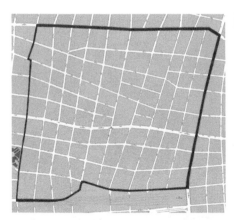

Fig. 5. The area of Abasto within the Autonomous City of Buenos Aires
Source: Adapted from www.mapa. buenosaires. gov.ar

1940   **2004**

Fig. 6. The transformation of Abasto 1940–2004
Source: Adapted from www.mapa. buenosaires. gov.ar

market', built in the 1890s, was demolished and replaced by a multiplex cinema complex, and the 'new market', built in the 1930s, was converted into a shopping centre. The most noticeable transformation is the appearance of the tower complex (in the centre of the image). The reduction of shed roofs (in the 2004 satellite image) indicates the gradual decrease of warehouse and industrial functions, which were mostly replaced by residential development. The photograph in Figure 7 was taken on the day of the inauguration of the shopping centre in 1998. In the background, the gated tower complex was still under construction, and between the shopping centre and the tower, the construction of the hypermarket was just starting.

If the transformation of the area of Abasto is a case of urban fragmentation, it is essential to analyse first in what sense the buildings of the *Abasto Project* are different from their surrounding built environment. In this way it may be possible to understand why it is widely perceived and referred to as being fragmented.

One of the first things to notice, particularly in the case of the gated tower block, is an alteration of the urban fabric: the aggregation of individual plots into a new single big plot (Fig. 8). If one considers that internal divisions within blocks 'are important mechanisms regulating city form' (Moudon, 1986, p. xviii), this alteration is bound to substantially influence the nature of this urban area. If plot

Fig. 7. Aerial view of the former market of Abasto at the day of the inauguration of Abasto de Buenos Aires shopping centre, 1998
Source: Sebastián Szyd

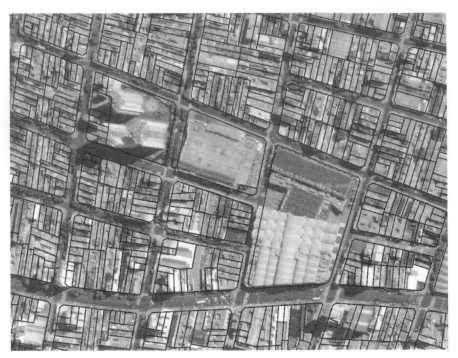

Fig. 8. Satellite image of the Abasto Project with plot divisions superimposed Source: Based on www.mapa. buenosaires. gov.ar

divisions are considered as smaller elements of the urban grid, then this alteration relates to Pope's (1996) concept of 'grid erosion'.

Although Pope also analyses central locations, particularly 'financial downtowns', his focus is mostly on suburban areas where closed systems of enclaves either replaced traditional open grids or were created *ex novo*. A transformation such as that of the block of the gated tower complex in Abasto was not included in Pope's conceptualisation of 'grid erosion'. Nevertheless it can be argued that this type of transformation, which has become a common practice in Buenos Aires over the last two decades, also erodes the geometry and character of the urban grid, albeit on a different scale.

The hypermarket and shopping centre can be associated with a concept mainly considered only in relation to the periphery, and which started with Dick and Rimmer's (1998) discussion of the 'rebundling of urban elements': the idea of the 'rebundled city'. This is the phenomenon of 'ever larger-scale buildings and complexes that encompass multiple uses and facilities under a single roof' (Graham and Marvin, 2001, p. 428). Comparing the examples given by Dick and Rimmer (1998) on the periphery with that of the Abasto in the centre, it is possible to conclude that they are similar but at a different scale. The difference between urban fragmentation on the periphery and in the centre seems to be primarily a matter of scale, meaning both: scale as a size problem and scale as a level of intensity of the phenomenon.

Unlike in the typical cases in the periphery, in Abasto no public streets were privatised and closed. However, during the process of construction of the Abasto Project the condition of 'public space' was repeatedly discussed and in some cases resignified. The process which finally led to the acquisition of the buildings of the market and two adjacent blocks, among other properties in the area, by IRSA (the investment group associated with Soros in Argentina) was a protracted one. In this process a myriad of projects were proposed and these properties changed hands twice. As a result of an agreement between the first company to own the market of Abasto (SAMAP) and the Municipality of Buenos Aires, a piece of land with an area of 300m^2 dividing the buildings of the 'new market' and the 'old market', was destined to become an open public square (Boletín Municipal de la Ciudad de Buenos Aires, 1984, pp. 63,462–63,462 and appendix).[6] All the projects proposed since the closure of the market in 1984, including the first schemes developed by IRSA, had this open square (Kozak, Forthcoming). The final and existing project also included the square, but elevated 4.5m above the level of the sidewalks, with closed accesses guarded by the shopping centre security and a glass roof (Fig. 9). Thus it became difficult to still recognise it as a 'public square'. Despite its official 'public status', it is obviously part of the private shopping centre. The developers also made a request to build a skywalk between the shopping centre and hypermarket, seen in the first drawings published to market the new development (La Nación Newspaper, 1997, p. 1). This was denied by the Secretary of Urban Planning.

Despite the example of the 'public square' in the shopping centre, it is noticeable that the absence of private closed streets makes a clear distinction between the Abasto Project and many other private developments on the periphery. The open-grid street system of Buenos Aires City seems to be so consolidated that it still functions as a reassuring form of integration. Nonetheless, the buildings of the Abasto Project and the 'unity' that they form are still referred to and perceived as being fragmented from the city.[7] Thus, an analysis of the boundaries, the interface between the private buildings and the public city, is the key to this discussion.

Fig. 9. Indoor 'public square' Plaza del Zorzal and accesses at the Abasto de Buenos Aires Shopping Centre Source: Daniel Kozak

Boundaries in Abasto

A quick comparison between the physical boundaries of the buildings of the Abasto Project and its neighbours confirms their exceptionality in this urban area. A more detailed analysis shows the magnitude of this difference and its implications. Figure

10 measures the difference between the building's heights, illustrating the extent to which the towers supersede the average height of its closest surrounding and Fig. 11 shows the difference between the number of plots per block. Taking the four average blocks that unfold from the map in Figure 9 (those blocks divided into 33 plots) it is possible to contrast the typical boundaries in Abasto (Fig. 12) with the boundaries of the *Abasto Project* (Fig. 13). A set of physical-spatial indicators and variables designed to compare the integration–fragmentation potentiality of this borders, can then be analysed (Table 1).

Key

- 0
- 3–6
- 7–11
- 12–13
- 14–18
- 19–27
- 28–33
- 34–42
- 43–54
- 55–69
- 70–89
- 90

Fig. 10. Building's heights in the area of Abasto Source: Based on data provided by the Secretary of Urban Planning of the Autonomous City of Buenos Aires

Key

- 1
- 2–19
- 20–27
- 28–32
- 33
- 34–35
- 36–42
- 43–46
- 47

Fig. 11. Number of plots per block in the area of Abasto Source: Based on data provided by the Secretary of Urban Planning of the Autonomous City of Buenos Aires

Note: The blocks numbered are analysed below in Table 1. The block numeration corresponds with that used by the Autonomous Government of the City of Buenos Aires.

*Fig. 12. (above)
Typical
boundaries in the
area of Abasto
(block 20)
Source: Daniel
Kozak*

*Fig. 13. (right
and below)
Boundaries
of the Abasto
Project (blocks
60, 53 and 42)
Source: Daniel
Kozak*

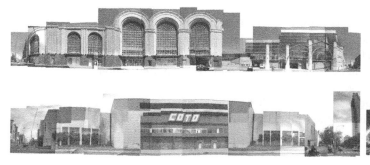 

	Average distance between pedestrian entrances [m]	Average distance between accesses to semi-public spaces [m]	Ground-floor windows [%]	Ground-floor fences [%]	Buildings aligned to the prevailing geometry [%]	Newest building in the block [built in...]	Oldest building in the block [built in...]	Dominant period of development [decade]	Plot-size average [m²]	Building's height average [m]	Floor Area Ratio (FAR)
Abasto Project boundaries											
Shopping Centre (block 60)	138.0	138.0	35	0	91	1998	1934	1990s	21135	18.0	6.0
Hyper-market (block 53)	52.0	184.5	20	0	60	2000	2000	1990s	9613	12.0	5.7
Gated tower complex (block 42)	137.2	N/A	0	90	0	1999	1999	1990s	8161	90.0	7.1
Typical boundaries in Abasto											
Block 20	7.0	47.7	36	3	97	1993	1922	1920s	229.7	7.6	1.8
Block 91	4.0	6.3	56	12	100	1996	1922	1950s	334.4	18.0	4.3
Block 39	3.9	14.7	48	5	100	1997	1922	1920s	279.6	13.7	3.0
Block 37	6.3	18.5	36	13	80	1999	1922	1920s	276.5	12.5	2.3

The variables in Table 1 are divided between two broad groups of indicators. The first group attempts to establish physical-spatial differences between the buildings of the Abasto Project and the typical buildings of this area (e.g. alignment to the prevailing geometry in the neighbourhood, plot size). The second group determines different levels of permeability (e.g. physical and visual interaction between buildings and sidewalks, type of accesses). The typical boundaries in Abasto, for example, have their pedestrian accesses between 7.5 and 34.8 times on average closer than those of the Abasto Project; and these are more evenly distributed. The buildings of the Abasto project have large parts of their boundaries completely opaque to the outside.

It is clear that the harder and less permeable the borders, the more integrated the buildings and activities in its inside and also the more fragmented from their immediate context. This is valid for the case of the gated tower complex, in which the internal towers are considerably more related among themselves than any random adjacent buildings in the area (e.g. they share pedestrian paths and a number of amenities such as a swimming pool), as well as in the case of the shopping centre, in which different stores are also substantially more linked than normal neighbouring stores at the street level.

Table 1. Physical-spatial indicators for an analysis of boundaries in the area of Abasto Source: Daniel Kozak

The buildings and complexes of the Abasto Project are examples of new units of scale appearing in cities that are adapted versions of typologies initially typical only on the peripheries; and they have kept many characteristics of the original models. They do not seem to relate well with their immediate context but are quite well connected to other populations and similar spaces not necessarily nearby. According to the *Argentine Chamber of Shopping Centres* (Cámara Argentina de Shopping Centers, 2006) approximately one million people visit *Abasto de Buenos Aires* each month, and almost one thousand people work there each day.

The separation that these buildings establish from their immediacy does not seem, however, inevitable. A simple variation in the design of the borders, which may include different uses connected to the street, should not necessarily interfere with the activities taking place in the core of the blocks. For example, hypermarkets in central urban areas could well have commercial stores facing the streets; and that was indeed one of the unrealised intentions of the architectural office in charge of the project of the hypermarket in Abasto (*Pfeiffer y Zurdo arquitectos*).[8] Nevertheless, this does not seem to be the trend. The rationale which these new units of scale appear to follow resembles the arguments reviewed above about new 'glocal scalar fixes'; albeit in a different context and scale. That is the argument, which in Castells' ([1996] 2000, p. 3) words, determines that 'global networks ... selectively switch on and off individuals, groups, regions, and even countries, according to their relevance in fulfilling the goals processed in the network'. These new units of scale in cities are indeed new 'scalar fixes' which, paraphrasing McCann (2003, p. 161), at the present time seem to have 'the power to persuade that [they] are imperative to the successful functioning of the city's economy and society'. In this specific case it is thus possible to conclude that the transformation of Abasto has indeed resulted in a combined process of increased integration at the larger scale and urban fragmentation at the local scale.

Conclusions

The 'impact of globalization' in cities is not just confined to the effects of the new buildings and developments built with FDI. There are new logics of clustering and connecting in current large cities that can be associated with a globalization rationale. The concept of urban fragmentation seems to be a useful heuristic tool to conceptualise some of the most current urban transformations associated with these new logics, including the 'importation' of typologies from the periphery to the centre.

The case of the Abasto Project also seems to be a good example of a current trend, not only in Buenos Aires but also in most large and mega-cities in the world, in which the public space recedes and its place is taken by 'semi-public' or 'semi-private' spaces such as that of the closed square inside the *Abasto de Buenos Aires* shopping centre. The boundaries that divide the public from the private are increasingly harder and less permeable; and the spaces of universal encounter which are not mediated by consumption, and consequently separated by social classes, are also becoming increasingly rare. In Buenos Aires, urban fragmentation in the centre has not reached the degree that has in the periphery, but it has significantly increased within the last decades. Likewise, as current urban literature seems to indicate, this is a largely generalised tendency.

Building up from current literature on urban fragmentation and this empirical study in Buenos Aires, it is possible to propose a definition of urban fragmentation probably applicable to many cities in the world currently undergoing this type of urban transformation:

> *Urban fragmentation implies an organisation of space – understood as both a process and a resulting spatial state – in which impermeable boundaries and enclosure have a central role. It is a state of disjointing and separation which is often coupled with socio-economic and/or ethnic divisions. A fragmented city is one in which the ability to use and traverse space is dominated by the principle of exclusivity and there is a reduction in the number of places of universal encounter.*

As argued in the first part of this chapter, the recurrence of the term fragmentation in contemporary urban discourse gives an idea of how significant this concept has become in the context of current urban transformations. As Lefebvre once affirmed, the fact that certain terms become 'so persistent is in itself an indication of something' (1968, pp. 120-121, author's translation). A parallel conclusion of this study suggests that the recurrence of a term under different interpretations indicates the emergence of a problematic that requires further study. The variety of understandings and uses that are applied to the idea of fragmentation in the urban realm especially encourages further research into this subject.

Notes

1. There are other uses of this term in urban texts. The most frequently recurring one, after the three general categories, is *urban fragmentation as an organisational state*; the extreme compartmentalisation of jurisdictions in a territory is the most common case of this type of use (e.g. Szajnberg, 2006, p. 8). One comes across terms such as *jurisdictional fragmentation* or *municipal fragmentation*. These refer to the lack of communication between adjacent governing territorial bodies (e.g. municipalities) or to the absence of an effective overall entity to coordinate them. Although this meaning of the term is used in urban texts, it is not directly analysed here because it is implicit in the other three main categories. It can be considered a particular form of the more general *fragmentation as a way of operating in the city*, and for the purposes of this limited study it is relevant only when it results in *fragmentation as a spatial state*.
2. The interviews were in-depth and semi-structured of an average time of 45min. They were conducted to thirty respondents equally divided between architects, planners and social scientists, from November 2004 to September 2005.
3. See note above.
4. These figures were provided by the Secretary of Urban Planning of the Autonomous City of Buenos Aires, and correspond to the National Census of Population, Households and Housing 2001 performed by the National Institute of Statistics and Censuses (INDEC).
5. The overall density of Buenos Aires City is 13,680 persons per km². (See Chapter 5, p. 78.)
6. Another result of this agreement was permission to build exceptional higher and larger buildings than what was allowed by the regulations for this area in the blocks where the towers and hypermarket now stand.
7. This is based on the interviews mentioned in note 2, as well as in a literature review conducted in Buenos Aires and from ninety elicitation interview questionnaires with users in Abasto, carried out from January to March 2006.
8. In an interview with one of the architects in charge of this project (included in a second series of in-depth semi-structure interviews with diverse actors involved in the case of Abasto, e.g architects, developers, civil servants) she thought that the addition of a line of commercial stores in one of the borders of the hypermarket might be feasible; but not in the near future. A few shopping centres in Buenos Aires, located in highly commercial areas, do have retail stores facing the streets in their boundaries. They constitute a more interesting hybrid between the typical suburban shopping centre and the traditional commercial arcades.

References

Banham, R. (1959) City as Scrambled Egg. *Cambridge Opinion*, **17**: 18–23.

Benjamin, W. (1999) *The Arcades Project*, Belknap Press, London [First published 1939].

Berjman, S. and Fiszelew, J. (1999) *El Abasto: Un barrio y un mercado*, Corregidor, Buenos Aires [First published 1984].

Boddy, T. (1992) Underground and overhead: building the analogous city, in *Variations on a theme park: the new American city and the end of public space* (ed. Sorkin, M.), Hill and Wang, New York, pp. 123–153.

Boletín Municipal de la Ciudad de Buenos Aires (1984) *Apruébese convenio. Ordenanzas, decretos y resoluciones. Vol. 17439*, Municipalidad de la Ciudad de Buenos Aires, Buenos Aires, pp. 63462–63463.

Borja, J. and Castells, M. (1997) *Local and Global: the Management of Cities in the Information Age*, Earthscan, London.

Bosma, K. and Hellinga, H. (1997) Mastering the city II, in *Mastering the city: North-European city planning, 1900–2000* (eds Bosma, K., Hellinga, H. and Nederlands Architectuurinstituut), Rotterdam: NAI Publishers, pp. 8–17.

Brenner, N. (1998) Between fixity and motion: accumulation, territorial organization and the historical geography of spatial scales. *Environment and Planning D: Society and Space*, **16**(4): 459–481.

Brenner, N. (2000) The urban question as a scale question: reflections on Henri Lefebvre, urban theory and the politics of scale. *International Journal of Urban and Regional Research*, **24**(2): 361–378.

Brenner, N. and Theodore, N. (2002) *Spaces of neoliberalism: urban restructuring in North America and Western Europe*, Blackwell, Oxford.

Bridge, G. and Watson, S. (2003) City differences, in *A companion to the city* (eds Bridge G. and Watson, S.), Blackwell, Oxford, pp. 251-260 [First published 2000].

Brown, L. (1993) *The New Shorter Oxford English Dictionary: on Historical Principles*, Clarendon, Oxford.

Burgess, R. (2005) Technological Determinism and Urban Fragmentation: A Critical Analysis, in *9th International Conference of the ALFA-IBIS Network on Urban Peripheries*, Pontificia Universidad Católica de Chile, Santiago de Chile, July 11th–13th 2005.

Burgess, R., Carmona, M. and Kolstee, T. (1997) *The Challenge of Sustainable Cities: Neoliberalism and Urban Strategies in Developing Countries*, Zed Books, London.

Cámara Argentina de Shopping Centers (2006) *Shopping Centers en la Argentina*. Retrieved in February 2006 from: http://www.casc.org.ar

Carmona, M. (2000) The regional dimension of the compact city debate: Latin America, in *Compact cities: sustainable urban forms for developing countries* (eds Jenks, M. and Burgess, R.), E. & F.N. Spon, London, pp. 53–62.

Castells, M. (2000) *The Rise of the Network Society*. Blackwell, Oxford [First published 1996].

Davis, M. (1998) *City of Quartz: Excavating the Future in Los Angeles*, Pimlico, London.

de Certeau, M. (1984) *The practice of everyday life*, University of California Press, Berkeley [First published 1980].

Debord, G. (1981) Toward a Situationist International, in *Situationist International anthology*, (ed. Knabb, K.), Bureau of Public Secrets, Berkeley, pp. 22–25. [First published 1957].

Delaney, D. and Leitner, H. (1997) The political construction of scale. *Political Geography*, **16**: 93–97.

Dick, H. and Rimmer, P. (1998) Beyond the third world city: the new urban geography of south-east Asia. *Urban Studies*, **35**(12): 2303–2321.

Fainstein, S. S., Gordon, I. and Harloe, M., eds (1992) *Divided cities: New York and London in the contemporary world*, Blackwell, Oxford.

GaWC (2006) *The GaWC Inventory of World Cities*. Retrieved in 15–10–2006 from: http://www.lboro.ac.uk/gawc/citylist.html

Graham, S. (2000) Constructing Premium Network Spaces: Reflections on Infrastructure Networks and Contemporary Urban Development. *International Journal of Urban and Regional Research*, **24**(1): 183–200.

Graham, S. and Marvin, S. (2001) *Splintering urbanism: networked infrastructures, technological mobilities and the urban condition*, Routledge, London.

Hall, S. (2006) Cosmopolitan promises, multicultural realities, in *Divided cities: the Oxford Amnesty lectures 2003* (ed. Scholar, R.), Oxford University Press, Oxford, New York, pp. 20–51.

Harvey, D. (2005) *A brief history of neoliberalism*, Oxford University Press, Oxford.

Harvey, D. (2006) The right to the city, in *Divided cities: the Oxford Amnesty lectures 2003* (ed. Scholar, R.), Oxford University Press, Oxford, New York, pp. 83–103

Kozak, D. (2004) *Urban Fragmentation in a cultural context*, in *Globalization, Urban Form and Governance* (eds Carmona, M. and Schoonraad, M.), Delft University Press, Delft, pp. 137–152.

Kozak, D. (2005) Entre la celebración del fragmento y la condena a la fragmentación, in: *9th International Conference of the ALFA-IBIS Network on Urban Peripheries*, Pontificia Universidad Católica de Chile, Santiago de Chile, July 11th–13th 2005.

Kozak, D. (Forthcoming) Otros 'Abastos' posibles: Veinte años de proyectos para el área en torno al ex–Mercado de Abasto, 1978–1998, in *Abasto a cielo abierto: estrategias de recuperación de áreas degradadas* (eds Tella, G. and Dieguez, G.), Nobuko, Buenos Aires.

Krier, L. (1984) The City Within the City. *Architectural Design*, 54(7–8): 70–105.

La Nación Newspaper (1997) *Abasto: el barrio que viene*, 30 April, pp. 1–8.

Lefebvre, H. (1968) *El derecho a la ciudad*, Ediciones Península, Buenos Aires.

Lefebvre, H. (1991) *The production of space*, Blackwell, Oxford [First published 1974].

Lyotard, J. F. (1979) *La condition postmoderne: rapport sur le savoir*, Editions de Minuit, Paris.

Marcuse, P. (1989) Dual city: a muddy metaphor for a quartered city. *International Journal of Urban and Regional Research*, 13(4): 697–708.

Marcuse, P. (1993) What's so new about divided cities? *International Journal of Urban and Regional Research*, 17(3): 355–365.

Marcuse, P. and van Kempen, R. (2000) *Globalizing cities: a new spatial order?*, Blackwell, Oxford.

Marcuse, P. and van Kempen, R., eds (2002) *Of states and cities: the partitioning of urban space*, Oxford University Press, Oxford.

McCann, E. J. (2003) Framing space and time in the city: urban policy and the politics of spatial and temporal scale. *Journal of Urban Affairs*, 25(2): 159–178.

Mollenkopf, J. H. and Castells, M., eds (1991) *Dual City: Restructuring New York*. Russell Sage Foundation, New York.

Moudon, A. V. (1986) *Built for change: neighborhood architecture in San Francisco*. MIT Press, Cambridge, Mass.

Pope, A. (1996) *Ladders*, Princeton Architectural Press, Princeton.

Rossi, A. (1982) *The Architecture of the City*, MIT Press, Cambridge, Mass. [First published 1966].

Rowe, C. and Koetter, F. (1978) *Collage City*, MIT Press, Cambridge, Mass.

Sassen, S. (1998) *Globalization and its discontents: [essays on the new mobility of people and money]*, New Press, New York.

Sassen, S. (2001) *The global city: New York, London, Tokyo*, Princeton University Press, Princeton [First published 1991].

Scholar, R. (2006) *Divided cities: the Oxford Amnesty Lectures 2003*, Oxford University Press, Oxford.

Soja, E. (1989) *Postmodern geographies: the reassertion of space in critical social theory*. Verso, London; New York.

Soja, E. (2000) *Postmetropolis: critical studies of cities and regions*. Blackwell, Malden, MA.

Soja, E. (2002) Six Discourses on the postmetropolis, in *The Blackwell City Reader* (eds Bridge, G. and Watson, S.), Blackwell, Oxford, pp. 188–196 [First published 1997].

Swyngedouw, E. (1992) The mammon quest: 'glocalisation', interspatial competition and the monetary order: the construction of new scales, in *Cities and Regions in the New Europe* (eds Dunford, M. and Kafkalas, G.), Belhaven Press, London, pp. 39–68.

Szajnberg, D. (2006) *La suburbanización: Partidarios y detractores del crecimiento urbano por derrame*, Ediciones FADU, Buenos Aires.

Welch Guerra, M., ed. (2005) *Buenos Aires a la deriva*, Biblos, Buenos Aires.

16

Perry Pei-Ju Yang

Tracking Sustainable Urban Forms and Material Flows in Singapore

Introduction

The ecological effect of urbanization is becoming an increasingly important research issue in the context of the rapid urban changes of Asian cities and regions. The urbanization processes usually create highly visible changes such as urban sprawl, or landscape fragmentation where large habitats or land areas get broken into small parcels and splintered shapes. The vegetated and agricultural landscapes are radically transformed by the built environment through the stock and flow of materials related to the construction industry.[1] How then do we minimize the utilization of land resources, diminish the environmental impact from the management of material flow, and maximize the ecological integrity in a context of urban growth and intensification of land development? How should urban planners and designers make better planning decisions of the extent, intensity and spatial configuration of the natural and built environments?

The relationship between urban form, material flow, and the impact on the natural surroundings is rarely explored. In the past decade, some research has attempted to quantify spatial form, city shape, and natural and landscape patterns; for example Batty and Longley (1994) measured the urban boundaries and edges, urban land use patterns and urban growth form by using fractal geometry. At much the same time, the quantification of landscape patterns has received considerable attention in landscape ecology, focusing on ecological processes and spatial configurations and forms (Forman, 1995; Farina, 1998; Turner, Gardner and O'Neill, 2001). However, the ecological meaning and implications of urban form are less well known. As some landscape ecologists indicated, most of the contemporary work on patterns has focused on the analysis or description of spatial geometry rather than on understanding the ecological consequences, significance or meanings of these patterns (Opdam, Foppen and Vos, 2002). In the area of urban and environmental planning, the concept of sustainable urban form has been a core issue within the academic community, for example in the International Urban Planning and Environment Association through a series of conferences and publications (e.g. Miller and de Roo, 1999; Jenks and Dempsey, 2005; Williams, 2005), and through other key publications (e.g. Jenks *et al.*, 1996; Jenks and Burgess, 2000; Williams *et al.*, 2000). In these, the normative proposition of sustainable urban form was tested and debated, based on extensive empirical studies and analyses.

Research design

In order to understand the impact of urbanization on the landscape, there is a need first to measure urban spatial form and track material flows, and second to ensure that the ecosystem is integrated into the analysis of an urban system, composed of materials, energy, water and ecology across different spatial scales. This leads to a series of questions. How do we measure the material stock of urban environment? How do we track the urban transformation and the material flow that lie behind it? How do we measure the material stock and resource use in different types of urban development (given a similar population density)? What are the effects of scale? How do we conduct multiple-scale assessment and analysis? The answers to these questions may lead to ways of minimizing the material uses and increasing the resource efficiency of the urban construction process. Such quantitative measures will be needed to understand how spatial form influences the ecosystem in general and the material stock and flow in particular.

Some research has focused on measuring material flows at the city or regional level (Beers and Graedel, 2003, 2004; Hendriks *et al.*, 2000; Huang and Hsu, 2003; Tanikawa *et al.*, 2002). These studies raise the issue of efficiency in material usage in the process of urban development, in which city becomes a reservoir of material flow. For example, Müller *et al.* (2006) proposed a procedure and matrix for a multi-scale life-cycle assessment, in which products categorised into subsystems and discrete components, and then aggregated into larger and more complex products. These studies, in summary, indicate that there are two approaches to measuring and tracking the material stock and flow. One is top-down, tracking the input and output flows over time, and then computing the difference. The second is a bottom-up measure based on mapping land use types in the city, and deriving the material stock by estimating the material consumption of different land-use types.

Singapore City can be considered as a system with spatial characteristics over a range of spatial scales comprised of different urban forms and distribution of the material flow. The research reported here mapped changes of land coverage patterns using the geographic information system (GIS) in grids of 20km to 40km across the entire Singapore City. The mapping on GIS provided a platform that integrates data from remote sensing image processing. Some landscape ecological indices were applied to the measures of land cover patterns that enable the data to be interpreted, and to assess the implications of changes in the patterns. It is noted that urban growth and land cover transformation are closely related to the changing spatial distribution of the material stock and flow. The data for material stocks were extracted from mapping development density and land use types based on the city-wide development plan. Two approaches to the material flow analysis were conducted: tracking overall input and output flows; and, estimating material consumption, based on measurement of density and land use types.

Space image processing

Within the context of some general questions: is Singapore City becoming more compact or more sprawling? What constitutes sustainable urban form? The case study research addresses the debate about sustainable urban form, landscape patterns, compact forms and urban sprawl. In particular, urban and landscape form and changes in them over a twenty-year period of Singapore's urban development

Table 1. Data sources of Singapore's satellite images Source: Perry Pei-Ju Yang

Sensor	Year	Type / band(s)	Resolution
SPOT	1986	RGB/3 (1-NIR*, 2-Red, 3-Green)	20 metres
SPOT	1994	RGB/3 (1-NIR*, 2-Red, 3-Green)	20 metres
Landsat ETM	2002	RGB/8, only 3 are used (4-NIR*, 3-Red, 2-Green)	30 metres

*NIR = Near Infra Red

are measured using image processing and computation within the GIS platform. The landscape ecology is measured through the use of ecological indices such as the patch area, patch perimeter, patch number, and the compactness of forest shape and city form.[2]

To measure urban form and its changes, satellite images of Singapore from 1986 to 2002 were analyzed. Some details of the images are described in Table 1. The RGB images are useful for the comparison between 1986, 1994, and 2002 because they have similar specifications. The image processing, including NDVI (Normalized Difference Vegetation Index) transformation, ROI (Region of Interest) sampling, and supervised classification were conducted using RSI ENVI 4.1 image processing software. The vegetation classification requires only a single NDVI band, which provides a unilateral spectrum of NDVI values clearly separated by two classes, 'light vegetation' and 'dense vegetation'. Urban classification involves a multilateral spectrum of RGB values of SPOT and LANDSAT images, in which ROI sampling was applied. The classifications of 1986, 1994, and 2002 were performed using the Parallelepiped method. The geo-referencing and the raster calculation of images were conducted on GIS platform. The classification results from ENVI were exported to the GIS platform for data processing.[3]

Land cover transformation

Changes in three different land classifications – dense vegetation, light vegetation and urban land cover – were analysed across the years 1986, 1994, and 2002. The analysis of land cover changes, using the ArcGIS Raster Calculator tool, was conducted by selecting patches that had observable changes over the years. The results were derived from the pairing of two different years: 1986–1994 and 1994–2002.

Urban expansion: conversion from the dense and light vegetation to urban land cover

The change from vegetation to urban land cover indicates urban growth from 1986 to 1994 and 1994 to 2002. The result reveals that urban growth was dispersed all over the island, but concentrated in new developments in several new towns: Jurong West, Chua Chu Kang, Bukit Panjang, Bukit Timah, the northern new town Sengkang, and the eastern new towns Pasir Ris and Tampines (for locations see Fig. 1). The new town developments revealed by the satellite image are consistent with Singapore planning policy from the 1970s. Most concentrated growth is located along the mass-rapid transit (MRT) lines. The East–West Line was completed in 1987, followed by the North–South Line in 1995 and the North–East Line in 2003. There were also minor developments on the smaller islands, such as the resort and

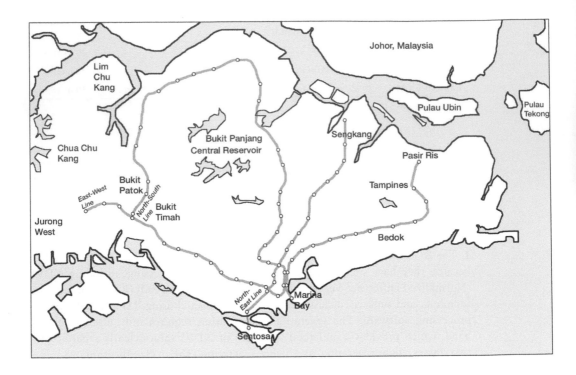

Fig. 1. Singapore: showing key locations referred to in the text

recreational facilities at Sentosa, facilities for natural reserve at Pulau Ubin, and training facilities at Pulau Tekong.

Deforestation: conversion from the dense vegetation to light vegetation

The conversion from dense to light vegetation land cover indicates a process of deforestation or decreasing vegetation intensity. This may have been caused by the opening up of new grass fields for different purposes, such as the increase of sports training grounds from 1986 to 1994. From 1994 to 2002, small vegetation patches around the central reservoir catchment area were cleared up for its associated facilities.

Reforestation and rezoning

The next type of change, the conversion from light to dense vegetation, reveals that some open fields that were cleared from 1986 to 1994 were replanted with trees in 1994 to 2002. This suggests that some areas of natural forest were replaced by man-made forest areas. This change indicates a reforestation process, including the environmental management of new parks and the central reservoir catchment area. This process has been driven mainly by the government's tree-planting campaign, initiated in 1963.

The conversion of urban land to vegetation indicates the impact of a rezoning process. The rezoning policy since the late 1960s included rezoning from kampongs or informal urban settlements, agricultural fields and farms into green fields and forests. The mapping of the period 1986 to 1994 shows that many traditional kampong settlements had been rezoned to be cleared green patches at Jurong West, Bukit Timah, Bukit Batok, Bukit Panjang, and the northern area of Sengkang. Some of these cleared green patches were later redeveloped into new towns and public housing areas over the period 1994 to 2002. In the same period, patches of reforestation through the tree planting campaign occurred at Lim Chu Kang, Sentosa Island, Bedok, and new green fields on reclaimed Marina Bay.

Landscape ecological analysis of urban and natural form

Fig. 2.
Landscape
ecological
analyses:
2a (top–left)
Sum of Area;
2b (top–
right) Sum of
Perimeter;
2c (bottom–left)
Mean of Area;
2d (bottom–
right) Number of
patches
Source: Perry
Pei-Ju Yang

Key

— Dense
  Vegetation

– – Light
  Vegetation

━ Urban

The following analyses show the landscape changes in the three types of land coverage using the landscape ecological indices, including the number of patches, patch area, and patch perimeter (Fig. 2). The changes of urban patch area indicate increasing urban expansion from 1986 to 1994, but the trend is shown to slow down from 1994 to 2002 (Fig. 2a). The increasing patch perimeter of the vegetation implies more fragmented green space and longer interface between the urban area and nature (Fig. 2b). When the number of patches increases and the mean of patch area decreases, this indicates a developing pattern of small-scale scattered urban development and small green areas – representing the phenomenon of new town developments in Singapore's urban fringes from 1980s (Figs 2c and 2d).

Urban growth in Singapore increased dramatically by 27% from 337km^2 to 429km^2 of the development area in 1986–1994, but slowed down afterwards. This does not mean that there was strictly limited urban development in 1994–2002, but rather implies a balance in urban growth, redevelopment and natural conservation during that time. However, a significant change of urban area over

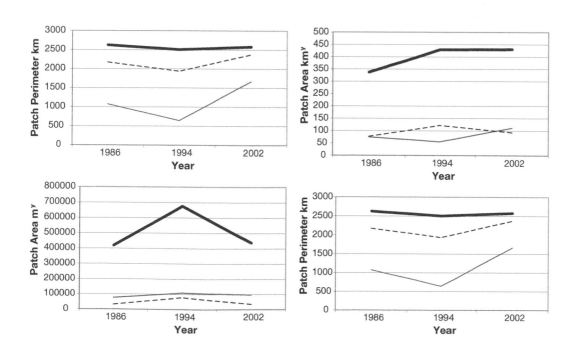

1986–2002 is associated with only a minor change of urban patch perimeter ranging from 2,505km in 1986 to 2,624km in 2002. This suggests that Singapore's urban form was becoming more compact and less convoluted during the later urbanization process.

Concerning natural form, dense vegetation areas decreased from 74km^2 to 55km^2 in 1986–1994, but increased dramatically to 111km^2 in 1994–2002, possibly caused by maturing of the tree canopy since the tree-planting campaign in 1963. Light vegetation areas increased greatly by 58% to 122 km^2 in 1986–1994, probably induced by the 'kampong' rezoning policy and the clearing of dense vegetation, i.e. forest or mangroves were mostly converted into light vegetation areas during the period. However, the light vegetation areas decreased to 93km^2 in 1994–2002, with the increase of dense vegetation areas in the same period by a maturation process into dense vegetation patches. The changing number and mean areas of vegetation patches have interesting implications, showing a pattern of concentrated green growth in 1986–1994, followed by another pattern of dispersed green growth in 1994–2002.

Material flow and spatial distribution

Urban growth and land cover changes relate closely to the spatial distribution of the material flow that generates the physical urban features of Singapore City. The research has used the activity of construction industry to indicate the extent of the urbanization process. By tracking the amount of materials consumed by the construction industry within a certain period of time, the material flows are measured and their spatial distributions are identified. These, as noted above, involves two measures, the first being the input and output flows over time and computation of the difference between them. The data for Singapore's building construction from 1994 to 2004 is shown in Table 2, and it provides a basis for deriving the material stock of cement, steel, coarse and fine aggregates (Singapore BCA, 2006).

The second measure is based on mapping the land use types in the city. It is difficult to compile data on all building materials over time, since data on all the end uses are not available. Instead a small number of key materials in the making of the material stock assessment were selected. The criteria for selection include major materials used in the construction industry and those potentially recyclable materials. The case study assessed the main building materials – steel, cement and aggregates; and the following types of land uses – residential, commercial, industrial and institutional. Commercial buildings include shops and stores and multi-storey shopping complexes including car park structures. Industrial uses include all the major establishments where the manufacturing process, storage and service activity takes place. The average material stock of the individual land use type was then estimated.

The density and distribution of urban development is derived from the density indicator Gross Plot Ratio (GPR). The GPR is obtained from the Singapore development guide plans of the 1998 and 2003, which contain the information on density and land use types of individual land plots (Singapore URA, 2006). GIS-based density maps of Singapore were drawn up in both 1998 and 2003, representing the density distribution of development as well as the change between the plans in the 5-year period.

	Manufacturing			Building and Construction				Building materials (tons/m²)			
Year	Net Investment Commitments	Total Output	Index of Industrial Production (2003=100)	Written Permission	Building Plan Approval	Building Commencement	Building Completion	Steel	Cement	Coarse aggregate	Fine aggregate
	Million Dollars			1,000 m² of gross floor area							
1994	5,765	103,013	64.9	na	na	na	na	na	na	na	na
1999	8,037	136,937	87.8	2,862	1,845	2,087	3,348	137,268	331,452	950,832	759,996
2000	9,209	163,721	101.3	3,594	2,415	2,725	3,457	141,737	342,243	981,788	784,739
2001	9,172	138,323	89.5	2,981	2,339	2,799	2,589	106,149	256,311	735,276	587,703
2002	9,009	147,296	97.1	2,624	2,602	2,183	3,446	141,286	341,154	978,664	782,242
2003	7,511	158,697	100.0	1,482	1,683	1,446	2,192	89,872	217,008	622,528	497,584
2004	8,258	189,789	113.9	2,301	1,672	1,404	2,439	99,999	241,461	692,676	553,653
Percentage change over previous year											
1994	47.3	15	12.9	na	na	na	na	na	na	na	na
1999	2.7	9.9	13.9	-29.6	-52.4	4.2	-32.5	-33.0	-33.0	-33.0	-33.0
2000	14.6	19.6	15.4	25.6	30.9	30.6	3.3	3.0	3.0	3.0	3.0
2001	-0.4	-15.5	-11.6	-17.1	-3.1	2.7	-25.1	-25.0	-25.0	-25.0	-25.0
2002	-1.8	6.5	8.5	-12.0	11.2	-22.0	33.1	33.0	33.0	33.0	33.0
2003	-16.6	7.7	3.0	-43.5	-35.3	-33.8	-36.4	-36.0	-36.0	-36.0	-36.0
2004	9.9	19.6	13.9	55.3	-0.7	-2.9	11.3	11.0	11.0	11.0	11.0

*Table 2.
Permitted,
approved,
commenced,
and completed
building
construction
from 1994–2004
Source: Perry
Pei-Ju Yang*

The Figure 3a shows a process of intensification appearing in the downtown area and in some newly developed areas such as Punggol and Sengkang. Density is lower in the industrial areas. To obtain a fine-grain distribution of development density, raster datasets were interpolated using an 'inverse-distance-weighted' procedure from the point dataset containing GPR values in the 1998 and 2003 development guide plans. The estimated values of Gross Floor Area (GFA) are 375km² in 1998 and 390km² in 2003, showing an estimated increase of 15 million m² GFA over the 5-year period.

The amount of materials consumed from 1998 to 2003 was calculated using a raster technique from the difference between the two datasets. It shows a total land area of approximately 60 km² where new material stock has been added. The areas with positive changes are locations that contain new building construction, including new residential and commercial developments sporadically distributed all over the whole Singapore. It also shows the increase of new industrial development in Jurong Island, the south-western part of Singapore. The areas with negative changes are the locations for redevelopment or rezoning, such as the kampongs, the traditional settlements cleared for new types of development in the eastern and north-eastern part of Singapore (Fig. 3b).

The density maps of different land uses types show the spatial distribution of urban development which provide the basis for the measure of material flows over the entire city. The spatial distribution of the five land use types in 1998 and 2003

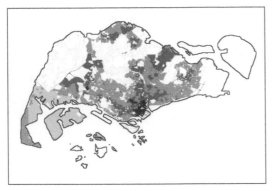 

were converted to material stock, and the material consumption of steel, cement, coarse and fine aggregates during that period were calculated from the land use types, development density and GFA (Fig. 4).

Conclusion

Measuring the urban transformation of Singapore City provides a potential approach to planning a more sustainable urban form in an intensively developed urban environment. The research shows the spatial distribution of the construction materials used and the amount of material added to the city. Construction follows the major public infrastructures such as MRT lines and avoids the interruption to the critical green area such as the central reservoir catchment of Singapore Island. The intensification of construction material use happened not only in the central city area, but also in some peripheral areas through building new towns.

The results of image processing of the satellite data indicate an increase in the green area over a sixteen-year period of Singapore's urban development. It provides evidence of some success in governmental efforts to manage and plan the park system, and of their policy of green buildings such as the greening of rooftops. However, it was observed that there were more smaller-sized green patches as well as smaller-sized urban development areas scattered around Singapore Island. The results of landscape ecological analyses show that the urban form of Singapore City is becoming more intensive and developed in a more compact and less convoluted form.

Is Singapore's urban form becoming more sustainable, and what constitutes sustainable urban form? The analyses of Singapore's urban form and its linkage to the spatial distribution of construction materials used imply a well connected 'polycentric' urban form. However, it may not be sufficient for making a strong argument for an urban sustainability model. The measurement of urban environmental performance should be expanded from the land cover pattern and material stock to other dimensions like water conservation and solar availability. The mapping and analysis should cover multiple spatial scales: from the city, the districts and reaching to development clusters. Urban transformation occurs in parallel to the usage of material, water and energy flows across different spatial scales. In future, the tracking of sustainable urban forms and material flows should be conducted

Fig. 3a: (left) Density mapping of urban development (by Gross Plot Ratio) in 2003; 3b: (right) The change of development density between 1998 and 2003 Source: Perry Pei-Ju Yang

Note: the darker the tone the higher the density

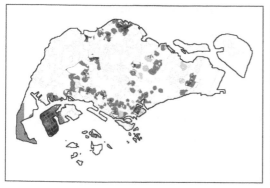

at both the coarse and fine spatial scales, to allow an integrated assessment of the urban environmental performance to be developed.

Notes

1. 1) The material stock refers to the materials such as steel, concrete and aggregates that already exist within the built environment. 2) The material flow means the materials that 'flow' through or change in the built environment or the construction industry during the urbanization process.
2. The 'patch' refers to a landscape feature containing the characteristics of size and shape of the land coverage such as forest, agricultural land and urban area. According to landscape ecology, the computation of landscape ecological indices, e.g. the patch number, patch size and patch shape, provides a way to measure the ecological quality of landscapes and regions (Forman, 1995; Farina, 1998).
3. RSI ENVI: RSI (Research Systems) is the name of the company which developed the ENVI 4.1 (Environment for Visualizing Images) software for processing of remotely sensed image. (http://www.tec.army.mil/td/tvd/survey/ENVI.html).
   SPOT (Satellite Pour l'Observation de la Terre) is a high-resolution, optical imaging earth observation satellite system operating from space. It is run by Spot Image based in Toulouse, France. It has been designed to improve the knowledge and management of the earth by exploring the earth's resources, detecting and forecasting phenomena involving climatology and oceanography, and monitoring human activities and natural phenomena. (http://en.wikipedia.org/wiki/SPOT_%28satellites%29).
   LANDSAT: The Landsat program is the longest running enterprise for acquisition of imagery of Earth from space. The first Landsat satellite was launched in 1972; the most recent, Landsat 7, was launched on April 15, 1999. The instruments on the Landsat satellites have acquired millions of images. The images are a unique resource for global change research and applications in agriculture, cartography, geology, forestry, regional planning, surveillance, education and national security. (http://en.wikipedia.org/wiki/Landsat).

References

Batty, M. and Longley, P.A. (1994) *Fractal Cities: A Geometry of Form and Function*, Academic Press, London.

Beers, D.V. and Graedel, T.E. (2003) The Magnitude and Spatial Distribution of In-Use Copper Stocks in Cape Town, South Africa. *South African Journal of Science*, 99(1/2):61–69.

Beers, D.V. and Graedel, T.E. (2004) The Magnitude and Spatial Distribution of In-Use Zinc Stocks in Cape Town, South Africa. *AJEAM-RAGEE*, 9:18–36.

Farina, A. (1998) *Principles and Methods in Landscape Ecology*, Chapman & Hall Press, London.

Frey, H. (1999) *Designing the City: Toward a More Sustainable Urban Form*, E & FN SPON.

Forman, R.T.T. (1995) Horizontal processes, roads, suburbs, social objectives and landscape ecology, in *Landscape Ecological Analysis: Issues and Applications* (ed. J. Klopatek and R. Gardner), Springer Press, New York.

Hendriks, C. *et al.* (2000) Material Flow Analysis: a Tool to Support Environmental Policy Decision Making: Case Studies on the City of Vienna and the Swiss Lowlands. *Local Environment*, 5(3): 311–328.

Huang, S.L. and Hsu, W.L. (2003) Material Flow Analysis and Emergy Evaluation of Taipei's Urban Construction. *Landscape and Urban Planning*, 63: 61–74.

Jenks, M., Burton, E. and Williams, K. eds (1996) *The compact city: a sustainable urban form*, E & FN Spon, London.

Jenks, M. and Burgess, R. eds (2000) *Compact cities: sustainable urban forms for developing countries*, Spon Press, London.

Jenks, M. and Dempsey, N. eds (2005) *Future forms and design for sustainable cities*, Elsevier, Boston.

Miller, D. and de Roo, G. (1999) *Integrating City Planning and Environmental Improvement: Practical strategies for sustainable urban development*, Ashgate, Aldershot.

Müller, D.B., Wang T., Duval B. and Graedel T.E. (2006) Exploring the engine of anthropogenic iron cycles. *Proceedings of the National Academy of Sciences of the United States of America*, 103(44):16111-16116.

Opdam, P., Foppen, R. and Vos, C. (2002) Bridging the gap between ecology and spatial planning in landscape ecology. *Landscape Ecology*, 16: 767–779.

Singapore BCA (2006) *Singapore Building and Construction Authority* official website: http://www.bca.gov.sg/

Singapore URA (2006) *Singapore Urban Redevelopment Authority* official website: http://www.ura.gov.sg/

Tanikawa, H., Hashimoto, S. and Moriguchi, Y. (2002) *Estimation of Material Stock in Urban Civil Infrastructure and Buildings for the Prediction of Waste Generation*, in The Fifth International Conference on Ecobalance, Nov 2002, Tsukuba, Japan.

Turner, M. G., Gardner, R. H. and O'Neill, R. V. (2001) *Landscape Ecology in Theory and Practice: Pattern and Process*, Springer, New York.

Williams, K., Burton, E. and Jenks, M. eds (2000) *Achieving Sustainable Urban Form*, E & FN Spon, New York.

Williams, K. ed. (2005) *Spatial Planning, Urban Form and Sustainable Transport*, Ashgate, Aldershot.

17

Wael Salah Fahmi

The Right to the City:
Stakeholder perspectives of Greater Cairo Metropolitan communities

Introduction

This chapter reinterprets the Greater Cairo metropolitan area as one of multiple layers with diverse fragmented communities. This seemingly deconstructed social morphology is characterised by the proliferation of fragmented enclaves scattered throughout the city landscape, from ruralisation of urban fringes and the emergence of spontaneous peripheral and informal settlements (poverty belts), to the establishment of suburban gated communities within the eastern desert settlement of New Cairo City. Therefore Cairo's urban scenery incorporated both the poor and dispossessed who were forced to move out of historical core areas (for land development and speculations) to urban fringes in spontaneous settlements (poverty belts) and the rich affluent groups who chose to leave the city with its high density, traffic congestion, and environmental pollution and move to new suburban houses (gated communities) (Kuppinger, 2004; El Araby, 2002).

The impact of such population mobility is examined through case studies within the Greater Cairo region, employing qualitative data and focusing especially on the new towns and on New Cairo City, to the east of downtown Cairo. New Cairo City emulates and encapsulates features of Cairo's historical inner city, with its urban contradictions ranging from a hybrid mixture of new town, overspill estate with low cost resettlement housing scheme. This new private suburban development has contributed to the process of splitting the city, with a transition from compact inner city model to diffused suburban sprawl with no city centre (*heteropolis*). New Cairo City is made up of contrasting urban elements, from decayed public housing for ghettoised resettled urban poor to unplanned but up-market private gated communities (Katameya Heights and Golf City).

The empirical study adopts a qualitative analysis employing ethnographic techniques of informal discussions with secondary stakeholder agencies (local municipality, planners, NGOs activists, real estate agents and land developers), and in-depth interviews with primary stakeholder groups (ranging from urban poor households within resettlement housing to affluent residents within gated communities). Combining information from interviews with activists and policy-makers, newspaper articles, official reports, and secondary sources, the analysis explores narratives expressed by individuals, whilst examining the impact of Greater Cairo Master Plan and New Towns Policy on urban poor's right to the city and their forced relocation to the eastern desert of Katameya (New Cairo City). The chapter tackles the on-going contestation between the urban poor's right to the city, and between real estate and property speculations and between Greater Cairo urban policy, whilst calling for a stakeholder approach (integrated matrix) to the development of the built environment which would involve public–private partnership and grass-root co-operation between tenants, resettled people, NGOs, housing experts and local authorities.

Cairo – a world class city or fragmented metropolis

The built-up conurbation of the Greater Cairo Metropolitan area comprises three sectors. The main sector is the city of Cairo located on the east bank of the River Nile, extending south to Helwan, north to Shubra, and north east to Heliopolis and Nasr city. The other two sectors are Giza, a governorate on the west bank of the River Nile, effectively part of Cairo city, and Qalubya, a governorate north of Cairo. The Greater Cairo area – core, suburbs and exurbs – has an estimated population of 17.6 million. There has been an increase of nearly 3 million people since the 1996 census (CAPMAS, 1996). The city itself has a population of 7.8 million, Giza contributes nearly 5.8 million and around 3.9 million live in the Qaliyubia governate. This makes Cairo the seventh largest urban area in the world, with very high urban densities at 11,000 persons per square kilometre, with pockets of density as high as 193,000 persons per square kilometre.[1]

Since the 1970s post-infitah period (Sadat's open door economic era), capital investment has moved away from downtown Cairo and the old suburbs to newer ones. A new class of *nouveux riche* millionaires have abandoned the city for the newly emerging exclusive desert enclaves and condominiums (e.g. Al Rehab City and Mirage City in New Cairo City, and Dreamland and Gardenia in 6th October New Town). Recently several fashionable new compounds and gated communities have mushroomed around the city, readily accessible via ring-roads, bridges and tunnels. These gated communities were part of newer planned urban extensions and new towns located within the eastern and western deserts of the Cairo metropolitan area. According to Raafat (2003) whilst Cairo's city centre is being abandoned, characterised by urban decay, overcrowding, environmental and acoustic pollution, it is difficult to grasp that modern Cairo was once an architecturally attractive world class city.

Greater Cairo Master Plans – 1980s and 1990s

As a result of Cairo's urban problems evident in its overcrowding, proliferation of small environmentally polluting enterprises, housing shortages, poor infrastructure and deteriorated public services, a Master Plan was launched in 1981 and approved by the Egyptian authorities in 1983 (Sutton and Fahmi, 2001). The 1983 Master Plan originally planned for ten new settlements to house two to three million people (GOPP/IAURIF, 1988). To create a more integrated urban network, these new settlements were incorporated in a series of eastern and western development corridors, aiming to promote transportation linkages and to funnel growth away from Cairo (Stewart, 1996, p. 464). Concern was expressed that these new settlements were sited in the 'hostile' desert and would probably be unpopular with people.

The Greater Cairo Master Plan's attempt to restructure the existing metropolitan area was promoted through a scheme of '16 Homogeneous Sectors', seeking to break up the 'mononuclear' arrangement of the city (Fig. 1). Each sector was to be an

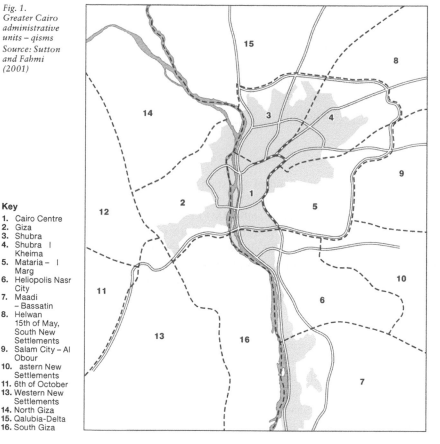

Fig. 1.
Greater Cairo
administrative
units – qisms
Source: Sutton
and Fahmi
(2001)

Key
1. Cairo Centre
2. Giza
3. Shubra
4. Shubra I
   Kheima
5. Mataria – I
   Marg
6. Heliopolis Nasr
   City
7. Maadi
   – Bassatin
8. Helwan
   15th of May,
   South New
   Settlements
9. Salam City – Al
   Obour
10. astern New
   Settlements
11. 6th of October
13. Western New
   Settlements
14. North Giza
15. Qalubia-Delta
16. South Giza

autonomous urban unit with a population size ranging from 500,000 to 2 million people and to be relatively self-sufficient in jobs and services. The 1983 Master Plans and subsequent modifications increasingly sought to direct Greater Cairo's urban growth and sprawl into the desert areas to the east and west of the Nile Valley. Several new secondary centres were to be created in these homogeneous sectors.

Some modifications to the 1983 Master Plan were made in 1991–1992 (GOPP)/ IAURIF 1991). The ring road route was re-routed to the east through the desert for military reasons, with a western arc through arable land on Giza's outer fringes. While representing the more successful aspect of the plans, the ring road served to segregate inner Cairo from the expanding outer fringe areas of higher class residences. Such modifications affected the location of some new settlements as well as contributed to the cancellation of a green belt between Cairo and the eastern New Settlements. This led to the emergence of 'New Cairo City', effectively combining New Settlements 1, 3, and 5 (Fig. 2, p. 274).

Several factors can be suggested as being behind the relative failure of the 1983 Master Plan and its modifications, especially the unplanned, illegal nature of much 'spontaneous urbanisation' often ignoring the exhortations to preserve the precious green land. Another major factor stems from the failure of the New Towns (Table

	Target Population	1996 Population
10th Ramadhan	500,000	47,839
6th October	500,000	35,477
15th May	100,000	65,865
Badr Town	250,000	248
Sadat City	500,000	16,312
El Obour	250,000	no data

Table 1. Lack of progress of Cairo's New Towns
Source: Stewart (1996); CAPMAS (1996) in Sutton and Fahmi (2001)

1) and New Settlements to attract people, and of the state to provide adequate funding for public housing for lower classes in them. This was attributed to financial constraints due to the International Monetary Fund's (IMF) restructuring of the economic plan which contributed to changes in housing finance since 1991.

Delays occurred in the implementation of the concept of homogeneous sectors were also attributed to inefficient local urban planning and land development processes, violations of building regulations and inadequate infrastructural implementation associated with lack of coordination between Governorates, local councils and planning authorities (GOPP/IAURIF, 1990). This has led to the failure to achieve the goals related to the deconcentration of population and activities away from the existing Greater Cairo agglomeration.

Despite the state being forced to re-house thousands of the 1992 earthquake victims within New Settlement 3 and Badr Town (Sutton and Fahmi, 2001; El Noshokaty, 2002), and despite the success in attracting factories and job opportunities, these new towns did not achieve the self-sufficiency aim of the original plan (Stewart, 1996; Denis, 1997). According to Stewart (1996, p. 475), poor basic services and lack of social and educational infrastructure have also discouraged families from settling in the New Towns. Rather than relocate to remote desert cities with unattractive monotonous architecture, low income families preferred to remain in familiar if crowded environments in Greater Cairo, experiencing a new form of encroachment – the informal addition of rooms, balconies and extra space within buildings. Those housed in public projects built by the state, illegally redesign and rearrange their space to suit their needs by erecting partitions, and by adding new rooms. Consequent overcrowding has resulted in intense struggles and negotiations within inner city poverty areas (Sutton and Fahmi, 2001, p. 143). Instead of the planned movement out, centrality and polarisation of functions in Central Cairo has continued, albeit with a more poly-centred spatial pattern. As new town housing proved to be too expensive for the urban poor, it eventually attracted speculators rather than residents.

In addition to the ring road, a new '26th July Axis' road was created on Giza's western outer fringes, which has boosted private development and land speculation within 6th October New Town and Sheikh Zayed Town (New Settlements 6a and 6b). By the early 1990s the state had handed over the management of some of these New Towns to private promoters and speculators who constructed villa complexes, enclosed élite compounds, and gated communities (Denis, 1997) – rather contradicting the original concept of low-cost dwellings and decent transport infrastructure in order to attract poorer people from the inner city areas.

Urban fragmentation and spatial contestation within the Greater Cairo Region

In the last decade Central Cairo has lost population, while the peripheral villages and towns have grown rapidly. This has not prevented Cairo's urban configuration from being transformed by informal communities into a 'rural city'. High land prices, increasing population density, shortage of affordable housing and the decay of existing housing stock, all contributed to low income migrants seeking peripheral districts and joining already marginalised groups in informal settlements around the metropolitan area of Cairo (Bayat and Denis, 2000).

Between 1981 and 1991, rural poverty doubled and urban poverty increased more than 1.5 times. By the early 1990s, more than half of Cairo and adjacent Giza were classified either as 'poor' or 'ultra-poor'. Thus, to escape from high rents, millions of rural migrants and the urban poor in Cairo have quietly claimed state or public land and cemeteries on the outskirts of the city, creating largely autonomous communities.

With the deregulation of agriculture, villages expanded, with a new rural affluent class emerging out of their investment in real estate, construction activities and cash cropping. This urbanisation of large villages and small towns is a polarised one, with dynamic communities characterised by increasing mobility and a new pattern of social stratification.[2] Such diffused urbanisation has shifted from ruralisation of the city (the influx of rural populations into the city) and urbanisation of countryside (the densification of villages) giving way to the emergence of a 'post metropolitization' of the city (Bayat and Denis, 2000).

This new population concentration poses a challenge to the political economy and state authority, as the state regards such 'ruralised ' and 'urbanised' agglomerations as spontaneous settlements, and does not provide them with either urban services or public facilities. These environmentally degraded settlements are not only regarded as poverty belts, but also as localities for middle class professional urbanites. They consist of heterogeneous occupational and cultural groups; these 'spontaneous' settlements are stigmatized as 'rural', attracting both migrants and Cairo's urban youth (aged between 20–25 years) from inner city and urban core areas.[3]

In addition, spatial constraints have contributed to the densification of old established districts where old spacious high status villas were transformed into dense high rise and lower status apartment blocks. Older districts have been replaced by new private cities, with the ring road encircling the city and facilitating efficient means of private transport and communication. There is a duality of peripheral informal and marginal settlements on the one hand, and planned exclusive (often gated) suburbs on the other. According to Bayat and Denis (2000) this urban polarisation and social transformation has divided the city into six spatial patterns (Figs 2 and 3):

- Ruralisation of the urban fringes and the emergence of spontaneous informal settlements around the city peripheries
- Urbanisation of rural villages
- Densification of areas inside the ring road
- Saturation of the CBD and the decline of the inner old city as people seek to move out from core areas to the fringes
- Inefficiency of New Settlements in providing housing for low-income

people, except for emergency shelter schemes for victims of the 1992 earthquake

• Exclusive private suburban districts emerging within the New Settlements, inhabited by the new elite upper middle class.

Whilst Cairo's population growth has certainly slowed down in recent years, and despite the positive effect of the Ring Road on eastern and western private suburban development, the present diverse spatial patterns have not resulted from the Master Plan strategy. The Plan's aims of controlling urban growth and east–west expansion, reducing population concentration in the inner city, protecting arable land to the north and south of Cairo, and upgrading public facilities have in large part not been achieved. Indeed, it can be argued that, not only has Greater Cairo not been mastered or planned, but that the Master Plan of 1983, as revised in 1991–92, no longer really exists in an effective way. Instead Cairo's continued development has been carried out by private entrepreneurs and property speculators rather than by the city's planners.

The shift of emphasis from the public to private spatial development has been exemplified in newly emerging exclusive townships incorporating affluent private neighbourhoods. These are mostly located in the western desert in 6th October City and Sheikh Zayed Town, and in the eastern New Settlements of Al Shorouk and New Cairo City. With six out of ten New Settlements that include gated communities there has been a transition from a compact inner city model to a vast diffused spatial pattern (suburban sprawl) with no symbolic city centre and no identity. The remainder of this chapter presents the findings of research illustrating some of the key issues raised above.

Fig. 2. (left) Fragmentation of Greater Cairo Region Source: Sutton and Fahmi (2001)

Fig. 3. (right) Population movement within Greater Cairo Region Source: Sutton and Fahmi (2001)

Survey of four housing areas in Cairo and Giza

Between April and June 2005, a small area survey was carried out within a sample of housing schemes in four survey areas across Cairo and Giza. The four survey

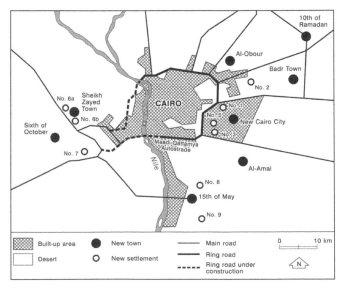

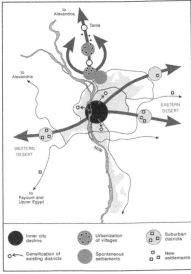

areas were selected to represent eastern and western locations, the period when they were constructed and their design criteria, and a sample of gated communities. The study areas were New Cairo City, Badr Town, Sheikh Zayed Town, and Maadi-Katemeya Autostrada development (a private housing scheme between Cairo's eastern suburb of Maadi and New Cairo City). Interviews were undertaken with a variety of primary and secondary stakeholders.[4]

New Cairo City

Gated communities

Initially the area was named Katameya, and was inhabited in 1992 by earthquake victims officially relocated by the government to public housing in New Settlement 3. In 1996 the low income tenement housing project, which overlooked the newly created *Katameya Heights*, was truncated. The official reason given was 'building irregularities' but incompatibility with nearby villa developments and with 'gated communities' appears a more likely cause. Since 2000 *New Cairo City* has been created as a result of merging the eastern New Settlements 1, 3 and 5 with newly established private gated communities in the desert to the east of the ring road (Fig. 4). The first phase of development included *Golf City*, which consisted of luxurious residential districts to encourage middle and high-income families to move out of Cairo to its 521 villas with prices ranging from US $258,000 to US $430,000 (e.g. Fig. 5). It is surrounded by a newly constructed 18-hole golf course with hundreds of planted palm trees. The second phase incorporated the *JW Marriott Hotel* and *Mirage City* with an additional 330 villas, ranging between 700–900 square metres, with landscaped gardens. In response to high demand for property

Fig. 4. New Cairo City and New Settlements 1, 3, 5
Source: Sutton and Fahmi (2001)

Key
1. New settlement no 1 (Mirage City – Gated Community)
2. New settlement no 5 (Youth Housing)
3. New settlement no 3 (Resettlement Housing)
4. Katameya Heights Golf City (Gated Community)
5. Nature conservation area
6. Green area
7. Al Rehab City
8. The American University Site
9. Universities district
10. mbassies district

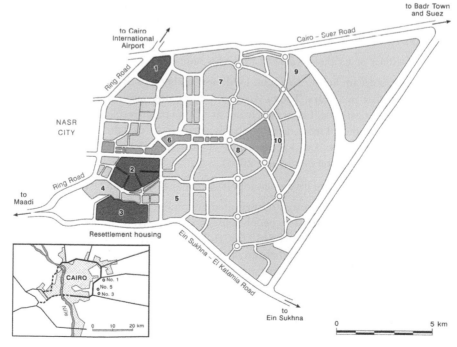

Fig. 5. Villa – Katameya Heights Golf and Tennis Resort – New Cairo City
Source: http://www.katameyaheights.com/gal4.htm

in *Katameya Heights*, more land was purchased in 2003, with a waiting list of over 400 interested buyers for 280 prototype villas (with prices ranging from US $340,000). Another two compounds are in the process of development: *Lakeview*, a 500-villa luxury development, and New Cairo Centre, a US $2 billion area which includes commercial, residential and tourism development (owned by Carrefour and Majid Al Futtaim of Dubai).

Regarding these gated communities in New Cairo City, an official at the Authority of New Urban Communities (ANUC)-Ministry of Housing, attempts to justify their policies:

> To provide housing for poor and limited-income families, the authority must raise funds elsewhere. The only way was to encourage businessmen to invest in real estate, and sell them land at high prices (Interview – April 2005).

A lawyer owner of a villa in *Katameya Heights* indicates:

> This is a dream house, no doubt. I will not be able to move into the villa for another two years. Then I will use it as an alternative to my North Western Coast villa, but definitely not as a permanent residence, since it is far from where I work. Most probably once I retire and do not need to commute to Cairo, I will permanently move here with my children and grandchildren (Interview – June 2005).

A managing director of an urban development area at *Katameya Heights*, argues:

> Our objective is to create an integrated community where people find everything. Although most people cannot afford a villa, we also offer small flats for LE 80,000 to LE180,000 [LE = Egyptian pounds, equivalent to US $13,000–$30,000] that can be purchased by instalments. At least 70% of all housing units have already been sold. Egyptian expatriates are among our main target groups. Now that we have these communities, we can encourage them to return, and offer an alternative to those who tend to buy holiday homes abroad (Interview – June 2005; Fig. 6).

In contrast, urban planners at the General Organisation for Physical Planning (GOPP) expressed opposition to these gated communities:

> The government has sold or leased vast areas of desert, supplied with infrastructure, to private investors at very low prices to be used for luxury housing. The investors received large loans from national banks to implement the projects, pouring people's savings into glossy dwellings that will serve only a fraction of society (Interview – April 2005).

More radical opinions were expressed by respondents at the Egyptian Centre for Housing Rights (ECHR):

> These resorts are symbols of social inequality, tangible manifestations of the rise of the *nouveaux riche* who are obsessed by the dream of the lavish lifestyle. Why did the government offer land at such low prices for luxury housing? This land should have been sold at double or triple the price and the profit could have subsidised limited-income and youth housing projects. Why should the government provide such projects with infrastructure when many impoverished areas are still deprived of these basic needs? (Interview – April 2005).

In addition to 11,322 housing units built by the private sector and by government agencies, 32,651 apartment units including 2,120 future housing units, and 19,717 youth housing units, were established by the New Urban Communities Authority in an area originally designated for accommodating the urban poor.

Youth Housing Project (New Settlement 5)

The Youth Housing Project was launched in 1995 for newly married couples and for limited income groups, with the aim to provide 2,120 units in New Cairo City. This was managed by the Future Foundation established in 1998 with a mission to provide affordable housing through public–private partnership using government land grants, and private entrepreneurs' contribution of LE 1 billion (US $167 million). Beneficiaries would be provided with loans, allowing them to pay in instalments over 40 years, at a 5% annual interest rate (Ministry of Housing and Reconstruction, 2002).

Despite official media propaganda regarding this project, negative attitudes were expressed by a married couple typical of many urban middle class couples searching for reasonable payment options to finance new homes:

> We have lived at (*the husband's*) grandfather's house in Giza since getting married two years ago, since we will not be able to move into our new flat for another two years. At the time of our marriage, both our parents put a down payment on the flat, and we have been struggling ever since to finance our 'future' home. We have been seeking a bank loan to pay an approximately LE 100,000 [US $17,000] final 'lump sum'. The problem is that our salaries do not cover the home prices, and unless you depend on our parents ... it's virtually impossible to rely on our personal savings (Interview – June 2005).

Resettlement housing (New Settlement 3: e.g. Fig. 7)

A survey by Sutton and Fahmi (2002b) illustrated case histories of various heads of households who described their lives after eviction from inner city poverty areas to the Eastern desert settlements, as 'Out here, there is nothing ... ' Some of them moved twice, choosing to live initially in temporary tin huts or tents on vacant land within fringes of eastern cemetery, and then forced again to resettle to Cairo's eastern desert communities (Fig. 8):

> Look what we have got ... nothing. There is no water, no electricity ... Nobody is going to help us. I worry about how I am going to feed my six children.

> The local officials for our area told us we had to get out. We asked where we could possibly go, and were told not to worry, as we would be provided with alternative plots. Some of the men did not agree to this, as there was nothing in this new place. And then. what about our livelihood? The local officials told us that in a week's time all facilities would be given to us. We were sceptical but what could we do? We are poor people; we have no money, no power. We had no options.

> The local authorities arrived with the police and a lot of other people and told us to pack our things, saying that they were going to move us to Cairo's New Eastern desert communities. What would happen to our children? How will we feed them? What about our homes?

> We have been given plots (areas = 60–100 square meters) in the new desert settlements. How can we build our homes there? The plots cost approximately LE 5,000 [US $835] which we have to pay in only two instalments. We can't afford LE 2,500. How can we save? Who will give us a loan?

According to official proposals at the Ministry of Housing and Urban Settlements, the new (re)settlement for the Zabaleen garbage collectors and tomb dwellers in Katameya would be equipped with complete piped networks of water supply and sanitation, roads network, open space, vocational training and heath care centres, libraries and schools (Fahmi and Sutton, 2006; Fahmi, 2005; Sutton and Fahmi, 2002b). Residents would be provided with a soft loan, 90% of which is required to be paid over 40 years, giving the residents a sense of ownership.

> The project is based on transparent dialogue with local inhabitants in both planning and management processes (Interview – April 2005).

In contrast, a more negative reaction to resettlement housing in Katameya was expressed by the Zabaleen who were to be evicted from their settlements in Muqattam:

> We have been assigned flats in buildings, housing which we are not accustomed to. I have a big family which will be obliged to live in an 80 square metre (or less) 'box'. These new housing units do not have courtyards and extra areas for rearing animals (Interview – May 2005).

Moreover residents of Gamalia district, who were forced to leave their decayed housing in Medieval Cairo area to resettle in the eastern new settlements, expressed their dissatisfaction,

> Those who had little money rented a place in Manshiet Nasser or somewhere else nearby. They said that at least they were close to their source of livelihood in the bazaar and Gamalia. In Katameya we are out there in the desert far from Cairo (Interview – April 2005).

Fig. 7. (below–left) Middle Income Housing – New Cairo City Source: http://www. rentalcartours. net/rac-cairo.pdf

Fig. 8. (right) Population eviction to New Settlement 3, New Cairo City Source: Fahmi and Sutton (2006)

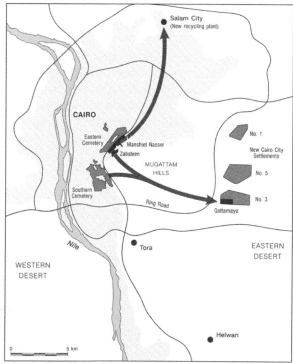

Similar opinions were confirmed by Bab al Nasr tomb dwellers who had already been evicted to the eastern settlements yet expressed their anxiety about being moved to the desert.

> We have to spend 6LE [US $1] a day on transportation taking us at least two hours going to and coming from Cairo. This is quite expensive for us. Katameya is so far from the city, where we seek work. Otherwise our families will starve. We are considering going back to Cairo even if it is difficult to find accommodation (Interviews – June 2005).

Suburban residential development (Maadi – Katameya Autostrada Axis: e.g. Fig. 9)

A few years ago, the former Cairo governor announced that the Nasr City–New Cairo City road with a length of 5km would go some way to solve housing problems through the help of private investors who had been allocated 80% of the land, encouraging residential and industrial development in the east of Cairo. However, local authorities have been in dispute since the new Cairo governor introduced a decree halting construction work, announcing that this residential development had violated building codes and heights restrictions, and had encroached on unregistered land. The residential blocks, which will be knocked down soon, had been earlier encouraged by officials at the Ministry of Housing as a means of linking Maadi's suburbs (Zahraa Maadi) and New Cairo City. An urban planner at GOPP elaborated:

> I expect the future demolition of these incomplete private middle class housing blocks where land will eventually be sold at a very high price (Interviews – April 2005).

The gate-keeper (*bawab*) pinpointed problems:

> We do not know the owners since they used to pay their due instalments through their agents. They mainly live and work abroad and have delegated some real estate agents and lawyers to deal with their cases. The problem is that no one knows who is currently in charge of managing these blocks. Overnight I anticipate that the whole area will

Fig. 9. Vacant Housing Development – Suburban Maadi – Katameya Autostrada Axis Source: Wael Fahmi

be bulldozed and next day a wall will be built surrounding 'another dreamland' (Interviews – May 2005).

Sheikh Zayed Town

The town was established in 1996 as an agglomeration of New Settlements 6a and 6b, located 38km from the centre of Cairo just outside the western arc of the Greater Cairo ring road. The New Urban Communities Authority provided 9,938 public housing units including 4,733 units for high-medium-low income levels and 5,205 youth housing units with nearly 5,000 inhabitants living in the city as compared to a planned target population of 450,000 residents to be achieved by 2017. In addition 5,354 housing units were built by the private sector and by other governmental agencies. Recently, the government encouraged private investment in housing development with nearly 38% of the planned residential area being implemented mainly for upper-income levels. Developing the town's infrastructure, public services and low-income dwellings is often regarded as the government's responsibility, while the private sector is involved in the production of luxury housing and entertainment, as the town became a resort for high-class gated communities. According to an owner of a privately-upgraded public housing flat in Sheikh Zayed Town:

> Some of the unoccupied low income housing stock is decayed even before being allocated to people. The government is encouraging private developers to buy these units for upward filtering to middle class standards. Since the opening of 26 July axis road, I was encouraged to move here permanently, especially with other friends already considering settling in Sheikh Zayed Town. We might even seek bank loans to buy land collectively and build a condominium for our grown-up children. The future is here since owning your home has many advantages. It is a great investment (Interviews – May 2005).

A more positive view was expressed by an upper class retired diplomat who resided within one of Sheikh Zayed Town's gated communities:

> It takes only 25 minutes to get here from Mohandessin, where I live, which means I can spend all my weekends there. Resorts like this one are essential for people to refresh themselves before starting another working week, and for children to play and spend some time away from Cairo's pollution (Interviews – June 2005).

Badr Town

Badr Town was created in 1983, about 47km east of Cairo, on the Cairo–Suez highway, as part of the 1983 First Master Plan. In 1995 additional industrial and vocational areas were introduced, expanding the whole built-up area to 73km^2, including the green belt. The city is characterised by decayed five-storey public middle and low-income housing which is mainly vacant, with the exception of a few inhabitants involved in construction work.

Despite the government's policy to develop an economy based on industry and a medical complex, the number of jobs was lower than anticipated. The 1996 census revealed that there were only 248 residents in Badr Town, despite a targeted population of 430,000 (CAPMAS, 1996). This was attributed to Badr Town's failure to achieve sufficient industrial development, and to poor transport links. Yet the vice-president of the Technical Affairs Department at the New Urban Communities declared:

> Every new city starts small then it grows and attracts residents. However, the availability of job opportunities plays a major role in this growth dynamic. Badr City has an industrial zone with 225 factories with a capacity to employ 20,000 people. We are planning to have more public transportation links between Badr City and Cairo (Interviews – April 2005).

The relocation of people affected by Cairo's 1992 earthquake contributed to an increase in population to 1,358 people by 2000. As Degg (1992, p. 227) observed, 'the effect of the serious 1992 earthquake has been to relocate some of the inner city's overcrowded and ill-housed population to the new settlements'.

Whilst the government sponsored public (low income and middle income) housing in Badr Town, other bodies such as the Dwellings Finance Organizations and the Construction Housing Bank financed upper income and private housing units. Whereas low-income earthquake victims occupied decayed dwellings, which needed maintenance, most residents in private housing units have introduced internal modifications and repair work. One low middle income resident in Badr Town complained:

> We have to cope with difficult terms of financial payment for housing units. We have to pay instalments subject to increasing compound interest rates. And there are cases where a government employee whose monthly income is LE150 [US $25], must make monthly payments of LE70 [US $12]. Whilst the houses need a lot of modifications, the inability to pay installments would lead to subsequent eviction. Most jobs go to people who do not live in Badr Town (Interview – June 2005).

Stakeholder perspectives

Empirical research attempts to examine qualitatively the complex reasons for the failure of various policies and project implementation (GCR master plan, New Settlements and New Town plans) in meeting the housing needs of middle and low income people – in particular the emergence of nearly empty new towns, and expansion of exclusive gated communities, a phenomenon which aggravated social injustice and housing inequality.

Therefore the four survey areas were divided into two subgroups, according to housing policies and type of development: the *Unplanned Eastern New Cairo City, and its Suburban Maadi–Katameya Autostrada development (55 cases)*; and the *Eastern Badr Town and Western Sheikh Zayed New Towns (35 cases)* (Table 2). The survey involved on-site interviews with 35 house occupiers, ranging from owners to tenants. Information about the other 55 vacant housing units was gathered through telephone contacts with non-occupier owners or tenants, and through informal discussions with neighbouring local residents. During related focus group discussions, questions were raised regarding reasons for the incidence of unfinished and unoccupied buildings (Tables 3 and 4), followed by a more qualitative examination of attitudes and expectations of various stakeholders with their housing environments (Tables 5a and 5b; Fig. 10).

*Table 2.
Stakeholder
Respondents
Source: Wael
Fahmi*

Primary Stakeholders (90 respondents)
Housing Tenants (35 respondents)
Resettlement Public Housing (New Cairo City – Eastern New Settlements) (25)   Low income households (Inner city urban poor: tomb dwellers – Zabaleen – Gamalia)
Public housing (Badr Town – Eastern New Towns) (10)   Middle and low middle income households
Housing Owners (45 respondents)
Youth housing (New Cairo City – Eastern New Settlements) (10)   Middle and low middle income households (Urban Youth)
Public-Private Development Housing (Sheikh Zayed Town – Western New Towns) (10)   Middle income households
Katameya Heights Gated Communities (New Cairo City – Eastern New Settlements) (10)   Upper middle income households
Sheikh Zayed Gated communities (Western New Towns) (10)   Upper middle income households
Incomplete/Unfinished Private Residential Blocks (Eastern Suburban Maadi – Katameya Autostrada Axis) (5)   Middle income owners (expatriates in Gulf countries)
Property Speculators (10 respondents)
Katameya Heights Gated Communities (New Cairo City – Eastern New Settlements) (5)   Upper middle income owners (land developers and investors)
Sheikh Zayed Gated Communities (Sheikh Zayed Town – Western New Towns) (5)   Upper middle income owners (land developers and investors)

Vacant dwellings within the survey areas

Generally, over a third of the total sampled houses (39%) was completed and occupied (Table 3). Nevertheless, within the two broad areas, there were specific variations amongst different housing types, with vacancies ranging from 100% within the eastern suburban Autostrada development; to 30% within public–private development housing in Sheikh Zayed Town (see Table 3 footnote).

Expectedly, the highest percentage of full and partial house occupancy was recorded amongst the housing compounds in the gated communities of both Sheikh Zayed Town (7 of 15 sampled cases) and New Cairo City's Katameya Heights (6 of 15 sampled cases). Whilst retired owners of these private compounds tend to live permanently in the gated communities, others, especially in Sheikh Zayed Town, travel to Cairo for work. This commuting pattern is greatly facilitated by the recent extension of the western section of the ring road.

Both a minimal and high incidence of unfinished and unoccupied buildings was more evident within Eastern New Cairo City and its suburban Maadi–Katameya Autostrada development, amounting to some two-thirds of the housing units. Such a high pattern of vacancy was prevalent within the public housing stock, particularly in Badr Town (7 out of 10 sampled cases). The vacancy rate dropped when residents invested in housing modifications, such as in Sheikh Zayed Town's public–private housing development, where only 3 of 10 sampled cases were partially unfinished and unoccupied.

A number of the vacant houses were finished but unoccupied by owners or tenants. Such a phenomenon, was evident mainly in the gated communities amongst nine sampled houses of New Cairo City's Katameya Heights and eight sampled houses in Sheikh Zayed Town. This was attributed to owners' speculation by purchasing in expectation of higher future house prices; others held on to housing for future occupancy by their married offspring.

Table 4 indicates various reasons given for their dwellings being left vacant by absent owners during telephone interviews. Other informal discussions with neighbouring local residents within New Cairo City's Resettlement Housing and with the gate-keeper in the Eastern Maadi–Katameya suburban Autostrada development were quite informative about housing vacancy. Inability to afford to develop the property was the main reason mentioned by 47% of respondents. Lack of utilities and infrastructure facilities, inconvenient location and inefficient transport system were mentioned by more than two-thirds of the respondents (these were mainly within resettlement housing areas), where people expressed satisfaction with their original inner city residence areas, but discontent with the resettlement housing because of limited development opportunities and substandard property. This opinion was further confirmed by tenants and owners of Public Housing Units in Badr Town and Youth Housing Units in New Cairo City, who could not afford to develop their property, and who expressed their need to sell their flats, currently regarded as frozen or dead assets.

Unfinished and Unoccupied Buildings	Eastern New Cairo City and Suburban Autostrade Development $n = 55$ (100%)		Eastern and Western New Towns $n = 35$ (100%)		Total $n = 90$ (100%)	
No incidence	18	33	17	49	35	39%
Minimal incidence	15	27	10	29	25	28%
Unfinished buildings	8		6		14	
Finished but unoccupied buildings	7		4		11	
High incidence	22	40	8	23	30	33%

Table 3. Incidence of vacant dwellings* Source: Wael Fahmi (Field Survey, April – June 2005)

Notes: * Types of housing (total number) – (number / % of vacant dwellings)
    Eastern New Cairo city +Suburban Autostrade Development (55) – (37 / 67%)
    1. Youth housing (10) – (6 / 60%)
    2. Resettlement public housing (25) – (17 / 68%)
    3. Katameya heights gated communities (15) – (9 / 60%)
    4. Private residential blocks (5) – (5 / 100%)

    Eastern +Western New Towns (35) (18 / 51%)
    5. Public housing (10) – (7 / 70%)
    6. Public – private development housing (10) – (3 / 30%)
    7. Sheikh Zayed Gated communities (15) – (8 / 53%)

Table 4. Reasons for the incidence of vacant dwellings *Source: Wael Fahmi (Field Survey, April – June 2005)*	Eastern New Cairo City and Suburban Autostrade Development *n = 37 (% responses)*		Eastern and Western New Towns *n = 18 (% responses)*		Total *n = 55 (% responses)*	
Owners' inability to afford to develop property	18	49	8	44	26	47
Property owners' speculations and investment	12	32	6	33	18	33
Bureaucratic (inheritance) arrangements/inaccessibility to property	6	16	3	17	9	16
Unofficial property allocation	4	11	0	0	4	7
Lack of utilities and infrastructure facilities	23	62	8	44	31	56
Limited development opportunities (decayed/derelict/substandard property)	22	60	9	50	31	56
Insecure property market	9	24	7	39	16	29
Lack of security and safety within area	16	43	7	39	23	42
Inconvenient location and inefficient transport system	19	51	6	33	25	46
Investment for future generations	12	32	8	44	20	36
Satisfaction with existing residence	16	43	3	17	19	35
Lack of housing mobility outside local area	10	27	3	17	13	24

Notes: Total numbers add up to more than 55 as respondents gave more than one answer.
The percentages are % of responses.

Factors related to Cairo's housing problems

Tables 5a and 5b and Figure 10 reveal the complex interrelated factors contributing to Cairo's dwelling crisis. These include: poor housing condition and management; type of occupancy; vacancy; funding; housing mobility and future housing expectations.

New Cairo City youth housing (10 cases) was initially a refurbishment of old public housing which was mainly owned, on mortgage and instalment schemes, by middle and lower middle income urban youth. Whilst housing conditions are well managed, despite housing being mostly vacant, the first home owners were partly optimistic about the future despite lack of affordability and their inability to pay instalments regularly.

Resettlement public housing (25 cases) was mainly assigned on a tenancy basis to inner city urban poor households. However, the lack of management and infrastructure has contributed to temporary occupancy and partial vacancies. The inclination is to return to inner city areas because of the lack of job opportunities.

In contrast to the other two housing groups in Katameya Heights, *the gated communities* in New Cairo City (10 cases) and in Sheikh Zayed Town (10 cases) represent unplanned speculative and private 'luxury' housing development and villa compounds owned by the upper middle income households. Despite the excellent conditions of these private housing developments, villa compounds are infrequently

Table 5a. Factors related to Cairo's housing problems – Eastern New Cairo City and Autostrade Development Source: Wael Fahmi (Field Survey, April–June 2005)

	Eastern New Cairo City and Suburban Autostrada Development (50 respondents)			Maadi–Katameya Eastern axis
	New Cairo City–Eastern Cairo		Katameya Heights gated communities *(n = 10)*	Private residential blocks *(n = 5)*
	Youth housing *(n = 10)*	Resettlement public housing *(n = 25)*		
Population characteristics	Middle and low middle income household owners (Urban Youth) (mortagage and instalments)	Low income household tenants (Inner city urban poor : tomb dwellers– Zabaleen– Gamalia)	Upper middle income household owners	Middle income owners (expatriates in Gulf countries)
Housing design characteristics	4–5 storey two– three bedroom units (100–150 square metres)	5–6 storey two bedroom units (60–80 square metres)	Villa compounds (300–600 square metres)	8–14 storey buildings (120–140 square metres)
Housing conditions/ management	Well managed housing (upward filtering of old public housing)	Decayed housing, poor management and lack of infrastructure	Excellent condition	Incomplete decayed blocks (Structural skeleton– no infrastructure) violation of building codes
Type of occupancy/ vacancy	Mostly vacant	Mainly vacant Temporarily occupied	Occasional occupancy Partial vacancy Weekend Commuters	Permanently Vacant
Funding agency	Private–public partnership and investment	Public funding	Private investment	Private investment
Housing history/ mobility	First home owners	Second home tenants (first formal home) mobility within informal housing	Second and third home owners	First and second home owners
Future housing expectations	Partially optimistic Concern about affordability and ability to pay instalments	Pessimistic Sees no future Lack of jobs and poor housing Tendency to return to inner city areas	Optimistic Home for young generation Property speculation	Mixed Feelings future property speculations for mixed use development

| Eastern + Western New Towns (30 respondents) | | |
| Badr Town Eastern Cairo–Suez road | Sheikh Zayed Town Western Cairo–Alexandra road | |
Public housing *(n = 10)*	Public–private development housing *(n = 10)*	Gated communities *(n = 10)*
Population characteristics Middle and low middle income household tenants	Middle income household owners long term instalments and future mortgages	Upper middle income household owners
Housing design characteristics 4–5 storey two bedroom units (80–100 square metres)	Originally 4–5 storeys Vertical extension and addition of new rooms (100–150 square metres)	Villa compounds (300–600 square metres)
Housing conditions/ management Decayed–monotonous architecture	Modified privately by owners/tenants	Excellent condition
Type of occupancy/ vacancy Partially occupied	Partially occupied	More occasional occupancy Commuters, weekenders
Funding agency Public funding	Public–private partnership Public funding and owners' private investment	Private investment
Housing history/ mobility First home tenants	First home owners	Second–third home owners
Future housing expectations Pessimistic Sees no future Lack of jobs and poor housing Tendency to return to inner city areas	Optimistic Property speculation	Optimistic Home for young generation Property speculation

Table 5b. Factors related to Cairo's housing problems – Eastern and Western New Towns Source: Wael Fahmi (Field Survey, April–June 2005)

occupied by owners who are either commuters or real estate property speculators. They are optimistic that their houses will be homes for the younger generation.

Vacant incomplete private residential blocks (5 cases) within New Cairo City's suburban eastern Maadi–Katameya axis are mainly owned by absent middle income expatriates, who are working in Gulf countries. The 8–14 storey tower blocks are considered to violate the building regulations in terms of height restrictions (5-storey height limit along the Autostrada area), encroachment on unregistered land, having neither a building licence or planning permission (despite the governor's decree). The owners expressed various opinions about future speculation for mixed residential and commercial development. They were concerned about the phenomenon of land

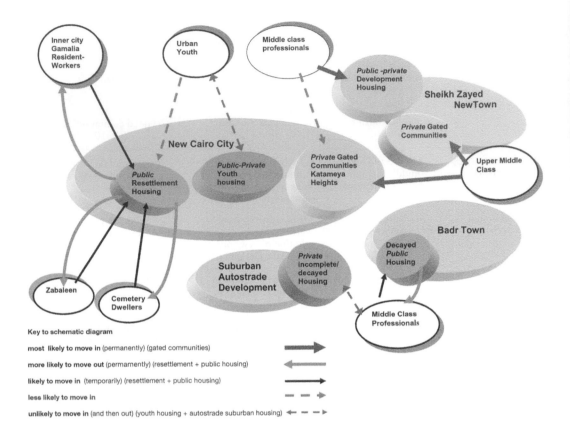

Fig. 10. Spatial contestation within survey areas
Source: Fahmi and Sutton, forthcoming

freezing, which occurs when speculators make large profits by doing nothing more than clearing the site and holding the empty land for speculation. As a result of inflation rates, land prices increase substantially where land plots can then be sold at great profit to commercial developers who will then construct 'luxury' accommodation.

Similar to New Cairo City's resettlement housing, *public housing* (10 cases) in the Eastern Badr Town is partially occupied by middle to lower-middle class tenants as part of the government's New Town Policy. Tenants expressed their dissatisfaction with the monotonous architecture and the lack of jobs and infrastructure. They expected no future within their decayed 4–5 storey two bedroom units, and they tend to return to inner city areas. Similarly, the *public–private housing* (10 cases) in Western Sheikh Zayed Town has been modified by the owner-occupiers' private investment. The original 4–5 storey *public housing units* have, in some cases, been extended upwards with the addition of new rooms for future occupancy by married offspring

Conclusion

The failure of official New Town policies and the Master Plans have contributed to Greater Cairo's housing crisis. Compared with official expectations, a much lower estimated resident population was found in Badr Town and in Sheikh Zayed Town.

Numerous vacant dwellings stand out as a glaring anomaly in the long-established housing crisis in Greater Cairo, and illustrate the inefficient public and private distribution of housing resources.

The shortage of affordable dwellings to purchase or to rent is reflected by overcrowded middle-class and lower-class inner city housing and teeming, illegal, if formal, constructions (El Batran and Arandel, 1998). 'Dwellings without dwellers and dwellers without dwellings' are to be found in practically all districts of Greater Cairo, though concentrations of both can be determined from census statistics or from local knowledge.

A further problem has arisen from certain recent official policies and practices aimed to assist poorer people acquire housing. Youth Housing dwellings, new town housing projects and other public housing schemes have allocated dwellings to low and middle income people hitherto caught up in the housing crisis. Some of these new residents return to their former inner city dwellings but keep hold of their new flat, creating an additional vacant dwelling. Other families cannot afford the rents or mortgage payments or resent the remoteness, distance or desert environment of the new settlement and abandon their allocated dwelling, which then proves hard to re-let. New towns in particular have failed to attract population away from inner Cairo locations.

Vacant dwellings

Official provision of dwellings by state or local government authorities has fallen well short of demand. Private and often illegal construction has met the shortfall, together with informal and squatter solutions to individual families' housing problems. Increasingly, private rather than public investment in housing has prevailed, following the restructuring of the Egyptian economy in line with the 1991 IMF structural adjustment agreement. The practice of keeping property vacant for future married sons has been further aggravated with property speculation encouraged by rising demand and the greater availability of mortgages. Speculative house building, for example within suburban eastern Maadi–Katameya axis, has resulted in empty dwellings and near complete but unoccupied flats and houses. There is a need for better regulation as well as better provision of housing for poorer families, with regulatory controls to limit its occupation by middle class families and speculators. The shortage of affordable dwellings to purchase or to rent is reflected in the overcrowded middle-class and lower-class inner city housing.

There are three broad categories of vacant dwellings. Firstly: there are the plots with finished but unoccupied dwellings, ranging from resettlement accommodation, completed but too expensive, too remote or too unsatisfactory for the anticipated occupants, to unoccupied and under-occupied gated communities built speculatively for commuters or second home dwellers in self-styled resort or golf complexes; other speculative, ostensibly low-income, dwellings acquired as a middle-class family's future investment. Secondly, a type recognised by the Central Agency for Population, Mobilization and Statistics (CAPMAS) housing census as plots with unfinished and unoccupied buildings at various stages of construction – the numerous multi-storey blocks on the Maadi Autostrada development illustrate this type. Thirdly, there are undeveloped plots of derelict land as within New Settlement No. 3 in New Cairo City. Such areas are treated by the census as vacant even though the plots may be occupied by illegal and informal housing.

Gated communities

The peripheral unplanned speculative and private luxury housing within the gated communities of New Cairo City and Sheikh Zayed Town has affected eastern and western new settlements in a way not predicted by the Master Plan nor by the GOPP. Most of this ring road-associated housing development has been private rather than public, and higher class rather than lower (Sutton and Fahmi, 2001, p. 145). To a certain extent New Cairo City, whilst replicating old social divisions, unresolved housing problems and ghettoisation, encapsulates most of the features and problems of Greater Cairo's housing situation. This stems from its nature as a hybrid mixture of low cost re-housing scheme and speculative middle class development. New Cairo City is made up of three contrasting urban elements which contribute to the vacant dwellings phenomenon: the unplanned but decidedly up-market private gated community of Katameya Heights incorporating upper middle class élite residents and/or property speculators; an unaffordable public–private Youth Housing Project; and a decaying and inadequate resettlement housing project for victims of the 1992 earthquake and other relocated urban poor who often prefer to return to their inner city informal housing.

Resettlement of urban poor population

The trauma of evictions of urban poor people (tomb dwellers, Zabaleen garbage collectors and residents of the Medieval city) is mainly attributed to the social disruption caused to those who had to move and forced from their inner city homes, in which they have lived for decades and in which they have often invested a considerable proportion of their income over the years (Fahmi, 2005; Fahmi and Sutton, 2006; Sutton and Fahmi, 2002a, 2002b). As no warning is given before the bulldozers destroy their settlement, they often lose sources of livelihood. Where provision is made for resettlement, this is almost on a distant site where the people are expected to build, once again, their homes but on land with little or no provision for infrastructure and services. Selected sites for relocation in Cairo's New Eastern desert Settlements are far from places of work and housing too costly for low-income households, with 'low-cost' housing projects often ending up in the hands of middle-class groups. Those evicted rarely receive any financial support for rebuilding.

The Government's programme aimed to stimulate upper and middle-class residential construction by clearing such strategically central areas of the inner city from the presence of the poor, in the name of government's concern for the welfare of 'less favoured' families, with legislation to protect 'the environment' as a justification for securing access to land for development. Recent evictions were clearly not motivated by purely aesthetic factors, where financial gains were made from reclaiming illegally occupied land, much of which has increased substantially in value in the recent past. Once cleared of urban poor, characterised as a major obstacle in the path of 'prosperity and development', this land can be sold at great profit to commercial developers who then construct 'luxury' accommodation and shopping centres. The eviction of inner city poverty belts was therefore attributed to the desire to use the cleared land more intensively, where developers could make large profits by doing nothing more than clear¬ing the site and holding the empty land for speculation.

Stakeholder approach

A stakeholder approach can be offered which would involve grass-root co-operation between tenants, resettled people, non-governmental organisations (NGOs), housing experts and local authorities to try and resolve the imbalance between vacant dwellings and families seeking accommodation (Fahmi and Sutton, forthcoming). Such a stakeholder approach would seek to introduce rental reform to meet the needs of the urban poor and young couples; legalise informal occupancy and spread security of housing tenure, so upgrading much inner city informal housing; and foster the rehabilitation of decayed housing stock for urban youth housing, and lease public land for low-income rent-controlled housing. Public–private partnerships of organisational stakeholders such as owner's associations, real estate agents, developers and planning authorities could contribute through better mortgage schemes to finance housing, improved infrastructure in and transport to new outlying settlements, and restrictions on property speculation. Such co-operation could aim to diversify housing production and tackle the phenomenon of the gated community and associated housing segregation through the encouragement of more mixed residential development. These alternative approaches could lead to a better provision of housing for poorer groups suffering from the housing crisis by providing dwellings in districts where they want to live, not by following a bureaucratic resettlement programme.

Acknowledgement

This chapter is based on collaborative work with Keith Sutton at the School of Environment and Development (SED) at the University of Manchester.

Notes

1. Estimated from original figures of 28,600 and 500,000 persons per square mile, respectively.
2. Urbanised villages with over 10,000 population in Egypt have increased from 463 in 1986 to 708 agglomerations in 1996 (Bayat and Denis, 2000).
3. As a result of marginalisation and lack of infrastructure and state control within these settlements, a new spatial division was created, excluding citizens from urban participation, leading to the emergence of Islamic activism. The early 1990s witnessed the intervention of the security forces in the district of Imbaba in Giza which confirmed the negative images of these spontaneous settlements, regarded as loci of extremism and poverty.
4. Narratives were recorded with various primary stakeholder respondents through informal discussions with local residents within New Cairo City, Badr Town and Sheikh Zayed Town, as well as with gate keepers (bawab) of vacant housing units in the Eastern Maadi–Katameya Autostrada development. Secondary stakeholders included housing experts and urban planners from the General Organisation for Physical Planning (GOPP), the Centre for Housing Research and Construction Building and the Housing and Development Bank, in addition to four real estate agents (Al Ahly Real Estate Development Company, El Ta'meer for Real Estate Financing, Coldwell Banker Middle East, E-dar Real Estate Agency). Local authorities and politicians from the Ministry of Housing, Utilities and Urban Communities (MHUUC) and the New Urban Communities Authority (NUCA) expressed the official view concerning New Settlements and New Town housing policies. The Egyptian Centre for Housing Rights, an NGO, provided a critical perspective on official housing policies, whilst addressing the need to consider the social dimension of housing provision.

References

Bayat, A. and Denis, E. (2000) Who is afraid of ashwaiyyat ? Urban change and politics in Egypt. *Environment & Urbanization*, **12**(2): 185–199.

CAPMAS (1996) *General Census of Population, Housing and Buildings 1996*, Cairo Governorate: Central Agency for Population, Mobilization and Statistics.

Degg, M.(1992) The 1992 Cairo Earthquake: Causes, Effects, and Response. *Disasters*, **17**(3): 226–238.

Denis, E. (1997) Urban Planning and Growth in Cairo. *Middle East Report*, **27**(1): 7–12.

El Araby, M. M. (2002) Urban growth and environmental Degradation; the case of Cairo, Egypt. *Cities*, **19**(6): 389–400.

El Batran , M. and Arandel, C. (1998) A shelter of their own: informal settlement expansion in Greater Cairo and government responses. *Environment and Urbanization*, **10**(1): 217–232.

El Noshokaty, A. (2002) *Promised Cities*, Al-Ahram Weekly Online, 5–11 September, 602.

Fahmi, W. S. (2005) The impact of privatization of solid waste management on the Zabaleen garbage collectors of Cairo. *Environment and Urbanization*, **17**(2): 155–170.

Fahmi, W. and Sutton, K. (forthcoming) Vacant Dwellings in Greater Cairo's Housing Crisis: An Anomaly in a World Mega-City (submitted to *Cities*).

Fahmi, W. S. and Sutton, K. (2006) 'Cairo's Zabaleen Garbage Recyclers: Multi-Nationals' Takeover and State Relocation Plans. *Habitat International*, **30**(4): 809–837.

General Organisation for Physical Planning (GOPP)/ Institut d'Amènagement et d'Urbanism de la Région d'île-de-France (IAURIF) (1990) *Upgrading and Enhancing Central Districts of Cairo- North Gamalia Project*, Implementation File, Final Report, Greater Cairo Region Long Range Urban Development Master Scheme. Cairo: Ministry of Development, New Communities, Housing and Public Utilities.

General Organisation for Physical Planning (GOPP)/ Institut d'Amènagement et d'Urbanism de la Région d'île-de-France (IAURIF) (1988) *The Implementation of the Homogeneous Sector Concept- Homogeneous Sector No 1*, Greater Cairo, Region Long Range Urban Development Master Scheme, Cairo: Ministry of Development, New Communities, Housing and Public Utilities.

General Organisation for Physical Planning (GOPP)/ Institut d'Amènagement et d'Urbanism de la Région d'île-de-France (IAURIF) (1991) *Greater Cairo Region Master Scheme*, Implementation Assessment. Updating Proposals, Cairo: Ministry of Development, New Communities, Housing and Public Utilities.

Kuppinger, P. (2004)Exclusive Greenery: New gated Communities in Cairo. *City & Society*, **16**(2): 35–61.

Mitchell, T. (1999) *Dreamland: The Neoliberalism of Your Desires*, Middle East Report, <http://www.merip.org/mer/mer210/mitch.htm> Accessed 17th October 2006.

Ministry of Housing and Reconstruction Report (2002) 21 years of Achievement, New Urban Communities Authority, Retrieved in September 2006 from: www.nuca.com.eg/root/general/map_of_cities.html.

Raafat, S. (2003) *Cairo, the glory years: Who built what, when, why and for whom*, Harpocrates Publishing, Alexandria.

Stewart, D. (1996) Cities in the Desert: the Egyptian New Towns Program. *Annals of the Association of American Geographers*, **86**(3): 460–479.

Sutton, K. and Fahmi, W. (2001) Cairo's Urban Growth and Strategic Master Plans in the Light of Egypt's 1996 Population Census Results. *Cities*, **18**(3): 135–149.

Sutton, K. and Fahmi, W. (2002a) Rehabilitation of Historical Cairo. *Habitat International*, **26**(1): 73–93.

Sutton, K. and Fahmi, W. (2002b) Cairo's 'Cities of the Dead': The Myths, Problems, and Future of a Unique Squatter Settlement. *The Arab World Geographer*, **5**(1): 1–21.

18

Nuttinee Karnchanaporn and Apiradee Kasemsook

'World-Class' Living?

Introduction

Bangkok's Ring Road probably conjures up an image of a ten-lane arterial road with speeding vehicles and muddy roadsides. However, with the encircling of Bangkok by the inner and outer ring roads, the positive prospect of the city's urban growth becomes apparent. Billboards advertising grandiose urban living utopias abound, offering frequent invitations to buy into a lifestyle that does not as yet exist. The rapid increase in the numbers of Thais moving to the suburbs and into secure residential enclaves has contributed to Bangkok's fragmented urban expansion. It is a form of urbanisation that runs counter to many of the aims of good urban design and social sustainability, yet is one that fulfils the ambitions and dreams of many Thais, and invites a more complex account of their motives and values (Fig. 1).

Fig. 1.
Photographs
showing a scene
of the western
ring road and
a close up view
on advertising
billboards of
some gated
communities
contrasted with a
melon seller
Source:
Karnchanaporn
and Kasemsook

Billboards

Housing estate companies have employed a number of strategies to persuade people to buy. And they add value to their products by offering a choice of design, decoration and furnishing. Although the range of choice is limited, these companies offer guaranteed contentment with a variety of full-scale mock-up houses, or what they have called 'sample houses.' Urban middle class Thais' obsession with 'the commodity-driven image of modernity' (Askew, 2002) manifests an obsession

with displaying affluence and well-being through consumption. Design, size, and private facilities (such as leisure centres or even private golf courses) become icons of 'exclusivity' to attract homebuyers. Private security arrangements become another selling point. Advertisements of these private housing estates tout a dream-like image of urban middle class living as much as they do their security features (Fig. 2).

The current trend of housing estates is graphically exposed on billboards that advertise these soon-to-be-built grandiose urban living utopias. They are there to suggest the future and a freedom offered through private consumption. The billboards display houses and their prices in private housing estates which could fetch up to one million USD (the average price is 100,000 USD), while the national population average annual income is approximately 1,215 USD (Department of Community Development, 2007), 82 times lower than the average house price.[1] The billboards are an overt symbol of their investment, intended to carefully nurture the idea (and selling point) of a 'hospitality-provided private home' that coexists with 'connections to vibrant urban activities' (Karnchanaporn, 2006; Fig.3).

The term 'urban town home' has been adopted as the language of the time. The billboards put private privilege (sports clubs, swimming pools, cafés, convenience stores, and even international schools) exclusively provided inside the housing estate ahead of any communal amenities. Indeed, 'connection to the BTS (Bangkok Transit System)' becomes almost the official line on billboards advertising housing estates along the Ring Roads, to ensure they appear to have the most convenient connection to 'urban' Bangkok. Several private housing estates guarantee that the buyers can buy into the 'real' living atmosphere without even having to bother viewing the sample houses. The economic logic is to sell the house as a finished product: – for example a home next to a lively park located in a secured estate with a sports club – which one could own and move into in as little as six days.

To an extent, the billboards and marketing begins to target a particular potential group of buyers: the working singles who are the main driving force of Bangkok's economy. Enticed by a promotional slogan 'You don't have to wait until turning 40 to own a house, you can own one now', here, the ultimate lifestyle goal need wait no longer (Fig. 4).

Fig. 2. Photographs showing several billboards of gated communities depicting family dreams, security, and house prices Source: Karnchanaporn and Kasemsook

Fig. 3.
A marketing
billboard
depicting
hospitality
provided in an
affluent housing
estate
Source:
Karnchanaporn
and Kasemsook

Living in Bangkok

Living in Bangkok was, and seems to continue to be, road-oriented, and urbanisation has always been totally dependent on road construction. The need to deal with Bangkok's urbanisation began around the 1960s, when the first city master plan was produced by Browne, Litchfield and Associates (Bangkok Metropolitan Authority, 1960). By that time the historical city (Rattanakosin) was the centre of Bangkok, and most of Bangkok's urban areas were in its immediate surroundings. The pattern of urban growth concentrated along two axes: towards the north with Donmuang International Airport as an anchor, and towards the east, with the port being another anchor (Fig. 5). The growth pattern was overlaid on the original river-and-canal-based spatial system, and marked a significant transition to a road-based spatial structure.

The speed of urbanisation in Bangkok accelerated in 1980s when the country's economy prospered and the elevated toll ways (the Expressways) were built. The

Fig. 4. A
marketing
billboard aiming
for young urban
worker
Source:
Karnchanaporn
and Kasemsook

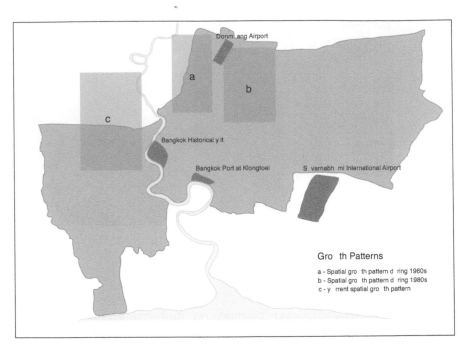

Fig. 5. A map showing urban growth patterns of Bangkok in three periods
Source: Karnchanaporn and Kasemsook

daytime population of Bangkok increased to 10 million, including commuters.[2] The urban middle class emerged as a new social class, and Bangkok appeared as the 'stellar' city of the whole country. With the large and growing population numbers, there came a huge demand for housing. As a consequence, urban areas expanded in all directions, and numerous housing estates were developed, privately or state-owned. The development of these housing estates has taken on a unique spatial form, as well as distinct social characteristics.

Spatial studies by Kasemsook (2003, 2007) on the Bangkok road network indicate some association patterns between the city's expansion and local grid development (Fig. 5). Originally, the spatial network of the city was laid out in a grid or quasi-grid pattern with various alignments. When the city grew, the local grid pattern became more fragmented and had a broken grid structure. This was a result of local area development in a linear fashion within the interstices between the major canals and arterial roads. In most areas, the local grid was made up of a series of short roads, often connected at right angles to an arterial road and which connected to very few other roads or were dead-ends (Fig. 6). This kind of development creates the local areas' spatial segregation. Despite the spatial segregation, inhabitants of the areas were socially integrated. The well-off and the poor, the educated and the illiterate, and the civil servant and the layman could live next to each other.

Spatial segregation became prominent when the expansion of the city intensified towards the periphery. But now it takes on a new form. Most of the local areas are made up of the quasi-grid structure, but they hardly ever connect to each other. Thus the local grid has an island-like form, with each island sequentially connected to the arterial roads by a small number of other roads. The islands themselves, generally, form the housing estates of the urban middle class.

In contrast to the emerging middle class, traditional Thai families were extended families where generations lived together, engaged predominantly in agricultural production. The traditional Thai domestic domain featured not only the house but also included other buildings related to agricultural production within a defined perimeter. Similar to other societies, changes from the extended family to the nuclear family often reflect changes in the economy, i.e., from agricultural production to industrial production (Tiptus, 1982). This is a result of the migration of rural people to the city seeking work opportunities and economic prosperity, and therefore leaving their extended families. Askew (2002, pp. 84–85) indicates that the 'Bangkok-centred trends of development' have given rise to a particular class of Thais: 'the urban middle class'.[3] The urban middle classes consist mainly of Thai, Chinese-Thai, and include a population that has migrated from rural to urban areas. They have benefited from development and have become a major element of Bangkok society. For them, the city is a site of 'better opportunities': higher income resources, conspicuous consumption, social mobility and improved social status (ibid.). Since the First National Development Plan, an increasing proportion of Thais have been able to buy their own home. The private housing estates appear to have a distinct spatial development – an allotted village, which

Fig. 6. Maps showing typical spatial structures in the recently developed urban area of Bangkok Source: Karnchanaporn and Kasemsook

was a product of changes in the building industry during the 1970s and increased land prices in the inner city and older suburban areas. Increasingly accessible through the expanding road network, land parcels in the cheaper farmland of the urban fringe were purchased by developers and subdivided into small blocks for ready-made housing. Boosted by the increased availability of financing at attractive terms through commercial banks and finance houses, many new companies took to the housing estate idea with zeal, responding to the demand for accommodation in the growing city (ibid., p. 64).

For the urban middle classes, a house in an allotted village is an ideal place. It represents a dream of the modernised Thai nuclear family: a new home in the Bangkok suburbs from which parents can commute to work and take children to schools (ibid.).

Within the allotted village, land parcels or houses within land parcels were offered for sale. The perimeter of the villages were not well-defined. The aim was that many allotted villages should be integrated within the city's spatial network, but this has not been realised: in practice they are not well connected to each other. Despite the spatial segregation of the villages, socially their occupants were still made up of a mix of occupations and income groups. Moreover, on the same sequentially linked road one could find the well-off mingled with the poor. This means accessibility between roads or public space and houses or private space was similar for all people. In these cases there is clearly a segregation or fragmentation of physical form, but this is not reflected, as yet, in social fragmentation. However, with a rise in the phenomenon of gated developments, there is a growing trend towards a social segregation that matches the physical fragmentation of urban form.

Fig. 7. A map showing the recently developed gated communities along ring roads Source: Bangkok Post, Property Focus, 23 Nov. 2006

Fig. 8.
Photographs
showing
entrances of
two gated
communitiess
Source:
Karnchanaporn
and Kasemsook

Gated communities

Bangkok's urban growth has accelerated with the extensive expressway construction programmes together with a series of ring road expansions intended to consolidate the city centre with the urban periphery. At the same time the mass transit system (Bangkok Transit System – BTS) was built to help ease the city's severe traffic problem. Due to these road construction programmes, and demand from a rapidly growing urban middle class, gated communities have become a significant part of the scene on the periphery. Figure 7 is a record of the locations of recently developed gated communities along the ring roads, showing 40 of them (whilst hundreds more remain unmarked). The roads provide accessibility to the previously underdeveloped land at the city edges whose prices are low enough for commercial housing development.

These gated communities are spatially designed with a cul-de-sac layout. A series of culs-de-sac branch from one to another through the main community roads. Very few connect directly to one another. This means that, topologically, the layouts lack permeability, resulting in limited accessibility between homes. Furthermore, two gated communities could be located next to each other with a high wall separating them – emphasising the characteristic of a spatial enclave. Within the high walls and guarded entrance and the spatially designed cul-de-sac layout, they suggest a spatial segregation of the gated communities that operates at two levels, within the community itself and between the community and the surroundings, even though they are built along the roads designed to integrate the city's areas (Fig. 8).

Recent interviews with 20 gated communities' residents in Bangkok and the vicinity area suggest numerous reasons for their decision to choose to live there, yet two of them stand out – namely security and social homogeneity (Karnchanaporn, 2005, 2006). A review of the house prices and household incomes clearly indicate that only the well-off can afford a house in these gated communities. A key characteristic of this group of people is that the majority of them are well-travelled, visiting many places in the world, or aspiring to go around the world. So why do they want to be excluded from their local context while buying into the global images of 'world class' living? What does 'world class' living mean to them? (Fig. 9).

Beyond the gates

Visiting, for example, a high-end housing estate to the north of the Ring Road, visitors pass through the guarded entrance – a check point with an elaborate security procedure – a request for a valid identification card in exchange for a visitor's pass,

Fig. 9.
A photograph
showing a
billboard
displaying a
world-class
living campaign
Source:
Karnchanaporn
and Kasemsook

asking for the house number to be visited, a check of the vehicle's rear boot (or trunk i.e. the rear compartment of the car), a radio check for every turn before reaching the destination, an acknowledgment by the house owner, a visit time-schedule, and a double-check at the vehicle's rear boot upon departure from the estate. What is found in the estate is a famous café, a supermarket carrying expensive imported products, an Olympic size swimming pool with club house and shopping arcade adjacent to a man-made lake, and a children's playground which has adult supervision all the time. All houses face a 'public' community space and at the same time maintain living privacy. Just a step outside the gate, high walls fence off intruders, or, to be exact, the locals: a sharp contrast between the local underdeveloped but green area and the well-maintained landscape, and between the lively local neighbourhood and the hygienic estate (Figs 10a and 10b).

In Bangkok the story of the housing estates and the gated communities raises several questions – what happens, when blinded by the images of 'world class' living, the other side of the local living experience can no longer be perceived? But most importantly to what extent can what is seen through the medium of the billboard and marketing dreams be made intelligible? The housing estate market operates, as reading the billboard advertisements suggest, in a similar way to other products aiming to sell 'self-improvement' by means of the quality of life and social status signified by a luxury living atmosphere. It is not only the ownership of the house that is achieved but also the totality of the experience created by a private and exclusive environment. The terms used for naming the housing estates such as Urban Villa, Park Royal, Elegance Villa, Golden Heritage, the Emperor, and Metro Park are all expressive statements, stressing the image of luxury urban living yet to happen (Karnchanaporn, 2006). These kinds of residence and lifestyle were unprecedented and completely at variance with the ways of living in traditional Thai society. Evidently, the will to resist western subjugation was not strong enough to maintain traditional ways, thus moving Thailand towards a perceived 'world class' status represented by a diverse iconography drawn from the west. In the case of Thai homebuyers, to consume and to fabricate the house that fits with a 'world-class' image is to appropriate a higher social status, hence making the image of world-class living marketable through advertisements.

Depicted by the billboards along the ring road, the current advertised trend emphasizes the sales pitch 'buy-into-the real' as a ready-to-move-into house. The 'house and its interiors' has become a product which is more real than ever: one which you can walk in and experience, one which is already tried and tested, and one which is part of the whole exclusive experience of living in a secured or gated community. Daily viewing of the billboard advertisements continually reinforces the 'dreamworld' which the housing estates create. The images of comfortable urban living presented on the billboards ensure that home-buyers should never be allowed to 'awake' from their dreams.

The dreamworld idea is consolidated by spatial and social segregation. On the surface, the gated community seems to pre-select a homogeneous social group whose members share similar values and personal characteristics. However, the spatial development reveals a deep preference. No longer is there a need to visit one's local neighbour – rather the preferred socialisation is within one's own enclave. The enclave takes an expansive form, from house to immediate neighbourhood and to the enclave's 'community'. The enclave itself is equipped with all the modern amenities and is security-provided. To some extent, its inhabitants are better connected with the global network than with the local one. It could be said that the convenient accessibility provided by the ring road and the expressway construction

Fig. 10a. (below upper-right) Location map of Nichada Thani

Fig. 10b (below) Nichada Thani, an affluent gated housing estate with an international school, a Starbuck coffee café, a high-class supermarket, an Olympic-size swimming pool, etc., located on a ring road of the greater metropolitan area of Bangkok Source: Karnchanaporn and Kasemsook

Key

1-20, 23-35 Single-house sub-compounds, some with private pool
12-13 Apartment units
21-22 Townhouses with common pool, gym, playground and park
Total number of units 868 units

A Nichada Thani head offices, Customer Relations & Security Office
B Dental and Family Health Clinics
C Nichada Club, Clark Hatch Nichada Club & Shopping Plaza
D Chaengwattana Community Church
E Samakee Gate Police Substation
F Post Box
G Rose-Marie Academy
H Nichada Sales & Rental Office
I International School Bangkok
J Nichada Plant Nursery
K Sunshine Police Substation

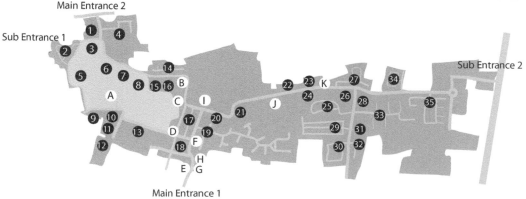

programmes turns to be a tool encouraging the urban middle class to gather in such enclaves. Thus the city becomes an illusion of 'world class' luxury living, without an interest to mingle with the locality.

In the other words, gated communities and their development are clearly physically and socially segregated. The form is self-reinforcing: the provision of infrastructure, roads and expressways will not help to integrate, but rather to connect 'world class' globalised housing with 'world-class' globalised areas on the city of Bangkok, and perhaps eventually to world cities through the new Bangkok airport. There is inherent in the form, in the illusions promoted by developers, and the dreams of the middle class, the creation (reinforcing) of unsustainable urban fragmentation both socially and physically.

Notes

1. The latest average annual individual income published by Department of Community Development, Ministry of Interior in June 2007 is of 43,737 Baht, which is equal to 1,215 USD or 101 USD/month – calculated by exchange rate of 36 Baht/ 1 USD.
2. Bangkok population statistical data was provided by Department of Provincial Administration , Ministry of Interior, on their website www.dopa.go.th (and an interview given by the Deputy Secretary, Office of the National Economic and Social Development Board on 28 June 2006, on 'Bangkok population sectors' in a report of Bangkok's Liveability: Who will answer?
3. Marc Askew (2002) names the social and economic changes that have been geared towards benefits of Bangkok as a capital rather than the rest of the country, whether in terms of the number of schools, universities, hospitals, medical practitioners, wages or salaries as the 'Bangkok-centred trends of development'.

References

Askew, M. (2002) *Bangkok: Place, Practice and Representation*, Routledge, New York and London.

Bangkok Metropolitan Authority (1960) *Bangkok Plan*, Bangkok Metropolitan Authority, Bangkok.

Bangkok Post (2007) *Economic Review*, Jan. 2007.

Department of Community Development (2007) *Average annual income survey of Thailand*, Ministry of Interior, Bangkok.

Karnchanaporn, N. (2006) *Cultivating the Home: A Study of Thai Dwelling Place with reference to Domestic Interior from the First National Development Plan (1961–66) to the Present, A Case Study of Bangkok*, Ongoing research (2006–2008) sponsored by Thailand research fund.

Karnchanaporn, N. (2005) *Fear as a Cultural Phenomenon in Thailand with Special Reference to the Spatial Relations of Domestic Architecture*, an Unpublished PhD Thesis, Architectural Association Graduate School, London.

Kasemsook, A. (2007) The Configuration Maps of Bangkok. *Nar-Jua: Journal of Architecture*, **22**, Faculty of Architecture, Silpakorn University, Bangkok

Kasemsook, A. (2003) Spatial Structure of the Bangkok Road Network, in *Spatial Layout and Functional Patterns of Urban Areas: A Case Study of Bangkok*, Unpublished PhD Thesis, University College London, London.

Tiptus, P. (1982) *Houses in Bangkok: Characters and Changes During the Last 200 Years*, Chulalongkorn University, Bangkok.

19

Wanpen Charoentrakulpeeti and Willi Zimmermann

Staunchly Middle-class Travel Behaviour:

Bangkok's struggle to achieve a successful transport system

Introduction

Different archetypes have been used in the past to rank cities, such as the City of God or the City of Satan, the City as a republic (Sternberger, 1985), to name but a few. Nowadays, metro-regions have been re-invented, and major cities categorised under labels such as 'Alpha-, Beta- or 'Gamma-cities'. Such concepts have become fashionable; they seem to exert a specific appeal to society and economy, to politicians and scientists. Bangkok is classified as one of the 'Gamma' world cities (Beaverstock *et al.*, 1999), but when it comes to a consideration of transport, these classifications in general and for Bangkok in particular, may not have much meaning.

World wide, road transport volumes are rising in tandem with economic growth; enthusiasm for spending on public transportation is not high and taxes on cars or gasoline which may curb traffic scare voters. Despite cleaner use of more environmentally friendly fuels and cars, emissions from transport are rising in nearly every European city, country and across the globe (EEA, 2006a). Concerning these problems, the OECD (2006a) speaks of nine dilemmas, several of which are also true for the Bangkok Metropolitan Area (BMA), for example:

- Passenger transport volumes have paralleled economic growth;
- Harmful emissions decline but air quality problems require continued attention;
- Price structures in OECD countries are increasingly aligned with, yet well below, external cost levels (this is particularly true for Bangkok where fuel costs have been kept 'artificially low').

The causes that lead to these dilemmas are multiple (see also EEA, 2006b). In the case of the BMA, we argue that one of the causes is to be found in the double 'structure and agency' (Giddens, 1984); i.e. in the transport system supply (mainly in form of large technical systems) in conjunction with the action orientation and behaviour of residents and commuters (Kluge and Scheele, 2003). In Bangkok the major drivers of the dynamics of the double structure are:

- *Supply structure* in the form of economic growth and rising standards of living, the supply of roads, low cost fuel and poor public transport;
- *Demand agencies* such as private car ownership, lifestyles and attitudes that fit with the current transport infrastructure and policy.

The remedies are many. OECD (2006b) recommends decoupling negative impacts of transportation from economic growth, and searching for the balanced city (OECD,

2006a). Bangkok has tried to find its own solution. The following sections examine the preferences of residents of three different density zones in Bangkok, and the transport infrastructure is mainly provided for private means of transportation. The analysis shows that structure and agency are largely complementary; both are based on a middle-class structure and its behaviour, and agency orientation. Last but not least, the chapter seeks to find the conditions for creating a 'balanced city' (see also Moenninger, 1999; OECD, 2006a; Ng and Hills, 2003; Donzelot, 2006).

Supply structure: favouring mobility

Several economic factors such as fiscal policy, fuel prices and express way subsidization highly favour (private) mobility. The supply of public mass transport has been comparatively poor. Public bus supply and ticket selling have decreased, tickets have become more expensive and the mass rail systems are too expensive for low income groups. To relieve the transport infrastructure and reduce mobility, sub-centres (polycentric urban forms) have been promoted, yet high mobility, congestion and pollution remain.

Economic growth and rising standards of living

Thailand has been doing well in the last 20 years. It is a middle income country with remarkable progress in human development (e.g. World Bank, 2007; UNEP, 2007). Life expectancy has steadily risen, on average people have 8.5 years of (compulsory) education, unemployment fell to 1.3% in 2005 and household incomes have risen in line with the economic recovery after the crisis of 1997, though pockets of poverty still exist. Three-quarters of Thai households live in their own house on their own land. The provision of transport infrastructure is good; the road network is extensive.

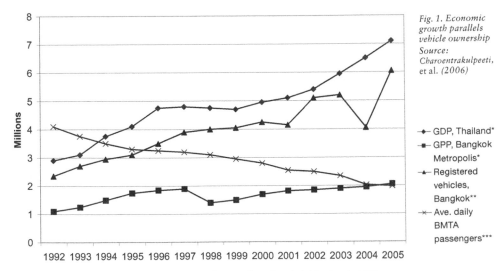

Fig. 1. Economic growth parallels vehicle ownership
Source: *Charoentrakulpeeti, et al. (2006)*

— GDP, Thailand*
— GPP, Bangkok Metropolis*
— Registered vehicles, Bangkok**
— Ave. daily BMTA passengers***

Note: *GDP refers to Gross Domestic Product and GPP to Gross Provincial Product in Million Baht at current market prices (NESDB, 2007)
**Data from DLT (Department of Land Transport, 2006)
***Air-condition and non-air-condition public buses operated by BMTA, 2006

Table 1. Annual public passenger volumes in Bangkok between 2003 and 2005
Source: Statistics of Traffic, Bangkok (2006)

Public transport mode	2003	2004	2005
Bus	871,113,380	753,842,530	713,990,735
BTS (Skytrain)	102,346,231	115,375,347	127,349,940
MRT (Metro)	–	–	59,618,300
Train	18,418,270	17,536,892	17,394,562

The GDP of the Kingdom and the GPP of the Bangkok Metropolis have steadily grown (see Fig. 1 above; all in current market prices). Economic growth has paralleled vehicle ownership in the Metropolis. However, the number of passengers transported by the Bangkok Metropolitan Transit Authority (BMTA) has declined. The bus is the dominant mode of public transport, carrying approximately 78% of total public transport passengers. Bus passenger numbers have declined while those using trains have increased, due to the relatively new mass rapid transit systems – the Skytrain and the Metro. In 2005, total annual public transport passengers numbered 918 million (see Table 1). Economic growth and increasing passenger transport need a corresponding infrastructure.

Supply of roads

Road length in Bangkok has increased from 2,785km in 1986 to 4,076km in 1999 (Table 2). While the road surface areas rose from 19.25km² in 1986 to 38.71km² in 1999, the number of vehicles went up significantly, to 4,162,846 in 1999, a threefold increase. Road capacity increased 1.5 times: from 71,990 vehicles per km² in 1986 to 107,539 vehicles per km² in 1999. At present (2007) the road length in Bangkok and the five adjacent provinces (including Samutsakorn, Samutprakarn, Pathumtani, Nontaburi, and Nakonpatom) is approximately 4,700km (BMA, 2007), covering a surface area of 34.52km² (BMA and UNEP, 2003, p. 15). Bangkok has only 11 % of urbanized land under roads, which compares 'poorly' with typical Western figures of around 20% (Orn, 2002, p. 9).

Table 2. Road length, road surface and numbers of vehicle registered in Bangkok compared between 1986 and 1999
Sources:
*Department of Public Work, BMA, 2000
**Department of Land Transport (DLT, 2004), Ministry of Transport

Bangkok's road network is not quite adequate to the demand. It is a radial system with insufficient circumferential roads. The growth of private car ownership (see below) and the comparatively low road capacity inevitably lead to chronic traffic congestion. In order to alleviate congestion the Expressway and Rapid Transit Authority of Thailand (ETA) established a Master Plan for expressways in Bangkok and its vicinity in 1981, consisting of five Expressway projects. They have been accomplished, with a total length of approximately 175.9km; they serve 1.2 million vehicles per day (ETA, 2005). The 'third stage expressway system' is under construction.

District	Road length (km)*		Road surface areas (km²)*		Numbers of vehicles registered (unit)**	
	1986	1999	1986	1999	1986	1999
Bangkok	2,785.01	4,076.13	19.25260	38.71266	1,385,801	4,162,846

Preference for roads

New transport investment – especially in motorways – provides a powerful stimulant to higher mobility as well as to urban sprawl. The budget for land transport (from 2001 to 2004) favours roads more than the railway systems – the budget for roads, bridges and expressways increased from 8,605 to 10,899 million Baht, whereas the budget for the rail mass transit budget increased from 3,195.8 to 4,975.8 million Baht.

Fiscal preferences for private transportation

In the period 1992–1996, the national economy grew rapidly and led to an increase of purchasing power and of numbers of vehicles enhanced by a reduction in the tax on vehicles in 1992. This modification of the tax allowed more people to buy new cars. The number of car licenses rose from 9.59 million in 1992 to 14.44 million in 1995 (Chanchaona *et al.*, 1997). A further incentive was provided in July 2004, when the Fiscal Policy Office (FPO, 2005) decreased taxes again, in essence lowering the registration fee, license plate and annual vehicle fee on small and fuel-saving car engines. Only drivers of cars with large engines (pick-ups and double cab vehicles) had to pay more. Taxes on motorcycles were slightly reduced. Tax reductions were not the only 'gift' that the government made to owners of private cars. They also benefited from fixed fuel prices until 2005.

Floating fuel prices and car ownership

The US–Iraq conflict led to higher oil prices which prompted the government into taking drastic measures. On 8 February 2003, the Energy Fund Administration started a price-stabilizing programme. Retail prices for gasoline, and especially diesel prices, were kept stable to 'lighten the burden' on the population, including farmers and small family enterprises (EPPO, 2005a; 2005b). In 2005, the government decided to open all gasoline to higher market prices. Floating the fuel price has had little impact on car ownership – 485 were registered per day in 2004: in 2005 the corresponding figure was 684 (OTP, 2005b). It seems that rising fuel prices have done little to reduce the demand for car ownership.

Express way subsidization

By 2005, many car friendly transport projects were operational; among them several new express and toll-ways, as well as the middle and outer ring roads of Bangkok. The supply of roads has increased and thus will further attract private car users due to a lack of other means of transport. To ease congestion and corresponding economic losses, as well as the burdens on car users, toll-way pricing has been subsidized for a while. Up to December 2004 motorists paid around 30 to 43 Baht per toll-way segment, a price already pegged below cost recovery. After that date, toll-way fees for the crucial Vibhavadee road were actually set at a flat rate of 20 Baht. The differential costs have been subsidized by the government, amounting to about 30 million Baht per month (Krungthep Turakij, 2005). This policy favours the middle and upper classes which could afford a car and the toll-way fees.

Inadequate public mass transport

There are several types of public bus in Bangkok, including regular, air-conditioned and joint service buses of the Bangkok Mass Transit Authority (BMTA) and joint service buses, minibuses, public vans and small buses plying lanes (bus routes within community and local roads i.e. in Thai 'soi'). BMTA serves bus routes in Bangkok and neighbouring provinces; in February 2005 it operated a total of 102 routes by approximately 1,673 regular and 1,905 air-conditioned buses. There are also privately-owned buses operated under the BMTA, totalling 3,415 regular and air-conditioned buses and 1,118 minibuses as well as 2,077 small buses providing services in the 'sois'.

The 'fleet' has undergone some changes in the recent past as Fig. 2 shows. BMTA has promoted a policy of privatizing bus services to companies operating either as joint ventures or to private companies using mini-vans. The number of minibuses decreased from 1997 to 2005 after public mini-vans were successfully introduced in 2000. Passengers' preferences have changed in favour of mini-vans due to their flexibility in routes and stops, although the price per ticket may be slightly higher than that of the bus. Public mini-vans can save travel time compared to buses, because almost all of them use the toll-ways, and thus they have many of the advantages of private cars. However, since travel costs of a public mini-van and private car are not very different (as indicated by the respondents of the survey, see below), many residents, especially the middle classes, prefer to buy private cars or motorcycles, thus generating more traffic and continuing to contribute to the seemingly 'endless cycle of traffic defeat'.

Public bus fares[1] were quite low before the fuel crisis. After the announcement of fuel price floatation BMTA increased ticket prices twice in 2005. Passengers faced a 10–100 percent increase; those using regular buses had to pay between six to eight Baht per trip;[2] those using air-conditioned buses paid between 12–22 Baht. Since Bangkok does not have a single ticket for all public transport, the passengers need to pay for each connection which increases ticket costs; a trip often may cost as much as a litre of diesel (22.99 Baht) or a litre of petrol (25.74 Baht). In addition, during the rush hours, more air-conditioned than regular buses are operating, forcing

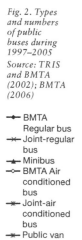

Fig. 2. Types and numbers of public buses during 1997–2005

Source: TRIS and BMTA (2002); BMTA (2006)

—◆— BMTA Regular bus
—✳— Joint-regular bus
—▲— Minibus
—◇— BMTA Air conditioned bus
—✳— Joint-air conditioned bus
—✳— Public van

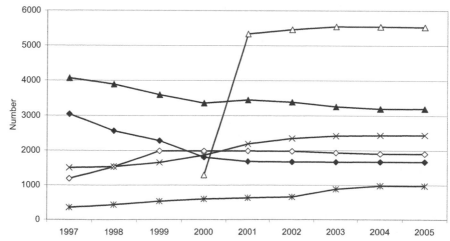

Note: 2005 data valid until June / Public vans were recorded for the first time in 2000

passengers who lack other alternatives to use the more expensive air conditioned buses. The shift of low income passengers to air-conditioned buses makes them 'crowded' and uncomfortable and the expense of 14 Baht per trip, rather than 6 Baht, may strain the budget of people with a monthly income of below 7,000 Baht. BMTA policy has caused a considerable decrease of ticket sales for regular buses; from almost 1,000 million in 1997 to 400 million in 2004. Tickets for air-conditioned buses rose from 250 million in 1997 to 300 million in 2004 (TRIS and BMTA, 2002, 16; BMTA, 2006).

To service road demand and relieve congestion, several transport projects have been proposed and implemented i.e. the elevated rail system (Skytrain) and the Underground Mass Rapid Transit System (MRT). These projects are the product of more concerted efforts by 'NESDB[3] technocrats' and by the BMA. The approaches have both been more professional and responsive to the growing demands of the middle-class Bangkok electoral constituency (Askew, 2002, 90); these projects also stimulate more mobility (Hirsch, 1977; Goodwin, 1996; BMA and UNEP, 2001, p. 21). Critics suggest that these approaches reflect a policy dominated by the élite, that favours middle-classes in relation to transport (see e.g. Tadiar, 1995; Christensen, 1993 in Askew, 2002, p. 84). Tickets are too expensive for low income groups and the number of passengers has been below target. In the face of the evidence presented above it seems that the projects deal with symptoms only. Mobility, congestion, the use of energy and pollution (noise, small particulate matter, Health Institute, 2004) continually increase. It is clear that the private car has no real competitor in the form of public transport, either by bus or the new rail mass transit systems (see also below). This situation may not improve significantly, since the means to balance the city, i.e. by the construction of sub-centres, has not been very successful.

Sub-centres are no alternative

The key strategy of the Bangkok Plan of 2002 for restructuring the city's urbanizing fringe is the construction of a system of metropolitan sub-centres. There are five metropolitan sub-centres in Bangkok: Lad Krabang in the east served the new international airport and industrial estate, Taling Chan in the west for a combination of office, light industry, commerce and housing, Bang Khun Tian in the southwest has potential for light industrial and service uses along with high density housing, Minburee in the east, for commercial nodes and mixed land use, and Lam Lukka in the north has the potential to support a large centre integrating service and employment uses to serve the surrounding areas. These five sub-centres have been promoted through the development of housing projects and the manufacturing sector. High intensity of land utilization, high land prices, low accessibility, and the bad traffic conditions of Bangkok are factors that have contributed to the creation of these sub-centres. Yet the lack of a systematic and interconnected transport network makes the government bear the burden of transportation and public works investment. Moreover, those who live in the outer areas of Bangkok or in its vicinity have to cover (on average) longer distances to reach the inner city, often using private cars to save travel time and enjoy a higher level of comfort, due to the lack of effective and efficient public transport services (OECD, 1998; Suthiranart, 2001; Surasawadi, 2000; Poboon, 1997). Thus the sub-centres policy for Bangkok did not succeed in terms of balancing the city. This is confirmed by the following survey carried out with residents from three different density zones in Bangkok.[4]

Demand side: middle class oriented

While the supply side is car oriented, favouring the use of private means of transport, demand for private car ownership is not only important, but often a necessity. Car and vehicle ownership increased dramatically after the Second World War. Vehicle ownership and income per capita in the 1990s are quite revealing. Comparisons of these two variables in different cities in the Asia Pacific region, show that low and middle income cities such as Bangkok and Kuala Lumpur have similar numbers of car ownership per 1,000 persons as those of established industrialized cities like Tokyo (Marcotullio, 2005).

The number of vehicle licenses in Thailand and Bangkok issued from 1993 to 2004 increased from approximately 3 million in 1993 to 5.5 million in 2003, followed by a decrease to 4.3 million in 2004 due to the fuel crisis (OTP, 2005a). Motor-cycles have been one of the favourite means of transport in Bangkok. The number of registered motorcycles is higher than that of passenger cars (carrying up to seven persons) for the period 1989–2004. At the end of that period the City of Bangkok had 1,526,417 registered cars and 1,593,685 motorcycles. Both had grown since 1989, with some decline in times of economic crisis (DLT, 2004, 2006).

The following sections report on research carried out in August 2002 which collected activity-based data on a sample of 426 respondents. The research analyzed the travel patterns and attitudes and preferences of residents in zones of different densities in Bangkok.[5] The different density zones are selected to test whether or not the city is balanced. In this respect the project produced empirical data for the first time. The data reveal that the residents commute a lot, both within a zone or across a zone to go to work; thus, the zones and the city are not balanced in terms of providing both a working and a living location. Car ownership suits the dominant middle-class paradigm. Whatever the parameter chosen, attitudes and preferences related, for example, to residence or income, the answers about increasing mass transport or reducing private transport by means of taxes remain fairly homogeneous. Even the opinions of the lowest income group do not differ much from those of the highest.

Travel patterns

The results below refer to a typical workday, i.e. Monday, and demonstrate high use of the transportation infrastructure by means of private cars. Car ownership of the 426 respondents is high; each household consists on average of four members with one car.

Of the 426 respondents, 42% were female and 58% were male, on average 35 years old. Almost 66% of the respondents consider themselves to be head of the household; the others are either a parent of the head or a young adult, at least 18 years old. The majority of the respondents hold a bachelor's degree (54%). 85% of the respondents have jobs (91% of them are regular jobs), the others are studying, retired or unemployed. 50% of the respondents work for a private company in which they earn on average between 7,001–15,000 Baht per month;[6] the remaining are either self-employed (22%), government employees (19%) or others (4%) with earnings within the same average salary range as those of employees of a private company.

Transportation plays an important role in the daily lives of the residents. 90% of respondents make one trip per weekday for work or shopping or leisure, which lasts 45 minutes, covers 16km and costs 36 Baht on average. 57% use private means of transport (including motorcycle), the others public transport (30%) or para transit[7] (11%); only very few walk (2%). The most favoured means for those who use private transport is the car 92%, followed by motorcycles 8%. Most of those who use public transport take buses (87%), the sky train (7%) or the train and other (6%). 54% of respondents can access public transport within 500m or 7–8 minutes from their place of residence (on average).

Each form of transport has its advantages and disadvantages. On average, those using private transport take 44 minutes to cover 17km, with costs of 43 Baht per trip. Those who use public transport travel a significantly shorter distance (14km), take more time (47 minutes) and pay significantly less (20 Baht). Private transport is significantly faster (23km/h) than public transport (18km/h). This difference explains the preference of many residents for private transport which also provides more comfort and flexibility in terms of choosing less congested routes, including the toll-ways. The figures above reflect the well-known traffic problems in low-income Asian cities including Bangkok (Kenworthy, 2005; Pianuan *et al.*, 1994; NESDB and DANCED, 1996); the reasons are mainly congestion, low average speed and air and noise pollution caused by a non-balanced city that requires commuting and has strong preferences for private transport.

Density-related travelling

The ANOVA test (LSD comparison of paired areas of different densities) reveals some statistically significant differences between the residents of the three zones in relation to age, education, occupation, number of household members and household cars. The three zones are high density (HDZ), medium density (MDZ) and low density (LDZ). The (on average) characteristics of the residents of the three density zones are as follows:

- HDZ residents are older, live in a bigger household and have more cars per household than the large majority of the residents from the MDZ and the LDZ.
- A higher percentage of HDZ residents is self-employed;
- The respondents from the LDZ are the youngest, and live in the smallest households. The LDZ residents are better educated than the

Fig. 3. (below–left) Average distance, time, and cost of work and education trip (absolute figures)
Source: Charoentrakulpeeti, et al. (2006)

Fig. 4. (below–right) The percentage and average distance for work and educational trips of residents in the three density zones
Source: Charoentrakulpeeti, et al. (2006)

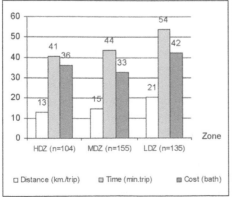

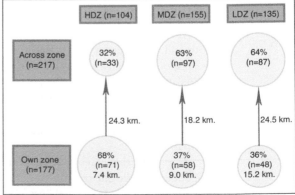

others. Thus the residents of the LDZ reflect the well educated, 'single-family settler' of the suburb.

The HDZ residents have several advantages in terms of a high density of workplaces and roads, and of public transport routes with a higher variety of public transport modes. These factors mean they spend significantly less time, and experience lower distances and costs for regular working and educational trips than those of the other two density zones, except for the travelling costs of respondents from the MDZ (Fig. 3).

Almost all of the residents (90%) of the three zones make regular trips for work and study. It takes residents of the HDZ on average 41 minutes to travel 13km to work, spending 36 Baht. It is less easy for the residents in the MDZ and the LDZ. Respectively, their trip to work takes them on average 44 and 54 minutes to cover a distance of 15/21km, and they pay on average 33/42 Baht.

a) Density zone and travel behaviour based on destination

Many respondents live and work in different places and need transport; this is particularly true for residents from the MDZ and the LDZ (see Fig. 4). There are significant differences between the three density zones and trip destinations at a 99% confidence level. More than two thirds (68%) of those living in the HDZ travel to a destination within their own zone, travelling on average 7.4km; 32% travel 24.3km out of the HDZ (Fig. 4). 37% from the MDZ travel 9km to their destination within their zone and 63% go to another zone, travelling 18.2km. Only 36% from the LDZ travel 15.2km within the LDZ; the other 64% travel 24.5km into another zone.

Fig. 5. Trip distribution of two sampled districts of LDZ Source: Charoentrakulpeeti, et al. (2006)

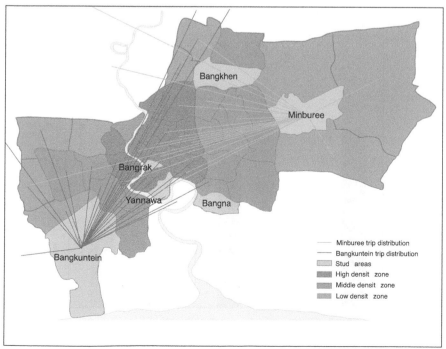

These results indicate a significant relationship between density and destination to work and education, i.e. the HDZ attracts respondents from the MDZ and LDZ. This is not surprising since the HDZ is a major centre of employment, service, telecommunication, commercial and industrial companies with both domestic and international business. However, the 'magnitude' has not been analysed so far; it is visualized in Fig. 5. The figure also illustrates that many travel to other zones, among them the industrial zones to the north and south of Bangkok, far from their residences. These exemplify an unbalanced city (region).

b) Density zone, travel behaviour and modal split

Private means of transport are dominant for the respondents from all three zones followed by public, para transit, and non-motorized transport. The chi-square test shows that there is no significant difference between density and transport mode.

Despite a good public transport system in the HDZ – in principle corresponding to a CBD – with public buses, skytrain, boat, and subways, 65% of the respondents from the HDZ use private vehicles, compared with 55% of the residents from the MDZ and 53% of the LDZ (Fig. 6). In contrast, public transportation is used by 22% of HDZ residents and by one third of the residents of the MDZ (34%) and LDZ (32%). Para transit plays a role for 15% of residents in the LDZ only, while non-motorized transport is negligible in all three zones.

Private transport is generally faster in all zones (except for those residents of the MDZ who use para transit), followed by para transit and public transport. However, given the traffic congestion in the HDZ, it is not surprising that the relative travel speed is slower in that zone for all modes of transport (around 19km/h) costing the

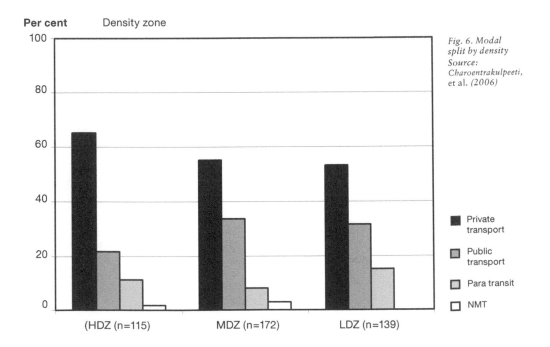

Fig. 6. Modal split by density Source: Charoentrakulpeeti, et al. (2006)

most per km (3.3 Baht in HDZ; 2.7 for the MDZ; 2.6 for the LDZ). Travel speed for public transport is the lowest in all three zones (around 18km/h) while those using private and para transit move faster in the MDZ and LDZ (between 22 and 25km/h).

Who carries the burden?

The travel behaviour of different income groups was analysed. For this purpose two groups have been constructed: the highest (>50,000 baht/month; n = 31 cases = 8%) and lowest income group (≤7,000 baht/month; n = 36 cases = 9%). In the highest income group two thirds are male; 68% are married, on average 42 years old and 81% have a university degree. Slightly more than half of the lowest income group are female; 31% of this group are married; on average 32 years old and 23% have a university degree.

There are no significant differences among these two groups in terms of residence and most respondents in both groups travel to work or for education (90% of highest group, 87% of the lowest). The travel destination differs significantly at 90% significance level: 55% of the highest income group travels outside its zone of residence compared to 33% of the lowest group.

Highly significant differences (99% level) exist also for car ownership, travel time for work or education, cost and distance:

- 87% of the highest income group own a car and 80% use private vehicles to go to work or for education, compared to 14% of car owners in the lowest income group – 31% of them use a private motor vehicle to go to work (14% by car; 17% a motorcycle). 56% of the lowest and 10% of the highest income group uses public transport.
- Respondents from the highest income group travel 24 km taking 53 minutes; the respondents from the lowest income group travel 8km taking 30 minutes (all figures per trip and on average).

The transport behaviour of the lowest income group is arguably more sustainable. They travel less beyond their zone, more use public transport and cover shorter distances. Their contribution to a balanced city is certainly higher. They also carry a heavier burden: They do not have the convenience of a private car and some travel up to two hours one way; many are exposed to air and noise pollution.

Homogeneity of attitudes and preferences?

It has been shown above that the transport infrastructure is conceived mainly for the motor vehicle; Bangkokians heavily depend on it. Many have a life-style (separated residence and work, car ownership) that needs transport infrastructure. Thus, the resulting (in)famous traffic problems of the Bangkok Metropolitan Region. Most governments have wanted to solve the urban traffic problems by trying to convince residents to use public transport. The government has launched several small and large transport projects to improve mass rapid transit, bus and boat feeder services, walking and biking streets. However, little is known about the attitudes of Bangkokians towards different strategies to reduce transport problems and transport-induced environmental pollution. The following sections analyse the attitudes towards public transport and policy change. They reveal that attitudes are fairly homogenous, just like the travel patterns.

Density related attitudes?

Residence and density do not seem to make much difference; respondents from different density zones not only have fairly similar travel patterns but also rank measures that are or could be introduced to reduce mobility in a rather similar way.

a) Density related preferences for using public transport

The Thai government (of the year 2005)[8] and many of those concerned for sustainable development in Thailand (e.g. NGOs, communities, green businesses and banking) propagate the use of public transport. The attitudes towards the use of public transport do not differ significantly between the respondents of the three density zones:

- A maximum of 16% of the MDZ respondents would agree to stop using a private car. Given the lower opportunities of public transport in the LDZ, only 7% would favour such a measure.
- More than 49% of respondents of the three zones might consider changing to public transport but want to keep the option of using the private cars open: between 19% and 30 % do not want any changes.
- Only a few remain undecided (6 to 11%).

b) Density related attitudes towards policy changes

In the effort to decrease environmental problems and energy consumption caused by increasing mobility, a number of measures have been debated in public and

Respondents from Strategy	Level of acceptance	HDZ	MDZ	LDZ
Control pollution emission**	high	0.84	0.98	0.93
Enhance fuel use awareness	high	0.83	0.82	0.81
Set up free pollution zone**	high	0.8	0.62	0.88
Promote clean technology*	high	0.85	0.79	0.67
Promote Bus lane**	high	0.74	0.89	0.88
Promote natural gas*	high	0.53	0.66	0.72
Promote short walk**	high	0.64	0.48	0.59
Introduce carbon tax**	medium	−0.01	0.46	0.38
Introduce road pricing**	medium	−0.31	0.09	−0.29
Increase license plate tax	very low	−0.12	−0.2	−0.17
Increase fuel tax*	very low	−0.14	−0.37	−0.41
Increase parking fee**	very low	−0.13	−0.49	−0.45

Table 3. Attitudes of respondents from different density zones towards strategic measures
Source: Charoentrakulpeeti, et al. (2006)

Note: *Sig=0.05, **Sig=0.01

by government. Results of an assessment of the attitudes towards energy and environmentally relevant measures differ in terms of acceptance and significance. In principle they can be classified into three groups:

- Most favoured measures are those with a value ranging from 0.48 to 0.98[9] (see Table 3). They are characteristically not directly related to costs, but have more to do with air pollution in Bangkok, which affects all residents and commuters. Thus measures like controlling pollution emissions, enhancing fuel use awareness, setting up pollution-free zones and promoting clean technology are very well received, with levels of acceptance of at least 0.62. These measures have been well publicized over the years and several campaigns to fight air pollution have been carried out. In a city that has been promoting private transport, little thought has been given to alternatives; thus the promotion of bus lanes, natural gas fuel and walking are accepted, though slightly less favourably.
- The introduction of the carbon tax is generally rejected by those from the HDZ while accepted by the respondents from the MDZ and LDZ. Similarly controversial is road pricing which is just accepted by respondents from the MDZ but decisively rejected by those from the HDZ and the LDZ.
- Measures that have monetary impacts on users of public and private transport such as an increase of the license fee, fuel tax and parking fees are clearly rejected by all three groups.

Though the classification into three groups of attitudes seems to be clear-cut, a finer analysis reveals that there are nevertheless significant differences between the three groups' respondents.

Respondents from the HDZ accept but are significantly (at 0.01 or 0.05 levels) less positive than those from the other zones about measures such as the control of pollution, the promotion of bus lanes and of natural gas. Since more residents of the HDZ own private cars and they use less public transport, they fear increases in costs as well as a possible reduction of their mobility and access to their residence due to bus lanes. Bus lanes, pollution-free zones and the promotion of natural gas fuel are very well received by residents from the LDZ, for they depend on public transport and are most likely also more sensitive to polluted air. Respondents from the MDZ strongly favour bus lanes and pollution control. However, HDZ residents more strongly accept the promotion of clean technology and walking than the other two groups. Both factors would make their place of residence more comfortable.

Respondents from the MDZ and LDZ are more accepting of the introduction of a carbon tax and road pricing. Both are rejected by the HDZ group – this group would lose most if these measures were implemented, given their high level of car ownership and car use.

Measures that may have cost effects for daily transport, like increased fuel tax and parking fees are rejected by all three groups apart from a higher license fee, although the former two more strongly by respondents from the LDZ and MDZ than from those living in the HDZ. There is a clear overall message: measures that 'directly' hit the pocket of the commuter are seen negatively.

Different income groups, fairly homogeneous preferences

The attitudes and preferences of the highest and lowest income group are compared. It has been noticed that the lowest income group has a more sustainable lifestyle than the well-to-do group. However, the members of these two groups have fairly similar attitudes in that they favor measures which do not hurt 'their pockets' while remaining sceptical or negative towards taxes and road pricing (see Fig. 7). Significant differences at 0.05 level exist for promoting bus free lanes. This is not surprising, given that many of the low income group depend on public transport while many of the high income respondents use their own car and fear losing 'lanes'. The low income group does not coherently speak for or against tax increases because a few are car owners and several use their own motorbike for commuting; others may fear increases of ticket prices for public transport.

Conclusions

Structure and agency are complementary. In the past road infrastructure and the promotion of private transport have been promoted in terms of budget allocation, construction of roads and highways, subsidized fuel and toll-ways. Public transport has been improved, and two, rather expensive, mass rail transit systems have been added. Economic growth and rising standards of living have enabled more residents

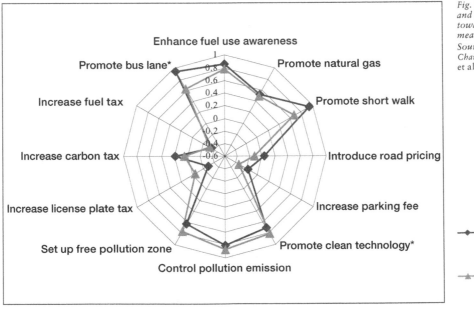

Fig. 7. Income and attitudes towards policy measures[9]
Source: Charoentrakulpeeti, et al. (2006)

Ver negative	egative	Slightly negative	Slightly Positive	Positive	Very Positive
(−1.0) to (−0.80)	(−0.79) to (−0.40)	(−0.39) to <0.00	0.00–0.39	0.40–0.79	0.80–1.0

*Note: *Sig=0.05, **Sig=0.01*

to own private cars. They are considered the primary cause of the expansion of cities and mobility. The suppliers of transportation infrastructure have contributed to the imbalance between public and private transport and to the 'imbalanced unsustainable Bangkok Metropolitan Region'. Policy makers and planners have confirmed in interviews that there is political intervention, lack of knowledge and skills in the major public organizations dealing with transport; furthermore, high organizational and policy fragmentation combined with little continuity as well as the lack of organizational cooperation between land use and transportation policies have hindered the implementation of coherent, sustainable transport and land-use policies and the creation of a polycentric city. Effective mechanisms of horizontal co-ordination between different government departments, between public and private stakeholders and between metropolitan region's municipalities, as well as vertical co-ordination between different levels of government are missing. Alternatives such as bicycle and bus lanes and pollution-free zones have not been offered.

However, the use of private means of transport is also a consequence of the dominant agency paradigm, which is the middle class life-style. Travel patterns, attitudes and preferences of respondents from three different density areas do differ. Whatever the differences, most preference is given to technical solutions, and transport information and planning measures; measures with a 'direct' cost such as increases in the license plate, fuel taxes and parking fees are not preferred. Controversial are the introduction of a carbon tax and road pricing. These middle class' attitudes and policy option preferences underscore a strong predisposition to protect its members' particular stake in car ownership and high dependence on the car for travel.

A structure and agency that is sustainable would need to reduce travel distances, travel time and make public transport more attractive by making car travel less attractive; an increase in urban density, mixed use of land and poly-centricity can only be achieved by a combination of land-use and transport policies. The latter have more direct effects than the former; however, land use policies are needed for the long term change in order to have a spatial organization that is not too dispersed. These policies are important in providing the preconditions for less car-dependent lifestyles in the future. (TRANSPLUS, 2003; Wegener and Fürst, 1999). How easy it would be to persuade the residents of Bangkok of these long-term benefits is another matter.

Notes

1. The Department of Land Transport (DOT) is in charge of ticket fares, bus routes or bus stop locations of both BMTA and joint service buses.
2. 1 US Dollar = 31.4 Baht; 1 Euro = 42.8 Baht; 1 Pound Sterling = 63.2 Baht on 8 July 2007 (www.xe.com/ucc/convert.cgi).
3. National Economic and Social Development Board.
4. The registered population of Bangkok was almost 7 millions in 2005 (BMA 2006), or about 10 millions for day time population (Thavisin and Suwarnarat 1995, 11). Evidence covering the period 1987–2000 shows that the population densities in the inner areas had decreased from 15.27 to 11.09 thousand per km^2 whereas the density in outer areas has increased from 0.77 to 1.28 thousand per km^2 (BMA and UNEP 2001, 14).
5. The research project focuses on the travel behaviour of residents of Bangkok living in areas of different density. The project was funded by the National Energy Policy and Planning Office of Thailand. Registered residents' density ranges from about 390 to almost 39,000 residents per km^2 and registered labour density varies from approximately 529 up to 35,000 per km^2. Accordingly, the authorities classified the districts into inner zone (high-density

zone: HDZ), middle zone (middle-density zone: MDZ), and outer zone (low-density zone: LDZ); this classification has been used in this research project.
6. 43.10 Baht approximately one US Dollar in 2002.
7. Para transit includes public motorcycle, public minivan, taxi and company bus.
8. This relates to PM Thaksin's régime before the military coup of 19th September 2006. Despite the régime change the policy of sustainable development is still valid in Thailand in 2007 since this concept has been promoted by the King himself; it has for example been contained in the current 10th National Plan (2007–2011) issued by the NESDB.
9. The Weight Average Index (WAI) is very often used by practitioners, for example in cases of measuring customer satisfaction. The averages have been scaled on a six-point Likert scale with the following intervals: (–1.0) to (–0.80) refers to very negative, (–0.79) to (–0.40) negative, (–0.39) to (<0.00) slightly negative, (0.00) to (0.39) slightly positive, (0.40) to (0.79) positive, and (0.80) to (1.0) very positive (see Fig. 7). However, in order to make the format better fit with A4 pages and keep the graphs 'readable', the vector will start at -0.6 and end at 1.0; these are respectively the lowest and the highest values that have been achieved in the analysis.

References

Askew, M. (2002) *Bangkok: place, practice and representation*, Routledge, London and New York.
Beaverstock, J.V., Taylor, P. J. and Smith, R. G. (1999) A roster of world cities. *Cities*, 16: 445–458.
BMA (2006) Retrieve from: http://203.155.220.239/en/01about/index.php?intContentTypeID=21&intContent ID=831.
BMA (2007) Retrieve from: http://www.bma.go.th/info/.
BMA and UNEP (2001) *Bangkok: State of the Environment 2001*, Bangkok Metropolitan Administration and United Nations Environment Programme Publications, Bangkok.
BMA and UNEP (2003) *Bangkok: State of the environment 2003*, Bangkok Metropolitan Administration & United Nations Environment Programme, Bangkok.
BMTA (2006) Statistics of Public BMTA bus, Bangkok Mass Transit Authority, Bangkok. Retrieve from: http://www.bmta.co.th/engversion/eng_version.htm.
Chanchaona, S. *et al.* (1997) *A study of strategies for energy conservation in vehicles*, A report for National Energy Policy Organization, King Mongkut's Institute of Technology Thonburi, Bangkok.
Charoentrakulpeeti, W., Sajor, E. and Zimmermann, W. (2006) Individual transport behaviour and attitudes: The case of the middle class. *Transport Reviews*, 26(96): 693–712.
Department of Public Work, BMA (2000) *Bangkok Statistics*, Bangkok Metropolitan Administration, Bangkok.
DLT (2006) *Statistics of Car License during 1982–2005*, Department of Land Transportation, Bangkok. Retrieve from: http://www.otp.go.th/pdf/Statistic/carregist36-48.pdf.
DLT (2004) *Statistics of Car License during 1982–2003*, Department of Land Transportation, Bangkok. Retrieve from: http://www.dlt.go.th/statistics_web/st1/bkk.xls.
Donzelot, J. (2006) *Quand la ville se défait: Quelle politique face à la crise des banlieues?* Paris, Seuil (When cities disintegrate: What policies for the crisis in the suburbs? Title translated by WZ).
EEA (2006a) *Transport and environment: Facing a dilemma*, EEA, Copenhagen, Report No. 3/2006.
EEA (2006b) *Urban sprawl in Europe. The ignored challenge*, EEA, Copenhagen, Report No. 10/2006.
EPPO (2005a) *Bangkok: Energy Policy and Planning Office*. Retrieve from: http://www.eppo.go.th/retail_prices.html.
EPPO (2005b) *Bangkok: Energy Policy and Planning Office*. Retrieve from: http://www.eppo.go.th/retail_changes.html.
ETA (2005) *The annual report the expressway and rapid transit authority of Thailand*, Expressway and Rapid Transit Authority, Bangkok.
FPO, Fiscal Policy Office (2005) Retrieved in May 2005 from Fiscal Policy Office Web Site: http://e-fpo.fpo.go.th/e-tax/Tax_update/data/27july-04.doc.
Giddens, A. (1984) *The constitution of society: Outline of the theory of structuration*, Polity Press, London.
Goodwin, P. (1996) Empirical evidence on induced traffic – a review and synthesis. *Transportation*, 23: 35–54.
Health Effects Institute (2004) *Health effects of outdoor air pollution in developing countries of Asia: A literature review*, Health Effects Institute, Washington.

Hirsch, F. (1977) *Social limits to growth*, Harvard, Cambridge, Mass.

Kenworthy, J. (2005) *Sustainable Urban Transport and Land Use Patterns for More Sustainable Cities in Australia: Some Key Policy Implications from an International Comparative Study.* Retrieved in January 2005 from: http:// www.aph.gov.au/house/committee/environ/cities/subs/sub107.pdf.

Kluge, Th. and Scheele, U. (2003) *Transformationsprozesse in netzgebundenen Infrastruk-turen* (Processes of transformation in infrastructure networks; title translated by WZ), ed. Forschungsverbund networks, Deutsches Institut fur Urbanistik, Berlin.

Krungthep, T. (2005) *The tollways subsidy, Krungthep Turakif newspaper*, Bangkok. Retrieved in July 2005 from: http://www.bangkokbiznews.com/2005/06/17/w010reg_15431.php?news_id=15431.

Marcotullio, P.J. (2005) Shifting drivers of change, time-space telescoping and urban environmental transitions in the Asia-Pacific Region, in *Managing urban futures. sustainability and urban growth in Developing Countries* (eds M. Keiner, M. Koll-Schretzenmayr, and W.A. Schmid), Aldershot, Ashgate.

Moenninger, M. ed. (1999) *Stadtgesellschaft* (Urban society; title translated by WZ), Suhrkamp Verlag, Frankfurt.

NESDB (2007) *GDP and GNP, National Economic and Social Development Board*, Bangkok. Retrieved from: http://www.nesdb.go.th.

NESDB and DANCED (1996) *Urban Environmental Management in Thailand: A Strategic Planning Process*, Kruger report, Bangkok.

Ng, M.K. and Hills, P. (2003) World cities or great cities? A comparative study of five Asian metropolitan. *Cities,* 20(3): 151–165.

OECD (1998) Land-use planning and sustainable urban travel: Overcoming barriers to effective co-ordination, in *OECD-ECMT Workshop on Land-Use for Sustainable Urban Transport: Implementing Change* on 23–24 September 1998, Linz, Austria.

OECD (2006a) *Competitive Cities in a global economy*, OECD, Paris.

OECD (2006b) *Decoupling the environmental impacts of transport from economic growth*, OECD, Paris.

Orn, H. (2002) *Urban Traffic and Transport: Building issues*, Lund University Press, The Netherlands.

OTP (2005a) *Vehicle licenses issued in Thailand and Bangkok during 1993–2004*, Office of Transportation and Traffic Policy and Planning, Bangkok. Retrieved from: http://www.otp.go.th/pdf/Statistic/carregist36-46.pdf.

OTP (2005b) *The daily average registration of passenger cars*, Office of Transportation and Traffic Policy and Planning, Bangkok. Retrieved from: http://www.otp.go.th/statdata_3.asp.

Pianuan K., Kaosa-ard M.S, and Pienchob P. (1994) Bangkok traffic congestion: Is there a solution?. *TDRI Quarterly Review*, 9: 20–23.

Poboon, C. (1997) *Anatomy of a traffic disaster: Toward a sustainable solution to Bangkok's transport problems*, Doctoral dissertation, Murdoch University, Austria.

Sternberger, D. (1985) *Die Stadt als Urbild* (The city as archetype; title translated by WZ), Frankfurt, Suhrkamp Verlag.

Suthiranart, Y. (2001) *The transport crisis in Bangkok: An exploratory evaluation*, Doctoral dissertation, University of Washington, USA.

Surasawadi, K. (2000) *Traffic solution in Bangkok and vicinity*, City Planning Department, Bangkok.

Statistics of Traffic (2006) *Statistic of Traffic*, Traffic and Transportation Department, Bangkok Metropolitan Administration, Bangkok.

Tadiar, N. X. M. (1995) Manila's new metropolitan form, in Discrepant Histories, in *Translocal essays on Fillipino cultures* (ed. C.L. Rafael), PA Temple University Press, Philadelphia.

Thavisin, N. and Suwarnarat, K. (1995) City Study of Bangkok, in *Magacity Management in the Asian and Pacific Region: Policy Issues and Innovative Approaches*, Volume II: City and Country Case Studies. Proceedings of the Regional Seminar on Mega cities Management in Asia and the Pacific, Manila, Philippines, October, pp. 3-25.

TRANSPLUS (2003) *Achieving Sustainable Transport and Land Use with Integrated Policies. Final Report.* Rome: ISIS – Istituto di Studi per l'Integrazione dei Sistemi.

TRIS and BMTA (2002) *A report of performance evaluation at fiscal year 2002*, Thai Rating and Information Services (TRIS) and Bangkok Mass Transit Authority (BMTA), Bangkok.

UNEP. (2007) *Sufficiency Economy and Human Development Thailand*. Human Development Report 2007, UNEP, Bangkok.

Wegener, M. Fürst, F. (1999) *Land-Use Transport Interaction: State of the Art. Deliverable 2a of the EU project TRANSLAND (Integration of Transport and Land Use Planning).* Institute of Spatial Planning, University of Dortmund.

World Bank (2007) *World Development Report 2007: Development and the Next Generation;* Washington, World Bank.

World Bank and ADB (2005) *Sustainable urban transport in Asia. Making the vision a reality;* Main report; Clean Air Initiative; Washington/Manila (A CAI-Asia Program).

20
Oana Liliana Pavel

To Be or Not To Be a 'World Class' City:

Poverty and urban form in Paris and Bucharest

Introduction

One of the key debates in urban planning today is centred on problems of social inequality and spatial segregation within large metropolises. While it can be argued that urban structure and form have an important role in determining both quality of life and the efficient distribution of resources, it can also be argued that a sustainable city is more than its mere physical form. As cities grow in size and the monocentric form of large metropolises evolves progressively into a polycentric structure over time, the problems of social exclusion and fragmentation may be exacerbated.

This chapter provides an insight into the social sustainability of Paris and Bucharest in relation to their spatial structure. Paris, traditionally considered one of the top ranking world cities, is an example of a European city which has its origin as a monocentric city, but has developed into a more polycentric city form and has expanded to become a polycentric metropolitan region. Bucharest, once known as the Paris of the East, is clearly a monocentric urban system, but its predicted growth over the next fifteen years is tending to move it towards a polycentric form. Despite their distinct differences, Paris and Bucharest can be seen as two cities searching for sustainability.

There are five sections in this chapter. First, the reason why Paris and Bucharest have been selected is discussed. Second, the possibilities of tackling social problems by changing spatial organisation are examined. Third, an analysis of the sociological problems of the two cities is made in order to reveal the social differences between 'world class' and non-'world class' cities. Fourth, the ways in which Paris and Bucharest transformed their spatial structures is explored, in particular the social characteristics of each period of their urban transformation, in order to build more complete images of them. And finally, as a conclusion, some suggestions for achieving more sustainable metropolises are proposed.

Why Paris and Bucharest?

Approximately 75% of the European population lives in urban areas today, and this is expected to rise to 80% by 2020 (EEA Report, 2006). Europe is highly urbanised, and it has many world cities of varying degrees of significance. Paris, according to the Globalization and World Cities Study Group and Network (Beaverstock *et al.*, 1999) is one of the top four 'alpha world cities', while Bucharest is categorised in the group as showing 'evidence of world city formation'. In addition to the ranking of cities by 'advanced producer services', there are many other definitions of world cities, such as a 'world city hierarchy based on air connections' (Smith and Timberlake, 2001), with Paris in third position and Bucharest 60th, or categorisations of global and sub-global cities in which Bucharest does not feature at all (Sassen, 1991). All these studies show that Paris is economically too developed and physically too large for a direct comparison with Bucharest. Yet there are a number of reasons why a study of the two cities can illuminate some issues of social sustainability and urban form.

First, the cities have a different urban form: Bucharest is an example of a monocentric city moving towards a polycentric form; Paris is an example of a polycentric city region that has evolved from a monocentric form. These differences offer some insight into the ways that a city and its form affect social urban life. Second, the selection of the two cities contrasts a western capitalist city with an eastern socialist city, in order to illustrate the role of history in the creation of an urban society. And third, comparing two cities at different levels of development provides a perspective on 'world class' city status through a consideration of the social urban development as an element of analysis.

These two cities have been selected as examples of metropolises, or more specifically as metropolitan areas – the cities and their directly and tightly polarized space – to take into account a possible and progressive disconnection between the cities and their surroundings. There is a conflict between urbanization and sustainable development, as the costs of urban concentration increase, and as socially and environmentally negative effects accumulate. It seems that looking for a sustainable metropolitan urban form has become the general aim of most metropolises.

In this study, monocentrism and polycentrism are used to describe urbanised areas, meaning towns, cities and metropolises. Precision is important as scale plays an important role; an area can be judged monocentric or polycentric at different scales. For instance, Paris is known as the largest monocentric city in Europe at the level of its core urbanised area, but on the larger scale it is considered an example of a polycentric metropolitan region.

Changing society by changing the space?

Europe's urban future is of concern because its traditional city model has been changed by the ever increasing expansion of urban sprawl that has created environmental, social and economic impacts not only for the city, but also for the regions. The European city has traditionally been much more compact, developed in a dense historical core shaped before the emergence of modern transport systems. But, as these cities have grown in size, urban form has evolved rapidly from the original monocentric structures into polycentric forms over time.

Urban sprawl generates large areas of residential development that often become segregated spatially and socially, especially in relation to income. In some cities, the social polarisation associated with urban sprawl is clearly apparent, with divisions between the inner city core and the suburban outskirts. A mix of unemployed people, the elderly poor, single young people and minority ethnic groups, often suffering from the impacts of the selective nature of migration and employment loss, is frequently concentrated in the inner city. However, these social problems are not unique to city centres. In many cities, similar social problems have increasingly developed in more peripheral areas where post war re-housing schemes are today the location of some of the lowest quality environments and the home of the most disadvantaged urban

Fig. 1. Example of an urban environment in a suburb of Paris Source: Oana Pavel

groups (Fig. 1). It is probable that urban structure and form have an important role to play in reaching a measure of social sustainability; they can affect the quality of life and socio-economic equity. As the welfare of the poor may be affected by the type of urban spatial structures in a number of ways, and large areas of dereliction, empty quarters and poverty destroy a sense of the community, the debates about sustainable cities and their form are increasingly important.

The traditional monocentric city form was considered by its proponents in the in the 1960s and 1970s (e.g. Alonso, 1964; Muth, 1969; Mills, 1972) to be a model form, arguing that it was more favourable to low-income households because it reduced distance and allowed an efficient network of public transport, thus providing better access to job opportunities for the poor. However, in this type of city, the poor are more likely to live in higher density (possibly overcrowded) environments, consuming less land and floor space than in lower density polycentric cities, and the quality of their environment might be worse (Bertaud, 2004).

More recently, the 'compact city' model, essentially a high-density, mixed-use city, with clear boundaries (Jenks *et al.*, 1996; Williams *et al.*, 2000), is seen as displaying a sustainable use of land, preserving the countryside by reducing sprawl and recycling brownfield land in urban areas. In social terms, compactness and mixed uses are associated with diversity, social cohesion and cultural development as people can live near to their work place and leisure activities. The potential disadvantage of the compact city is that, like the traditional monocentric form, it can either create too high a density in city cores exacerbating existing rates of overcrowding for the poor, or can lead to regeneration by gentrification of the core, driving the poor to the peripheries.

With the growth of cities into large metropolises and urban regions, urban forms have evolved into polycentric ones. Such polycentric forms seem to be a promising way to achieve a balanced and sustainable territorial development, a reduction in territorial disparities of metropolitan areas and better social cohesion. However, most of the major metropolises now have to cope with worsening social inequity within the city, with forms and degrees of segregation that vary from place to place (Martens and Verwaeke, 1997).

It seems that the future of the city should lie in exploring new solutions to problems that have been repeatedly solved over the centuries – those of adapting urban forms to new functions (and not adapting new functions to urban form).

Paris and Bucharest: a social perspective

What makes the city real is not only its environment and urban form, but also its reflection of a social reality. The city is the image of the totality of its society, even if at times its social problems become magnified; some of the Parisian suburbs provide such an illustration. The problems that can prevail there create a paradox that reflects a more fundamental crisis of social fracture as a whole. Urban fragmentation is one function of this social disjunction, as suggested from the analysis of the two cities.

Cities have always had to face serious problems, but the paradox is those same cities can ameliorate problems such as poverty and violence through development, the concentration of facilities, and a symbiosis between them. This chapter provides a social analysis of Paris and Bucharest using information from literature, interviews and the analysis of statistics, census and archival documents.

Poverty and exclusion

Poverty is a relentless plague that runs through history and transcends spaces where people live. Although poverty has been considered a particularly rural phenomenon, more and more cities are experiencing it. Slums (Fig. 2) are not only a manifestation of poor housing standards and lack of basic services, but are also a symptom of dysfunctional urban societies, where inequalities are not only tolerated but also allowed to fester.

Paris is often described as the city of the rich, but it also accommodates many poor people, approximately 12% of its total population. Geographically, the most disadvantaged population is concentrated in the districts of the north-east: 40% of the poor households concerned reside in 18th, 19th and 20th arrondissements, as

Fig. 2. Roma slum in the northern part of the Paris metropolitan area
Source: Oana Pavel

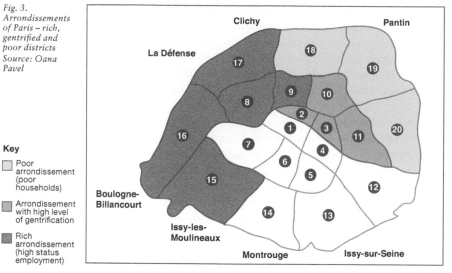

Fig. 3.
Arrondissements
of Paris – rich,
gentrified and
poor districts
Source: Oana
Pavel

Key

Poor arrondissement (poor households)

Arrondissement with high level of gentrification

Rich arrondissement (high status employment)

opposed to only 2% in the 4th and 6th (APUR, 2004; see Fig. 3). While there are many factors that explain the causes of poverty, people's occupations are a significant one. Analysis of the geography of higher status employment shows that, except for some specific extensions in the south-west, more than half such employment occurs in three contiguous central areas: that is, in western Paris, within the 2nd, 8th, 9th and 17th districts, the Defense and Boulogne/Issy-les-Moulineaux, and extending to the 15th and 16th districts (IAURIF,[1] 2002). These three areas form a triangle constituting the heart of the metropolis. By contrast, the eastern part of the city as well as its neighbour to the north (i.e. Saint-Denis) has little significant high status employment. This illustrates an imbalance between the west and the east of Paris in terms of its development.

While Paris is largely a rich city, Bucharest is predominantly poor. It is a legacy of delayed development because of its relatively recent communist past. The most visible social problem in the city is the marginalisation from mainstream society of the Roma minority.[2] Their marginalisation continues through their 'ghetto-isolation' in substandard housing complexes on the outskirts of the city. In many Roma families, five or more people live in one room, and in some cases even 12–21 people dwell in one room. This represents depths of poverty and exclusion more than of just a lack of income, and also of a situation of being outside society and without rights (Tourette, 2005).

There is a clear correlation between where people live and their chance of finding a job. In France, for example, it has been shown that job applicants residing in poor neighbourhoods were less likely to be called for interviews than those who live in middle- or high-income neighbourhoods (IAURIF, 2006).

Segregation and gentrification

Segregation and gentrification are important factors when poverty in the city is considered. The spatial concentration of a homogeneous population, generally the poor, leads to its designation as segregation, and often its stigmatization (Clavel,

2004). There are different types of segregation, such as residential, ethnic or racial, which are closely associated with marginalisation, racism, and social exclusion (Fig. 4). Gentrification is characterised by upper and middle income populations moving into older, centralized neighbourhoods, formerly occupied by the poor. Important researchers of the phenomenon of gentrification (Smith and Williams, 1986; Palen and London, 1984; Zukin, 1987; Sassen, 2000) perceive this process as an attack against the poor.

In the central area of Bucharest gentrification takes place in small pockets rather than uniformly. Some of the people responsible for the gentrification of the city centre are real estate agents, former owners, or politically powerful residents, who regained ownership of houses confiscated by the former communist state. The paradox of Bucharest as a social structure is that through household mobility, some ancient 'ghettoes' are changing into districts of luxury. This is the case of the north of Bucharest where villas are now built, and in the south where several companies have built prestigious stores. The centre of Bucharest is no exception; the same is happening in the peripheral area of the city. In the Titan district prices of apartments have increased by 30–40% after the construction of the first shopping mall in Romania. Thus in Bucharest, the general tendency is the creation of ghettoes for the rich and ghettoes for the poor that accentuate the contrast between populations and which lead to an open road to social exclusion.

Gentrification is observed in many cities or, more exactly in the centres of the great urban centres, and Paris is no exception (Pinçon-Charlot and Pinçon, 2004). For example, between the two Parisian censuses of 1954 and 1999, the percentage of resident workmen, employees and service staff fell from 65% to 35%, while that of the owners, senior executives and junior staff climbed from 35% to 65%, indicating

Fig. 4. Social exclusion in the centre of Bucharest
Source: Oana Pavel

a process of gentrification. Nevertheless, the 'Paris of the poor' remains a reality with, for instance, 10,000 homeless people eking out their existence on the streets.

The division of urban space is influenced by a set of various and antagonistic forces. Partly, the forms of segregation are explained by the cost of access to housing. The occupation of space and social evolution of the cities is explained by inequities of access to economic resources. The well-off households carefully choose their dwelling places and relegate the poorest to where they do not want to go. As such they create an evolving social structure and the geography of the city in which they reside. It seems, in large metropolises, that urban fracture and fragmentation is the price to pay for belonging to the global economy. A global city, such as Paris, gives economic power to the centre. And, even though Bucharest is not a world city, the phenomenon of segregation is a strongly perceptible reality in the urban landscape.

Constructing urban space towards sustainability

The significance of monocentric and polycentric urban forms varies at different scales. A monocentric city form has been described as 'a circular residential area surrounding a central business district in which all jobs are located' (Burgess, [1925] 1996). The polycentric city form consists of a centre and a number of concentrated sub-centres with relatively high population and employment density (Davoudi, 2002). Decentralized population and employment, extensive urbanization, the decline of central business districts and the emergence of employment concentrations outside the centre characterize contemporary metropolitan areas: there is an extensive literature on their evolution (e.g. Muller, 1981, 2004; Castells and Hall, 1994). Explanations for changing urban forms include public policies for transportation or housing, the dominance of the automobile, economic restructuring and technological change, social and racial segmentation and preferences for low density living environments.

Parallel to the research on the economic benefits of the current changes in urban form, there is a continuing discussion about their social costs. Today, more and more people in Europe regard a new house as the prime investment to be made in their lifetimes, ideally a semi-detached or detached house in the suburban areas outside the city. In contrast to the apparent attractions of the suburbs, the negative aspects of the inner city cores, including a poor environment, social problems and safety issues, create powerful drivers of urban sprawl. City cores are perceived by many as more polluted, noisy and unsafe than the suburbs. Unemployment, poverty, single parent households, drug abuse and minorities with integration problems are also often identified with inner-city areas. As families move out of the city, social segregation begins to intensify. The following sections on Bucharest and Paris indicate how urban forms and social problems occur in different city contexts.

Bucharest

The city of Bucharest is the result of natural growth, industrialisation and political centralisation. Since 1948, the city has more than doubled in population and land area, so that it has become congested and increasingly built-up. Beside increasing traffic congestion and air and noise pollution, the city's dispersed population and rising levels of car ownership have resulted in social problems. Bucharest faces the problems of population growth and dispersion from the centre to suburban areas, consequently developing from its strongly monocentric structure into a primitive form of a polycentric city (Fig. 5).

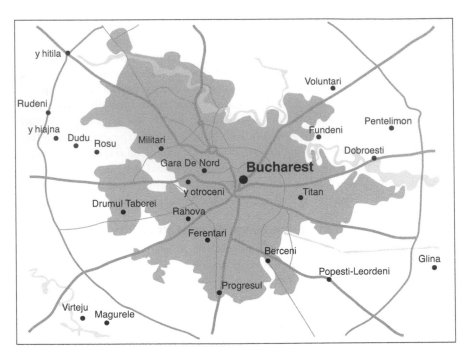

Fig. 5. Map of Bucharest Source: Adapted from www.mapquest. com

In the 1970s, Bucharest was a relatively compact city with large agricultural areas within and outside the city limits and the surrounding ring road, and large forests to its north and west. In the 1980s, during the presidency of Ceausescu, the townscape was dramatically restructured and transformed. Projects like the expansion of the airport and the building of the Bucharest–Danube Canal, which now divides the city into two sections, upset the traditional urban fabric. Socially, a number of villages on the outskirts of Bucharest were dismantled and their citizens were forced to move into the city. Meanwhile, about a quarter of the old city was demolished to construct massive new state buildings, such as the *House of People*. The city has become denser in the west, east and southeast. Urban growth is occurring both inside and outside the ring road, in its search to become a metropolis. The city is expanding on all sides, but in different ways. In the north, the richest district of Bucharest is expanding inside the city boundaries, while on the west side the city is expanding outside the ring road and beyond the city boundaries. In this transformation of its urban form, the small cities in the western Bucharest metropolitan area are advantaged by their proximity to the city centre (Fig. 6).

A classical theory of urban space is built on the premise that 'a city lives through the centre', 'a metropolis controls, presents, distributes and gathers by virtue of its centre' (Labasse, 1966). Bucharest is the only East European city where major sections of the historical city were torn down as part of a vast urban renewal scheme and not as part of war reconstruction (Angotti, 1993). The policy of 'systematization' of Ceausescu entailed the destruction of old central neighbourhoods, and in the place of the large sections of the demolished historical centre, large modern buildings followed. The low-rise traditional buildings that once characterized the heart of the metropolis were in stark contrast to large apartment blocks and big constructions like

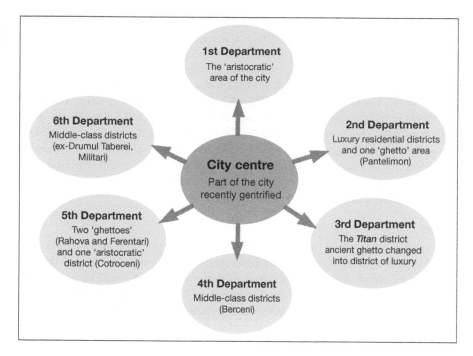

Fig. 6.
Segregation and
gentrification in
Bucharest
Source: Oana
Pavel

the Intercontinental Hotel (Fig. 7). Being aware of the impossibility of rebuilding the entire city, grand boulevards were constructed and lined with homogenised modern-looking flats. Some older individual homes surrounded by gardens remained almost intact behind the new blocks of flats. It is hard to imagine how all these changes could not have a disruptive effect on people's lives.

Fig. 7. Modernity
and poverty:
the dual face of
Bucharest
Source: Oana
Pavel

One of the key points of communist ideology was social harmonization, and the urban landscape of Bucharest offers an accurate translation of this social ideal into the materialisation of its urban form. To respond to a real deficiency in housing during the socialist period, new residential districts in the outskirts of Bucharest were constructed as a manifestation of the idea of social justice. Thus, the city centre, once a bourgeois monopoly, was rendered commonplace, by extending the same architecture and the same level of 'comfort' of the buildings in the centre to the new-built outskirts. This opening up of access by everybody to the centre attempted to eliminate the spatial hierarchy of the city and was considered to echo of principles of uniformity and equality. However, after the demise of the socialist period, this action of atrophying the centre of the monocentric Bucharest form activated the process of gentrification and segregation in the city.

Paris

The spatial organization of Greater Paris is transforming and metamorphosing rapidly, moving from planning that concentrated on the development of the old historic city centre, towards a more spread out, polycentric form. The polycentric organisation had its origin in the development of La Défense in the 1960s and 1970s, and was conceived to give an internationally visible business district to the nation as a whole. The symbolic architectural rupture embodied by the high rise buildings set against the historically horizontal skyline was a signal to competing nations that Paris was leading a strong national economy. At the same, polycentric policies evolved at regional level.

The long inherited monocentric Parisian form was compromised by rapid population growth. To prevent anarchic urban sprawl, all regional planning documents have favoured, at least since 1960s, polycentric spatial policies. In 1994, the SDRIF[3] set out a major objective: to plan the polycentric development of the metropolitan area of Paris. The principle of polycentrism not only aimed to reduce the pressure of development on Paris and its suburbs, but also to correct chronic spatial imbalances within the area.

While Paris and its region can be described as rich compared to other agglomerations in France, they also contain social inequalities and social exclusion that have a geography – divided between centre and periphery, and between east and west. The question is whether there are short- and long-term actions that can be put into place, so that the creation of wealth does not result in the social and territorial relegation of disadvantaged groups of population. Is rebalancing urban form through polycentric policies sufficient to create social cohesion?

Conclusion

The review of social problems in Paris and Bucharest, cities that have different scales of development and different urban forms, indicated that there is no particular urban form that either promotes or ameliorates poverty, segregation or gentrification. It suggests that attempts to change urban form do little to assist these features. The radical changes to Bucharest by demolition and attempts to homogenise and eradicate spatial differences simply failed when the regime was disbanded. The result was a fragmented geography of gentrification in the centre and of the displacement of the poor to the peripheries. At the same time, a process of change in urban form from monocentric to the beginnings of polycentrism served merely to concentrate poverty and exclusion at the city edge. In Paris and its region, there are clear areas of concentrated disadvantage within the monocentric urban core. The extension of the city into peripheral suburbs in many cases has concentrated poverty and exclusion in particular suburban areas. The plans to promote polycentric forms in the wider region have been overwhelmed by the spreading Parisian agglomeration (Berger, 2004). The picture here is still one of social fragmentation. So when it comes to poverty, segregation and gentrification, urban form appears to have little effect on promotion of a more equitable city, or to achieving a measure of social sustainability.

And what of Paris being a world city, and Bucharest lagging far behind: does it make a difference? As far as poverty and exclusion is concerned, seemingly very little. Clearly Paris has a lower proportion of the poor and excluded, but its global power and wealth merely serve to highlight such differences, as perhaps the violent unrest in the suburbs attests. Bucharest may have a higher proportion of the poor and excluded, and a relatively smaller proportion of the very rich, but the process of gentrification and displacement is growing fast as the city becomes part of the global economy.

If economic efficiency is the primary goal for cities, this chapter indicates that it is unlikely to succeed, or achieve a measure of sustainability, without considering the social element. From this social perspective of the two cities, the answer to the question of being or becoming 'world class' is found: 'world class' or not, social problems always exist. To be meaningful, a 'world class' city needs to be more than a world or global city: it needs to be a city that also achieves social sustainability, and is equitable for all its citizens.

Notes

1. Institute for Urban Planning and Development of the Paris Ile-de-France Region.
2. Roma, Romani or Romanies are otherwise known as gypsies. Romania has one of the largest Roma populations in Europe; the official census of 2002 estimates 535,250, whereas the UNDP suggests there are nearer 2 million (Wikipedia).
3. SDRIF is the Schema Directeur de la Region Ile-de-France – a prospective document concerning the planning and the development of the regional territory.

References

Alonso, W. (1964) *Localisation and land use*, Harvard University Press, Cambridge, MA.

Angotti, T. (1993) *Metropolis 2000: Planning, poverty and politics*, Routhledge, London.

APUR (2004) *La pauvrété à Paris*, Note de 4 pages.

Beaverstock, J.V., Smith, R.G. and Taylor, P.J. (1999) A roster of world cities. *Cities*, 16(6): 445–458.

Berger, M. (2004) *Les périurbains de Paris: De la ville dense à la métropole éclatée?*, Espace et milieux, Paris.

Bertaud, A. (2004) *The spatial organization of cities: deliberate outcome or unforeseen consequence?*, University of Califonia at Berkeley, Berkeley, CA.

Burgess, E. W. (1996) The growth of the city: An introduction to a research project, in *The inner city reader* (eds R. LeGates and F. Stouts), Routledge, New York [First published 1925].

Castells, M. and Hall, P. (1994) *Technopoles of the world: The Making of 21st Century Industrial Complexes*, Routledge, London.

Clavel, M. (2004) *Sociologie de l'urbain*, Anthropos, Paris.

Davoudi, S. (2002) Polycentricity-modelling or determining reality. *Town and Country Planning*, 7(4): 114–117.

EEA (2006) *Urban sprawl in Europe – the ignored challenge*, EEA Report No 10/2006, European Environment Agency, Copenhagen.

IAURIF (2002) Les emplois supérieurs en Ile-de-France. Les nouvelles polarités?. *Note rapide sur le bilan du SDRIF*, **305**, Paris.

IAURIF (2006) Les territoires de pauvreté en Ile-de-France: Etat des lieux. *Note rapide Population – Modes de vie*, **407**, Paris

IAURIF (2006) Les territoires de pauvreté en Ile-de-France: Typologie des territoires. *Note rapide Population – Modes de vie*, **408**, Paris.

Jenks, M., Burton, E. and Williams, K. (1996) *The Compact City: A Sustainable Urban Form?* E&FN Spon, London.

Labasse, J. (1966) *L'organisation de l'espace: Eléments de géographie volontaire*, Hermann, Paris.

Martens, A. and Vervaeke, M. (1997) *La polarisation sociale des villes européennes*, Anthropos Col. Villes, Paris.

Mills, E.S. (1972) *Studies in the structure of the urban economy*, Johns Hopkins Press, Baltimore.

Muller, P. (1981) *Contemporary suburban America*, Prentice-Hall, Englewood Cliffs.

Muller, P. (2004) Transportation and urban form: stages in the spatial evolution of the American metropolis, in *The geography of urban transportation* (eds S. Hanson and G. Giuliano), 3rd edition, Guildford Press, New York.

Muth, R.F. (1969) *Cities and housing*, University of Chicago Press, Chicago.

Palen, J.J. and London, B. (1984) *Gentrification, displacement and neighbourhood revitalization*, State University of New York.

Pinçon-Charlot, M. and Pinçon, M. (2004) *Sociologie de Paris*, La découverte, coll. Repères, Paris.

Sassen, S. (1991) *The Global City: New York, London, Tokyo*, Princeton University Press, Princeton, NJ.

Sassen, S. (2000) *Cities in a World Economy*, Pine Forge Press, Thousand Oaks.

Smith, D.A. and Timberlake, M. (2001) World city networks and hierarchies, 1977–1997: An empirical analysis of global air travel links. *American Behavioral Scientist*, **44(10)**: 1656–1678.

Smith, N. and Williams, P. (1986) *Gentrification of the city*, Allen and Unwin Edition, London.

Tourette, F (2005) *Développement social urbain et politique de la ville*, Gualino Editeur, Paris.

Williams K., Burton E. and Jenks M. (2000) *Achieving Sustainable Urban Form*, Routledge, London.

Zukin, S. (1987) Gentrification: culture and capital in the urban core. *Annual Review of Sociology*, **13**: 129–147.

21

Vijay Neekhra, Takashi Onishi and Tetsuo Kidokoro

The Inner Truth of Slums in Mega Cities:

A scenario from India

Introduction

Despite the arrival of the new millennium, it is increasingly apparent that the world is returning to some of its fundamental, unresolved issues: of equity, sustainability, poverty and social justice, among others. The relative paucity of knowledge of local and global forces shaping development and the production and reproduction of urban poverty, the complexity of the accompanying phenomena and the uncertainty of urban decision-making processes, call for more specific knowledge of inter- and intra-city differentials in poverty and inequality. This suggests the need for a better understanding of slums and those living in them.

The word 'slum' first appeared in the London Cant (UN-HABITAT, 2003b) at the beginning of the 19th century, designating initially 'a room of low repute' or 'low, unfrequented parts of the town'. The word then underwent a series of changes and it has many connotations. It often refers to settlements illegally occupying land and lacking in basic services. Slums can vary from high density, squalid central city tenements to spontaneous squatter settlements without legal recognition or rights. While some slums are more than 50 years old, others are land invasions just under way. What the word 'slum' covers is even more complex when one considers the variety of words it has generated in other languages; for example Colonias Populares (Mexico) or Villa Miseria (Argentina).[1] Although the various names given to slums emerge in different contexts, they describe the same poor quality of living conditions in their respective countries.

Slum dwellers experience multiple deprivations that are direct expressions of poverty. Houses in the slums are generally unfit for habitation and residents often lack adequate food, education, health and other basic services that the better-off take for granted. The places – neighbourhoods, residential areas and so on – where slums are located are generally not recognized by local and central authorities. However, in many parts of the world these 'invisible' areas are growing faster than the 'visible' ones. Slums are the products of failed policies, bad governance, corruption, inappropriate regulation, dysfunctional land markets, unresponsive financial systems, and a fundamental lack of political will. Each of these failures adds to the toll on people already deeply burdened by poverty and constrains the enormous opportunity for human development that urban life offers.

Slums in a global context

More than 920 million people lived in slums in 2001, constituting some 32% of the global urban population – in other words every third person of the world's urban population is a slum dweller. It has been estimated that four out of ten inhabitants in the developing world are informal settlers. In absolute numbers of slum dwellers, the majority are in Asia, but Sub Sahara Africa has the highest concentrations of people living in slums (UN HABITAT, 2003a).

While only 6% of the urban population in the Developed Regions of the world lived in slums, 43% of the urban population of all Developing Regions and 78.2% in the Least Developed Countries were slum dwellers (Table 1). In 2001, wthin the Developing Regions, the African continent had the largest proportion of its urban population resident in slums (60.9%). Asia and the Pacific Region had the second largest proportion of their urban population living in these precarious settlements (42.1%), while Latin America and the Caribbean slum dwellers population had the third largest proportion at 31.9%. In absolute numbers of urban slum dwellers, Asia and the Pacific Region dominate the global picture, having a total of 554 million informal settlers in 2001 (excluding China); Africa had a total of 187 million, and Latin America and the Caribbean had 128 million slum dwellers (Table 2).

Major reasons for the formation of slums

Researchers and scholars have stressed the value of many policies and approaches to help improve the quality of life of the slum dwellers. These range from passively ignoring or actively harassing men and women who live in slums to interventions aimed at protecting the rights of slum dwellers and helping them to improve their income and livelihood. There has been a lot of discussion about approaches/policies from negligence, eviction, relocation (Viratkapan and Perera, 2006), resettlement, infrastructure provision (Abott, 2002a; Choguill, 1999), *in situ* upgrading (Abbott, 2002b; Mukhija, 2001) to method-based re-planning of slums (Abbott, 2002a). Although individual issues and reasons for the formation of slums have been discussed by numerous researchers, the basic root causes of slums in the cities have been less discussed. Issues such as migration (Srivastava, 2003a; Mukherji, 2001; McDonald, 1999; Connel, 1987; Eke, 1982; Davis, 2005), poverty (Kapoor *et al.*, 2004; Daniere and Takahashi, 1999; Lipton, 1980), globalization and urbanization (Birdi, 1995, Deshpande, 1996, Davis, 2005), rural–urban partnership (Epstein,

Table 1.
Population of
urban slum areas
2001
Source: UN-
HABITAT
(2003b)

	Total pop. (in millions)		Total urban pop. (in millions)		Urban pop. as % of total pop.		Slum population as % of total urban pop.	Urban slum pop. (millions)
	1990	2001	1990	2001	1990	2001	1990	2001
World	5255	6134	2286	2923	43.5	47.7	31.6	924
Developed regions	1148	1194	846	902	73.7	75.5	6.0	54
Developing regions	4106	4940	1439	2022	35.0	40.9	43.0	870
Least developed countries	515	685	107	179	20.8	26.2	78.2	140

	Total pop. (in millions)		Total urban pop. (in millions)		Urban pop. as % of total pop.		Slum population as % of total urban pop.	Urban slum pop. (millions)
	1990	2001	1990	2001	1990	2001	1990	2001
World	5,255	6,134	2,286	2923	43.5	47.7	31.6	924
Developing regions	4,106	4,940	1,439	2022	35.0	40.9	43.0	874
Africa	619	683	198	307	31.9	44.9	60.9	187
Latin America and Caribbean	440	527	313	399	71.7	75.8	31.9	128
Asia (excluding China)	3,040	3,593	928	1,313	30.5	36.5	42.1	554
Oceania	6	8	1	2	23.5	26.7	24.1	5

Table 2. Population of slum areas in developing regions 2001 Source: UN-HABITAT (2003b)

2001), housing (Mukhija, 2001; Bijlani and Roy, 1991), property/tenure rights (Mukhija, 2001, 2002; Payne, 2001) have been given attention in the past. Some of researchers have stayed in slums over time in order to understand the minute details of the slum dwellers' day-to day activities and observe and feel their problems (Neuwirth, 2005).

This chapter attempts to highlight the major factors responsible for the formation of slums in India, and then considers some of the characteristics of slums in India's 35 mega cities. It finally questions the potential status of Mumbai, India's major city on the world stage.

Slums in India

Although India's economy is one of the world's fastest growing, with a GDP growth rate of 9.2% per annum (2005) (World Bank, 2007)[2] and is in the top 20 of the world's countries, it is still a poor country. Its GNI per capita in 2006 is $880 (World Ranking – 161) and PPP (Purchasing Power Parity) is $3,800 (World Ranking –145) (ibid.), placing it far below comparable developing nations in the world and just above the least developing nations.[3] A large proportion of India's population is poor and live in slums due to a number of factors including: family poverty and little education; regional inequities and urbanization; migration; a low-wage economy and unemployment; and shortages of housing. The following sections explore these factors, before considering the key problem of slums in India's mega cities.

Poverty and Education

The Planning Commission (2001a) estimates that 26.1% of India's population live below the poverty line. In 1999–2000, the states with the highest poverty levels were: Orissa (47.2%), Bihar (41.2%), Madhya Pradesh (37.4%), Assam (36.1%) and Uttar Pradesh (31.2%): for administrative purposes, India is divided into 28 states and 7 Union territories (Fig. 1). The literacy level (a person who can read and write irrespective of formal education) in India also varies across the states. In most of the region/states in rural parts of India the level is low and varies from 40% to 60%, although in urban areas, the literacy level is high and is around 80% in most states. However, a formal education of 10 years or more is vital for getting a reasonable job. The percentage of those who are illiterate, or with less than 10 years of education, rises to 89.8% and 71.1% in rural and urban areas respectively (Table 3). With such a mass of population with low levels of formal education, it is inevitable that most will end up in an insecure low wage job in the informal sector. This is one of the major causes of poverty in India.

India	Population (in millions)	Above 10 years education or tech. diploma (in millions)	% above 10 years education or diploma	Illiterate or below 10 years education (in millions)	% Illiterate or below 10 years education
Rural	742.5	76.1	10.2	666.4	89.8
Urban	286.1	82.8	28.9	203.3	71.1
Total	1028.6	158.9	15.4	869.7	84.6

Table 3. Education level in urban and rural India Source : Census of India (2001b)

The states with a large below-the-poverty-line population (BPL) also reported an equally large slum population. Indeed, there is a significant correlation between the percentage of the slum population and the BPL in urban areas in many regions in India, indicated by the regression line with an R square value of 0.79 (Fig. 2).

Regional inequality and unbalanced urbanization

In a country of India's size, the existence of significant regional disparities should not come as a surprise, but the scale and growth of these disparities is of concern. The ratio between the highest to lowest per capita income in all states increased from 2.6 in 1980–83 to 3.5 in 1997–2000 (Srivastava, 2003b). Regional differences in population, pressure on land, inequality of infrastructure, industrial development, and modernization of agriculture have led to large scale migration. Regional inequalities between the different states and regions can be established by comparing the Gini ratio for per capita consumption expenditure of various states over 1999–2000. It varies from 0.149 (Meghalya) to 0.292 (Arunachal Pradesh) in rural areas, and in urban areas from 0.205 (Meghalya) to 0.398 (Tamil Nadu) (Planning Commission, 2001b: Fig. 3). Regional inequalities in the various states also exist in terms of health facilities, education facilities, the economy, job opportunities, poverty level, gender, and so on.

Another important aspect of inequality is the imbalance of urbanization in different states of India. Variation in the level of urbanization rises from 12% in Assam to 93% in Delhi, depicting the large regional gap in different parts of India (Fig. 4). With wide gaps in urbanization, it is natural that the poor both from small urban areas and rural areas are pulled or pushed to highly urbanized regions in search of livelihood and survival.

Migration

The link between migration and slums is a key reason for slum formation, and has been of concern to development planners. Early studies have shown that

Fig. 1. India: showing key locations referred to in the text
Source: Adapted from www. censusindia.net

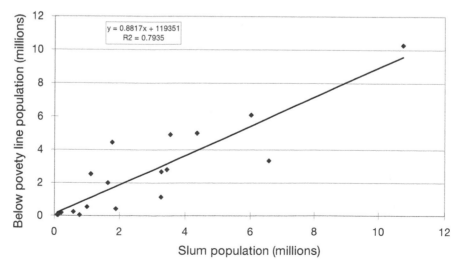

Fig. 2. Slum population against below poverty line population

Source : Planning Commission, Government of India, 2001b; Census of India (2001c)

337

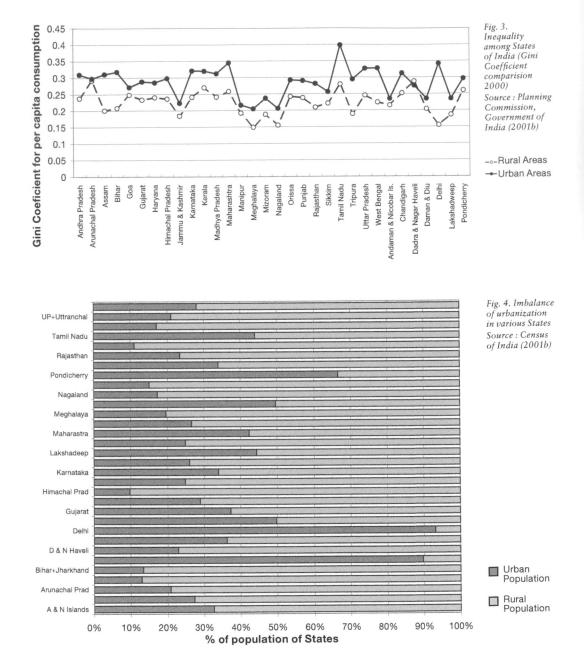

Fig. 3. Inequality among States of India (Gini Coefficient comparision 2000)

Source : Planning Commission, Government of India (2001b)

Fig. 4. Imbalance of urbanization in various States

Source : Census of India (2001b)

poor households participate extensively in migration (Connell *et al.*, 1976). More recent studies have confirmed migration as a significant livelihood strategy for poor households in diverse regions of India (PRAXIS, 2002; Mosse *et al.*, 2002; Hirway *et al.*, 2002; Haberfeld *et al.*, 1999; Rogaly *et al.*, 2001; Srivastava, 1998). The causes and consequences of rural to urban migration are of growing concern in developing country like India. Rural to urban migration, driven by inequalities, involves processes of change, adjustment, adaptation and assimilation by migrants.

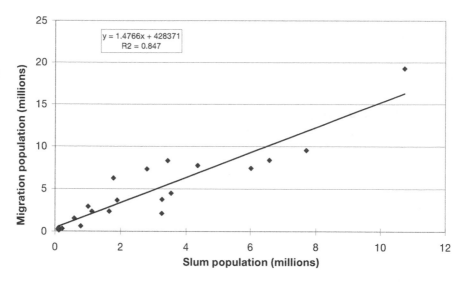

Fig. 5. Slum population and migration to urban areas
Source : Census of India (2001a; 2001c)

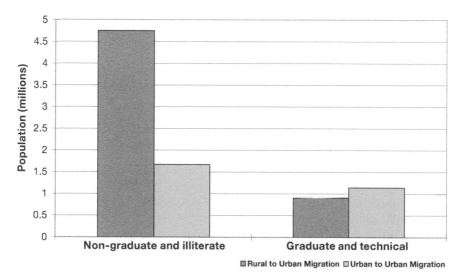

Fig. 6. Educational level of population migrating for employment
Source : Census of India (2001a)

The majority of migrants are absorbed into low grade and low productivity work in the mega cities, but virtually never into professional, administrative or even clerical work. The masses of illiterate and unskilled labour migrants have no alternative but somehow to eke a living in urban centres by performing odd jobs, petty sales or service work. As their income is low, migrants cannot afford house rent and are forced to live on vacant and hazardous land, thus forming slum settlements with very high population density, high room density and poor housing, and with inadequate access to basic civic amenities. Housing generally comprises shacks, earthen structures, flimsy structures or very old dilapidated houses. Analysis of the slum population and migration in different states shows a very close correlation (0.94); regression analysis shows an R-square value of 0.847 (Fig. 5).

On further analysis (Fig. 6) it is found that the majority of people moving from the rural areas are illiterate or have very little education; that is, some 4.8 million of the migrants moving to urban centres for jobs from rural areas. Around 1.7 million of those from urban areas are likewise with no or very low literacy (Census of India, 2001a).

Economy and employment

The economy and availability of employment have always been reasons to move from a rural to an urban area. In the majority of the cases, a person moves for employment. 26.1% of population migration reported in India is due to the search for employment and, as a family generally accompanies the earning member, a further 37.3% move with their households. Taken together, this indicates that more than 63% of migrants are moving to find jobs (Table 4).

Reasons for migration	Persons (in millions)	%	Males (in millions)	Females (in millions)
Work/Employment	14.4	26.1	12.3	2.1
Business	1.1	2.0	0.9	0.2
Education	2.9	5.3	2.0	0.9
Moved after birth	6.6	12.0	3.4	3.2
Moved with households	20.6	37.3	8.3	12.3
Other	9.5	17.2	5.2	4.3
Total	55.1	100.0	32.1	23.0

Table 4. Reasons for migration in India (other than marriage) Source: Census of India (2001a)

A wide gap between the various states is observed in terms of employment opportunities provided by the state domestic product (SDP) of each region. The slum population and SDP are clearly related and there is a strong correlation shown in the regression line, with an R-square value of 0.91 (Fig. 7). The rich states with high employment attract a large population from the poorer regions. However, these rich states fail to provide formal employment for the illiterate poor who moved to these regions in the hope of good job and better quality of life. These poor are ultimately absorbed into the informal low wage economy.

Housing shortages

There is a housing shortage of 24.68 million units in the country, including a shortage of 10.56 million housing units in urban areas (Ministry of Housing and Urban Poverty Alleviation, 2001).[4] For different reasons, central government and local bodies have not been able to meet the demands of urban citizens. With limited financial resources, various half-hearted efforts were made to meet the housing needs of the poor, but these have not produced the desired result. The rich or financially well off citizens certainly are able to afford housing, but the poor cannot. Their lack of money and the shortage of low cost or affordable housing mean that they have no choice except to occupy vacant land or slums. Indeed there is a clear relationship between the proportion of slum population in a state and its housing shortages, as shown in the regression line with an R-square value of 0.81 (Fig. 8).

Fig. 7. State Domestic Product (SDP) and slum population
Source: Planning Commission, Government of India, 2001b; Census of India (2001c)

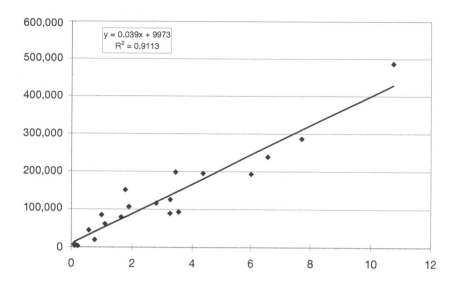

Fig. 8. Slum population and housing shortage
Source: National Building Organization based on Census of India (2001c); Town and Country Planning Organization (1996)

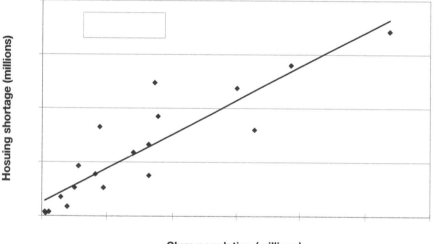

Urbanization in India

While the reasons for the formation of slums are a combination of social and economic deprivation and spatial inequalities, the location of slums is generally an urban phenomenon, as noted by UN-HABITAT (2003b, p. 11): 'The combination of high population density amid poverty and limited resources makes the developing world's mega-city an environment which favours the rapid growth of slum areas.' India experiences this same global trend towards increasing urbanization. In 2001, 27.8% of India's population (285 million) lived in urban areas and 72.2% in rural areas (Census of India, 2001b). Although the net addition of population in rural areas

(113 million) over 1991–2001 was much higher than in urban areas (67 million), the percentage of the total population living in rural areas declined from 74.3% in 1991 to 72.2% in 2001, and increased in urban areas from 25.7 to 27.8% over the same period (Table 5). Though rapid urbanization is welcomed for its positive effects, the high levels of migration to big cities from rural and small cities has imposed increasing pressures on the level of services in the urban centres and resulted in the formation of slum settlements in the major cities.

Year	Total population (in millions)	Rural pop. (in millions)	Rural pop. %	Urban population (in millions)	Urban pop. %
1981	682.9	523.5	76.7	159.4	23.3
1991	846.4	628.8	74.3	217.6	25.7
2001	1027.4	742.0	72.2	285.4	27.8

Table 5. Rural and urban population growth in India Source: National Institute of Urban Affairs (2001)

India has a total of 4,378 towns and cities (Table 6). Of these, only 35 cities have a population of more than 1 million – the mega cities. The mega cities, along with the 358 other big cities (with a population of more than 100,000), account for 9% of India's urbanization. 91% of cities fall into the small and medium class categories (Census of India, 2001b). The majority of the population live either in these small and medium size cities or in rural areas. Unfortunately most of the government-provided or assisted resources are utilized for the development of the mega cities. Little effort has been made towards the development of rural, small and medium size cities, which, compared to the mega cities, have relatively a very poor quality of life. Due also to their poor level of services and lack of job opportunities, large numbers of people migrate from these small and medium size cities and rural areas to the mega cities, exerting more pressure on an infrastructure already exceeding its carrying capacity. Most are forced to survive in slums, where the cost of living is much cheaper.

Class	Population size	No. of urban areas/towns	Percentage
Class I	1,00,000 and above	393	9.0
Class II	50,000 – 99,999	401	9.2
Class III	20,000 – 49,999	1,151	26.3
Class IV	10,000 – 19,999	1,344	30.7
Class V	5,000 – 9,999	888	20.3
Class VI	Less than 5,000	191	4.4
Unclassified		10	0.2
Total		4378	100.0

Table 6. City classification in India Source: Population Census (2001) and Ministry of Urban Development, India http://urbanindia.nic.in/moud/urbanscene/urbanmorpho/urbanmorph.htm

A total of 42.6 million people live in slums in some 640 cities spread across the states and union territories of India. Table 7 gives an overview of distribution of the slum population of cities of various sizes, and it can be seen that around 42% (17 million) of the total slum population lives in the mega cities. Rapid growth of Indian

Table 7. City population and slum population Source: Census of India (2001c)

City population (in millions)	Slum population (in millions)	Slum population %	Ratio: slum/city population
4 +	11.0	26.0	2.20
2–4	3.76	8.8	0.47
1–2	2.88	6.8	0.21
5–1	5.81	13.7	0.14
1–0.5	13.9	32.7	0.04
0.5–0.1	5.1	12.0	0.02
Total	42.6	100.0	0.07

cities has been associated with emergence and growth of slum and squatter settlements, characterized by overcrowding and a lack of sanitation and basic infrastructure.

Yet the large urban cities are centres of economic growth and contribute significantly to the GDP of the country. For instance, in 1950–51 the contribution of the urban sector to India's GDP was estimated at only 29%; this increased to 47% in 1980–81 and to 60% by the turn of the century (Planning Commission, 1992; Ministry of Housing and Urban Poverty Alleviation, 2002[5]). Cities hold tremendous potential as engines of economic and social development, generating wealth through economies of scale; they need to be sustained and augmented through high urban productivity for the country's economic growth and prosperity. But unfortunately this economic growth is limited to a few mega cities only. Small and medium size cities, along with rural development, need effective policies to enable them to contribute a major share to national economic growth. This could also help to alleviate the pressures of migration to the mega cities, and thus help reduce the growth of slums in the biggest cities.

Slums in mega cities in India

About 41.6% of the total slum population, around 17.7 million people, live in slums in India's mega cities (Census of India, 2001c). In absolute numbers, Greater Mumbai has the highest slum population at about 6.5 million, followed by Delhi (1.9 million) and Kolkata (1.5 million) (Table 8). It is disheartening that in the economic hub of

Table 8. Slum population in mega cities Source: Census of India (2001c)

	Total city population (in millions)	Slum population (in millions)	% of slum to city population
Greater Mumbai	12.0	6.5	54.1
Delhi	9.9	1.9	18.7
Kolkata	4.6	1.5	32.5
Chennai	4.3	0.8	18.9
Meerut	1.1	0.5	44.1
Faridabad	1.1	0.5	46.5
Hyderabad	3.6	0.6	17.2
Surat	2.4	0.5	20.9
Other Million Plus cities	34.4	5.0	14.4
Total	73.3	17.7	24.1

India, Mumbai, more than 54% of the city's population live in slums. The situation is not so different in other mega cities in other states of India; this varies from 20% to 50% of city population. The slums can be found all around these cities; along wastewater drains, railway tracks and next to strategic locations like airports, and are usually deprived of sufficient quantities of drinking water (Fig. 9).The size of the slum population in these cities is an indicator of the socioeconomic gap between the citizens, the scale of poverty, regional inequality, and incapacity of the local bodies to accommodate poor migrants in the city.

Caste and slums

In the past, Indian people were classified into castes, based on occupational status. The same caste system still exists, and socio-economically the most backward castes are the scheduled caste (SC) and scheduled tribes (ST). In spite of special government policies and efforts, the majority of them live below the poverty line. The majority of the slum dwellers in mega cities belong to these castes. Delhi and Mumbai have a combined population from these castes of 0.48 million and 0.44 million respectively. In most of the mega cities the population of the SC and ST exceeds 100,000. In the cities like Nashik, Bangalore, Nagpur, Chennai, Bhopal, more than 30% of the total slum dwellers belong to these two caste categories (Table 9).

Fig. 9.
Typical slum environments
Source: Vijay Neekhra

*Table 9.
Scheduled
caste (SC) and
scheduled tribe
(ST) population
in slums of
million plus
cities
Source: Census
of India (2001c)*

	SC population (in millions)	ST population (in millions)	SC + ST population (in millions)	Total slum pop. (in millions)	SC + ST population in slums %
Nashik	0.04	0.02	0.06	0.14	43.9
Nagpur	0.15	0.11	0.26	0.74	35.1
Bangalore	0.14	0.01	0.15	0.43	34.1
Bhopal	0.04	0.01	0.04	0.13	33.3
Chennai	0.27	0.00	0.27	0.82	33.1
Delhi	0.48	NST	0.48	1.85	26.1
Greater Mumbai	0.39	0.06	0.44	6.48	6.8

Child population in the slums

*Fig. 10.
Percentage
of slum child
population (0–6
yrs) to total child
population in
'million plus'
cities
Source: Census
of India (2001)*

In the majority of the cities a large percentage of the child population lives in unhygienic and unhealthy conditions, exposed to environmental and social hazards. More than 6 million children live in slums and they constitute 16.4% of the total child population of Indian urban areas. In other words, every sixth urban child in the age group 0–6 is a slum dweller. Around 2.5 million children aged 0–6 live in the slum areas of the mega cities, constituting 27.3% of the total child population of these cities. In Greater Mumbai alone there are 0.86 million slum dwelling children aged 0–6, and another 0.3 million in Delhi and 0.15 million in Kolkata (Fig. 10). More than half the child population of Greater Mumbai (62.8%) and Faridabad (50.6%) live in slum areas, and just below half in Meerut (48.5%) and Kolkata (38.3%). Such a situation raises the question: what will be fate of the city when the majority of the next generation is living in such deprivation?

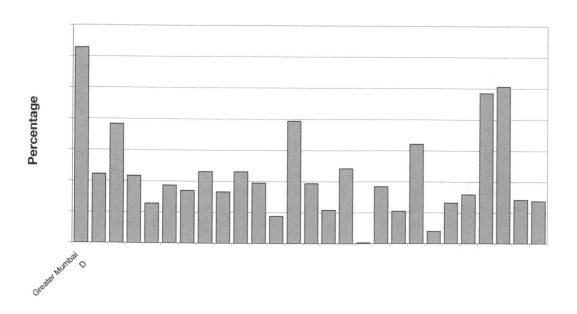

Employment in the slums

About one third of the slum dwellers in India are in work, close to the urban work participation rate (WPR) of 32.3%. Male WPR is high in the mega cities, suggestive of excessive male migration for employment. But despite relatively high employment, especially in slum areas, slum dwellers still live in poverty. The reasons are that most of them have a low literacy level and few technical skills, belong to a disadvantaged caste, and so end up in low wage jobs. The low wage jobs include daily labour, domestic help, sweepers, dish cleaners, rickshaw/cart pullers, and hawkers and so on. Slums dwellers are the service providers to non-slum citizens and support them in their daily life.

'World class' city – Mumbai, India?

Mumbai is India's premier industrial and commercial centre. The city accounts for about one-tenth of factory employment and manufacturing value-added in India, three-fifths of jobs in the oil industry, two-fifths of domestic air traffic and one-third of all tax revenues collected nationally (Swaminathan, 1995). The port of Mumbai handles more than one-third of the total value of foreign trade. About one-third of foreign tourists travelling to India visit Mumbai and 90% of them arrive by air (Deshpande, 1996). As India's economy becomes more open, with market-led economic liberalization policies in place, the strategic location of Mumbai with respect to global markets and its capacity to encourage growth and development of financial and business services gives it the potential to emerge as a major international city of the 21st century (BMRDA, 1995). Thus the city is important, not only in generating economic growth for the nation, but also as a potential facilitator, helping to integrate India's economy with the rest of the world.

As a result of its industrial and commercial pre-eminence within India, Mumbai has become one of the world's 'mega cities'. The population of Greater Mumbai in 2001 was 12 million, making Mumbai the one of most populous cities in the world. Yet over one-half of Mumbai's population live in conditions of abject poverty, squalor and deprivation (Hardoy and Satterthwaite, 1989). The poor live in overcrowded slums and hutments, on pavements, along railway tracks, beside pipelines, under bridges, on ill-drained marshland and on other vacant spaces available to them. Although not strictly categorized as 'slums', many relatively old and dilapidated single room tenements (*chawls*) house poor people. The slums and hutments are located in highly polluted and unhealthy environments, the result of proximity to industrial emissions and effluents, and/or poor sewerage, drainage and irregular garbage clearance. Conditions in the slums are terrible. Slum inhabitants constantly have to deal with issues such as constant in-migration pressure, lack of water, no sewerage or solid waste collection, lack of public transport, pollution and housing shortages. Infant mortality is high. During the monsoon season some slums are submerged knee-deep in water. Many dwellers work as labourers, domestic help or rag-pickers. It is against this background that the claim that ' ... Indian cities have become bulging overgrown villages and slums ... ' has been made (Desai, 1988; p 67) and Mumbai itself described as ' ... the unintended city' (Da Cunha, 1996; p 80).

Slums and shantytowns have been a part of Mumbai's landscape for a long time. The first official census of slums was carried out in 1976 by the Government of Maharashtra in Mumbai and its survey recorded 2.8 million people living in some 1,680 slum settlements when Mumbai's total population was 5.9 million (Government of Maharashtra, 1995; BMRDA, 1995). Today slum dwellers make up 55% of Mumbai's population, approximately 6.5 million people of the city's total of 12 million inhabitants. Slums are densely packed and more than 50% of Mumbai's population who are living in slums occupy just 3.5% of its space; that is, approximately 400,000 slum dwellers per square kilometre.[6] Most of these people are migrants from poorer regions or states. Some of the slums are very big and seem like cities in themselves. For example, Dharavi, Asia's largest slum, is spread over an area of 1.75 square kilometres in central Mumbai, with a population of over 1 million people living there.

In the city, where 55–60% of the people live in slums, it is the slum dwellers who dominate it. It seems that slums are the city; the legal areas are the small enclaves. The slum growth rate is actually greater than the general urban growth rate. The city is gaining the name 'Slumbay'. Slums in many world cities are seen as relatively small fragmented areas, but in Mumbai they are a dominating reality from which new urban forms are emerging, and which may need new conceptions of spatial planning to handle and to alleviate their inequalities.

The United Nations estimates that Mumbai's population will reach 26.1 million in the next 15 years (2015), making it the most populated city after Tokyo (United Nations Population Division, World Urbanization Prospects, 2002). However, considering only population numbers, but ignoring such a mass of slum dwellers, can Mumbai really be seen as a 'world class' city? To be seen as such, valiant efforts will be required to overcome the critical issue of its slums.

Conclusion

In India, the large metropolitan and mega cities are growing very rapidly, but unfortunately its slums are growing many times faster. Poverty, misery, exploitation, humiliation, insecurity, inequalities, and human unhappiness are also multiplying tremendously. These are indeed manifestations of India's inequitable society and faulty planning. Inequality, imbalance in levels of urbanization, low educational standards, poverty, migration (from both rural and neglected small and medium urban areas), low economic status, very limited job opportunities and lack of affordable housing are key components in the formation of slums. Such factors need urgent attention and immediate remedies. Overcoming one factor alone will not solve the slum and poverty issues. The factors are interrelated and need to be tackled in a holistic manner.

It could be argued that India is a fragmented society, with a wide gap between rich and poor. In the cities, fragmentation becomes extreme, with the trend of domination by the numbers of very poor slum dwellers, and the spatial and economic disparities that result. India has 35 mega cities, an economy that is growing fast, and the potential to attract the international investment and businesses that could place them in the world city category. None of these cities are 'world-class' as yet, although both Mumbai and Delhi have been classified as showing relatively strong evidence of world city formation (Beaverstock *et al.*, 1999). But even if the international businesses and producer companies appear and the cities make it onto the 'list', the predominance of slums and their poor would compromise any claim to be truly 'world class'. International businesses and producer companies perhaps would represent a 'world class' fragment, or enclaves of privilege in cities of disadvantage and poverty.

The problems of the cities occur in a wide regional and national context. The syndrome of poverty, distressed migration and urban degradation in India needs a novel approach. Realistic plans and effective implementation for the benefit of humankind and national growth are absolutely necessary. Effective policies need to be framed for development, not only in rural areas but also in small and medium sized cities. There is an urgent need to generate job opportunities in all regions of India to reduce the gap of inequalities. This will assist in combating poverty and human misery, and may alleviate some of the pressures that lead to the widespread formation of slums.

Notes

1. Examples include in French: Bidonvilles, Taudis, Habitat Précaire, Habitat Spontané; in Spanish: Asentimientos Irregulares, Barrio Marginal (Barcelona), Barraca (Barcelona), Conventillos (Quito), Colonias Populares (Mexico), Tugurio and Solares (Lima), Bohios, Cuarterias or Solar (Cuba), Villa Miseria (Argentina); in Arabic: Mudun Safi, Lahbach, Brarek, Medina Achouaia, Foundouks and Karyan (Rabat-Sale), Carton, Safeih, Ishash, Galoos and Shammasa (Khartoum), Tanake (Beirut), Aashwa'i and Baladi (Cairo); in Russian: Hrushebi, Baraks (Moscow); in Portuguese: Favela, Morro, Cortiço, Comunidade, Loteamento ; in American English: Hood (Los Angeles), Blight areas; in Indian languages 'chawls'/chalis (Ahmedabad, Mumbai), Ahatas (Kanpur), Katras (Delhi), Bustee (Kolkata), Zopadpattis (Maharashtra), 'cheris' (Chennai), and in other languages Katchi Abadis (Karachi), Iskwater, Estero, Eskinita, Looban and Dagat-dagatan (Manila), Umjondolo (Zulu, Durban), Watta, Pelpath, Udukku or Pelli Gewal (Colombo); Museques (Angola), Chereka Bete (Ethiopia) (UN-HABITAT (2003a), 'The Challenges of Slums – Global Report on Human Settlements', pp. 9–10).
2. http://siteresources.worldbank.org/DATASTATISTICS/Resources/GNIPC.pdf
3. Brazil: GNI- $8,800(rank-91), PPP-$4730 (rank-92); Russian Federation: GNI- $11,620 (rank-78), PPP-$5,780 (rank-79); China: GNI-$7730 (rank-102), PPP-$2,010 (rank-129).
4. http://nbo.nic.in/housing%20data%20table.pdf
5. http://mhupa.gov.in/pdf/annual-reports/2001-2002/4.pdf
6. http://news.bbc.co.uk/1/hi/world/south_asia/831833.stm
7. http://urbanindia.nic.in/moud/urbanscene/urbanmorpho/urbanmorph.htm

References

Abbott, J. (2002a) An analysis of informal settlement upgrading and critique of existing methodological approaches. *Habitat International*, 26(3): 303–315.
Abbott, J. (2002b) A method-based planning framework for informal settlement upgrading. *Habitat International*, 26(3): 317–333.
Beaverstock, J.V., Smith, R.G. and Taylor, P.J. (1999) A roster of world cities. *Cities*, 16(6): 445–458.
Bijlani, H.U. and Roy, P. (1991) *Slum Habitat*, Har-Anand Publication, Delhi.
Birdi, H.D. (1995) *Slums and Urbanisation*, Vidhi Publication, New Delhi.
BMRDA (1995) *Draft Regional Plan for Mumbai Metropolitan Region, 1996–2011*, Bombay Metropolitan Region Development Authority (BMRDA), Mumbai.
Census of India (2001a) *Migration tables, India and States*, Tables D-1, D-2, D-3, D-4, Department of Publication, Govt. of India.
Census of India (2001b) *Population Tables*, Department of Publication, Govt. of India.
Census of India (2001c) *Slum Population*, Department of Publication, Govt. of India.
Choguill, C.L. (1999) Community Infrastructure for Low-income Cities: The potential for Progressive Improvement. *Habitat International*, 23(2): 289–301.
Connell, J. *et al.* (1976) *Migration from Rural Areas: The Evidence from Village Studies*, Oxford University Press, Delhi.
Connell, J. (1987) Migration, Rural Development and Policy Formation in the South Pacific. *Journal of Rural Studies*, 3(2): 105–121.
Da Cunha, G. (1996) Bombay and the International Experience, in *Cities and Structural Adjustment* (eds N. Harris, and I. Fabricus), pp. 80–92.

Daniere, A. G. and Takahashi, L. M. (1999) Poverty and Access: Differences and Commonalities across slum communities in Bangkok. *Habitat International*, 23(2): 271–288.

Davis, M. (2005) *The Planet of Slums*, Verso Publication, New York.

Desai, V. (1988) Dharavi, the largest slum in Asia. *Habitat International*, 12: 67–74.

Deshpande, L. (1996) *Impact of Globalization on Mumbai in Globalization and Mega City
Development in Pacific Asia*, UNU-UNESCO Workshop, United Nations University,
Institute of Advanced Studies, Hong Kong and Tokyo, 13–15 October 1996.

Eke, E.F. (1982) Changing Views on Urbanisation, Migration and Squatters. *Habitat
International*, 6(1/2): 143–163.

Epstein, T.S. (2001) Development – There is another way: Rural-Urban Partnership
Development paradigm. *World Development*, 29(8): 1443–1454.

Government of Maharashtra (1995) *Programme for Reconstruction and Rehabilitation
of slum and Hutment Dwellers in Greater Mumbai*, Afzulpurkar Committee Report,
Mumbai.

Haberfeld, Y. *et al.* (1999) Seasonal migration of rural labour in India. *Population Research
and Policy Review*, 18(6): 471–87.

Hardoy, J. E. and Satterthwaite, D. (1989) *Squatter Citizen*, Earthscan, London.

Hirway, I. *et al.* (2002) *How Far Can Poverty Alleviation Programmes Go? An Assessment
of PAPs in Gujarat Ahmedabad*, Centre for Development Alternatives, India.

Kapoor, M. *et al.* (2004) *Location and Welfare in Cities: Impacts of Policy Interventions on
the Urban Poor*, World Bank Policy Research Working Paper 3318.

Lipton, M. (1980) Migration from Rural Areas of Poor Countries: The Impact on Rural
Productivity and Income Distribution. *World Development*, 8: 1–24.

McDonald, D. A. (1999) Lest the rhetoric begin: migration, population and the environment
in Southern Africa. *Geoforum*, 30: 13–25.

Ministry of Housing and Urban Poverty Alleviation (2001) *National Building Organization*,
Govt. of India.

Ministry of Housing and Urban Poverty Alleviation (2002) *Annual Report 2001–2002*,
Chapter 4, p. 17.

Ministry of Urban Development (2001) *Population Census 2001*, Urban Morphology.

Mosse, D. *et al.* (2002) Brokered livelihoods: Debt, labour migration and development in
tribal western India. *Journal of Development Studies*, 38(5): 59–88.

Mukherji, S. (2001) *Low Quality Migration in India: The Phenomena Of Distressed Migration
And Acute Urban Decay*, 24th IUSSP Conference, Session 80: Internal Migration – Social
Processes And National Patterns, Salvador, Brazil.

Mukhija, V. (2001) Upgrading Housing Settlements in developing Countries: The Impact of
Existing Physical Conditions. *Habitat International*, 18(4): 213–222.

Mukhija, V. (2002) An analytical framework for urban upgrading : property rights, property
value and physical attributes. *Habitat International*, 26: 553–570.

National Institute of Urban Affairs (2001) *Urban Statistics: Handbook 2000*, National
Institute of Urban Affairs (NIUA), Delhi.

Neuwirth, R. (2005) *Shadow Cities, a billion squatters: a new urban world*, Routledge,
Oxford, UK.

Payne, G. (2001) Urban land tenure policy options: titles or rights? *Habitat International*,
25: 415–429.

Planning Commission, Government of India (1992) *8th Five Year Plan*, Vol. 2, Department
of Publication, Govt. of India.

Planning Commission, Government of India (2001a) *Indian Planning Experience: A
Statistical Profile*, Department of Publication, Govt. of India.

Planning Commission, Government of India (2001b) *National Human Development Report
2001*, Department of Publication, Govt. of India.

PRAXIS (Institute for Participatory Practices) (2002) *MP Participatory Poverty Assessment*,
Report, prepared for the Asia Development Bank.

Rogaly, B. *et al.* (2001) Seasonal migration, social change and migrants rights, lessons from
West Bengal. *Economic and Political Weekly*, 36(49): 4547–59.

Srivastava, R.S. (1998) Migration and the labour market in India. *Indian Journal of Labour
Economics*, 41(4): 583–616.

Srivastava, R. (2003a) *An overview of migration in India, its impacts and key issues*, Regional
Conference on Migration, Development and Pro-Poor Policy Choices in Asia, Dhaka,
Bangladesh.

Srivastava, R.S. (2003b) Regional growth and disparities in Alternative Economic Survey
2001–2, in *Economic Reform: Development Denied* (eds Alternative Survey Group),
New Delhi, Rainbow Publishers, India.

Swaminathan, M. (1995) Aspects of urban poverty in Bombay. *Environment and Urbanization,* **17**: 133–143.

Town and Country Planning Organisation (1996) *A Compendium on Indian Slums 1996,* Department of Publication, Govt. of India.

Viratkapan, V. and R. Perera (2006) Slum relocation projects in Bangkok: What has contributed to their success or failure. *Habitat International,* **30**: 157–174.

UN-HABITAT (2003a) *The Challenges of Slums – Global Report on Human Settlements,* Earthscan, London.

UN-HABITAT (2003b) *Slums of the World: The face of Urban Poverty in the New Millennium?,* United Nations Human Settlement Programme.

United Nations Population Division (2002) *World Urbanization Prospects 2002,* United Nations.

World Bank (2007) *World Development Indicators database,* World Bank.

Conclusion

Mike Jenks, Daniel Kozak and Pattaranan Takkanon

Conclusion:
The Form of Cities to Come?

Introduction

The inspiration for this book came from the wide range of debates, problems and issues raised at the 7th International Urban Planning and Environment Association's Conference (UPE7, 2007) entitled 'World Class Cities: Environmental Impacts and Planning Opportunities'. These were important issues to tackle and this book has set out some challenging questions: when urban forms are clearly polycentric and cities becoming more fragmented, how significant are these concepts, can anything be learned from practice and are there any commonalities that help guide ways forward? Not only is this broad in scope, both in terms of the concepts discussed and cities reviewed, but also it is clear that the size, complexity, and differences between these cities make generalisations difficult. A further complication is that many of the debates and issues are separated in academic research and publication. That is particularly true for the two debates which are primarily discussed in this book: the world city debate and that of the sustainable urban form. To this end this book attempts to link the concepts of polycentrism and fragmentation to the question of whether there are any pathways or processes in the context of world, or even 'world class', cities in achieving more sustainable urban forms. Dcrudder and Witlox clearly define the world city debate and other chapters, such as those of Marcuse and Radović, extend the debate into issues associated with globalisation. Many of the contributions to this book come from research fields that delve into the relationships between urban planning and urban form, and with sustainability. This combination of research agendas is not without conflicting positions and even contradictory uses and understandings of terms. One of the key starting points is from the standpoint of world cities, and the identification of a gap in the debate, namely about the physical form these cities take. This is the first aspect of inquiry in the book.

Fig. (opposite page). It is not only planning that will change the form of cities, but also sustainable design and technologies – a zero carbon emissions mixed use development in Penryn (Bill Dunster, Architects)
Source: Mike Jenks

World city as a category for urban form inquiry

Throughout the book 'world class' has been put in parenthesis to reflect the range of, and ambiguities in, the meaning of the term. However there is an underlying consistency in its use, largely reflecting a qualitative definition. Global and world cities, as usually defined, clearly relate to the realms of economic power and global 'weight' – there are few ambiguities there. Many of the chapters draw on the GaWC 1999 roster of world cites. This has been updated in 2005 to categorise the 'leading' world cities, bringing in other criteria including measurements of economic, cultural, political and social aspects. A resulting list which supplements the 1999 roster has changed and been reduced in size. It is based on rigorous analysis and measurement drawing on a range of data sets in order to help overcome the 'evidential deficit' in world cities research (Taylor, 2005). However, there are inevitably dangers in this approach. What is possible to measure and include, also opens up the debate about what is excluded (Radović). Many aspects may be difficult to measure, many may have problematic data sources, and some may be purely qualitative and subjective. Wikipedia (2007) – while far from a peer reviewed resource – conveniently draws together a range of sources that point to other criteria that might relate to world cities including for example: population; significant transport infrastructure, techno-logical capabilities, institutions and cultural facilities; sites of religious pilgrimage, and world heritage sites; tourism throughout; and, quality of life and cost of living. Most of these have data sources, but even in this extended list, sustainability is missing (although quality of life might be seen as a partial substitute), and environmental issues and the environment are not there at all.

Does this leave a genuine place for the classification of 'world class' – to include the less easy to measure and the qualitative – for quality of urban life, environment and sustainability? It is likely that many of the world's great cities will embrace all three terms, but there are many that could be termed 'world class' cities that will never be world or global cities – for example one could think of Venice in that respect. It is also probable, especially with the rapid urban and economic growth and power of Asia, that there will be world cities that would be hard pressed to achieve a quality that could be considered 'world class'. Certainly the ubiquity of the term, as used by authors of chapters in this book, has that qualitative dimension behind it, opening up consideration of the changes in urban forms that may or may not affect sustainability. Far from being exhausted the debate around these concepts is increasingly lively, although it still requires more explanation of how they are understood and used. One thing is for sure, these concepts cannot be taken for granted: they need to be challenged and continuously reassessed. For example Rogers, asking what is to be sustained and for whom, raises key questions. Many academic concepts and sustainability in particular, have entered the media and political discourses. Sustainability is sometimes used as a catch-all term and moulded to assist the most diverse arguments. Its capacity to function universally is questionable and what may be sustainable for some is not necessarily sustainable for all. When polycentric urban forms and urban fragmentation are placed then in the context of world cities, the need to define, debate and provide insight and guidance becomes pressing. It is here that this book attempts to expand the debate. We suggest few measures, but pose questions and suggestions as to how the debate can be extended.

Polycentrism and urban fragmentation: the unsustainable trend

When considered globally, it is easy to see a polycentric world of cities, interconnected through networks of communications, businesses, global companies, international airport hubs, and so on. Mostly these are the networks of cities, usually being the world cites as described, for example, by Derudder and Witlox. At a lesser scale, Hudec and Urbančíková discuss the concept of polycentrism at a regional, supra-national level between Slovakia and Austria, and consider its potential to increase regional competitiveness and access to knowledge. There are other positive perspectives on the growth of polycentric forms, such as the interconnected mega-city regions of Japan (Kaido and Kwon), or the recognition of the importance of transportation connecting the major cities of Taiwan in corridor with trans-national potential to connect with mainland China (Lo). Managing the growth of polycentric regions is also critical, through public transportation infrastructure within the Barcelona metropolitan region or the Randstad in Holland (Carné and Ivančić; Jenks and Kozak), or through growth management policies in the Helsinki region (Maijala and Sairinen), or policies, such as in Singapore, to enhance landscape and green spaces (Yang). The positive potential of polycentrism can be seen as partly a-geographical, not dependent on place but rather communications that may be ethereal, and partly related to location with physical links that allow for the movement of people in addition to the communications between them. The idea for, or hope of, integration and inclusion is a common theme.

A similar view of polycentrism occurs within cities themselves, connected by transport and electronic communications, and characterised as networked cities. The arguments are more formal, related to centralities within cities that are, or are hoped to become, compact urban forms. The issues are about the polycentric forms that have higher densities that enable easier access to essential facilities (Kaido and Kwon), that use high densities and high quality design in urban regeneration projects (Carné and Ivančić), or that use the centralities as a locus for creating high density liveable areas and that consider the need of local people (So, Maijala and Sairinen). Again the imperative to connect the centralities within cities with public transport is fully recognised. Given these potential and positive impacts, it should lead to networked cities that are approaching the sustainability of that claimed for a monocentric compact city: good transport, spatial proximity, less dependency on private transport, and economic and social diversity.

However, the trend is moving towards the break-up of cities and the loss of the coherent urban forms so beloved of urban design theorists. The emerging segregated and excluding urban forms such as ghettos and gentrified neighbourhoods, and forms compromised by the mores of security are well characterised by Marcuse. And the general trends towards urban fragmentation are illustrated through examples in Buenos Aires and Bangkok by Jenks and Kozak. Urban fragmentation takes many forms. It can be the result of planning policies such as the special zones created in Shanghai's peri-urban area (Wu) leading to dispersed, car-dependent disconnected urban areas, or areas of disadvantaged housing planned on the Paris periphery (Pavel). More likely, fragmentation will occur through lack of planning, either by deliberately turning away from it in favour of the private sector, as in Houston (Qian), or by poverty and the spread of informal settlements and slums as in Mumbai (Neekhra *et al.*). The behaviour of the middle class and rich pushes the trend towards fragmentation even further through property speculation and the displacement of the poor (Fahmi), through lifestyle aspirations and desires to be part of a global style or gather with the like-minded in gated communities (Sintusingha; Karnchanaporn and Kasemsook). Further fragmentation and exclusion occurs through the privatisation of public space and of private shopping malls – such forms that once were the preserve of the suburbs have also appeared in main urban areas, as in Buenos Aires (Kozak). Surprisingly transport can also add to fragmentation, not just from the barriers caused by large heavily trafficked roads, but through pricing and behaviour – which again serve to isolate and exclude large sectors of the urban population (Charoentrakulpeeti and Zimmermann). The picture is of change and urban growth that is physically, socially and economically fragmented. Even for the highly 'connected' global and world cities, the exchange of people and cultures is amongst elite (Radović); for the majority of urban populations it is their city and region that will need to provide for their quality of life. Given these trends, how does sustainability enter the frame?

Towards more sustainable urban forms: polycentrism and sustainability

The positive view of polycentric forms is that they might be in effect small compact cities – or at least follow some of the principles embodied in that concept. Despite some uncertainties about compact cities, there is a fair amount of knowledge about sustainable design principles. There are some aspects where there is sufficient evidence to feel confident that they will contribute to sustainability and these include: land use and built form; environmental design; and communication and transport. It is important to preserve land through the intensification of its use, by consolidation and urban compaction policies, and to re-use existing urban land where building have become redundant – that is brownfield development. While these will increase density, attention needs to be paid to building in green spaces and green corridors, adding a further network to encourage biodiversity. Buildings should be designed with low embodied energy, and the recycling and reuse materials need to be considered. It goes without saying that all buildings and designs should be energy efficient, highly insulated and at least 'carbon neutral', but where possible even generators of energy. Systems need to be installed that monitor energy use and water use, so every resident can see the impact they have, the cost of their use of resources, and thus, hopefully modify their behaviour to consume less. But any energy used should be from renewable and local power generation or combined heat and power systems should have preference – systems that will be enabled by higher densities. Taking care of water, one of the most precious of resources, should be achieved through using grey water and recycled water for gardening, landscaping, and flushing WCs. The urban design needs to ensure that it is pedestrian friendly, and that distances between facilities or public transport stops are close enough (usually an average of 400m from dwellings) to encourage walking. Public transport also should be considered from a sustainability perspective: light rail or eco buses using bio-fuels should be considered. And to help make public transport more accessible and user-friendly real-time information needs to be available, not just at transport stops, but also in people's homes or places of work.[1]

Good urban design can help to create urban areas that are vibrant, lively and viable, and may overcome the negative aspects of urban development, by avoiding monotonous single use environments, the dead downtown office and commercial CBDs that close after working hours, and dead residential dormitory towns with vast areas of housing and few facilities to support it. A mixture of uses, and densities that are high enough to be culturally 'comfortable' and provide viable markets for local business and public transport should be provided. Such forms of design at the urban level are increasingly common in Europe and some other parts of the world, and are clearly related to their urban cultures and contexts[2]. However, they are often fairly small developments. There are questions about whether or not these provide a generalisable model for other parts of the world, particularly for world and mega-cities.

However, the physical environment is just one aspect; others need to be included, for example:

- Social sustainability – equity, social justice, poverty and social exclusion.
- Economic sustainability – income inequalities, employment, education and training, local business, services and facilities.
- Lifestyles and attitudes need to change – ever more consumption, with higher expectations, driven by a celebrity culture seem ultimately unsustainable.

The naïveté of assuming that what may work in one context and culture will work in another raises questions about what may seem obvious 'solutions' but which may be either myths, or concepts that are vague and imprecise. For example, at the heart of many arguments for sustainable urban form is the idea of higher densities – it is often repeated in guidance and policy documents throughout the world. And yet, it is not clear what high density means. It is not an absolute and clearly it is culturally specific – the table below shows the relative densities of some selected cities. Clearly what might be high density for London is likely to be seen as low density in Hong Kong.

Table 1. Relative densities, selected cities and metropolitan regions
Source: Mike Jenks

	Inner City	Metro Region
Hong Kong	7.5	10
Cairo	4.5	8
Bangkok	2	2
New York	1.5	1
London	1 (7250 p/km^2)	1 (600 p/km^2)
Los Angeles	−2.5	−3.5

But, even with the best intentions, and the best practice from a European context, it seems unlikely that a 'solution' can be easily transferred, perhaps this will be demonstrated in the much heralded 'world's first sustainable city', Dongtan within the Shanghai metropolitan region. Dongtan is being aggressively promoted by its overall designers ARUPs as a model for China, and yet it is clear that as a 'model' it is not greatly different in form and density, and other technicalities, to schemes such as Greenwich Millennium Village in London. For China and Shanghai, for all its technical wizardry, it may become a relatively low density, upper-middle class rich suburb for commuters to Shanghai – anything but socially and economically sustainable.

A process of defragmentation: a pathway to urban sustainability

What can be done to *defragment* the fractured cities? Polycentric forms and their tendency to lead to urban fragmentation pose new problems in achieving a measure of sustainability. A new paradigm is needed that may incorporate relevant aspects of compact city theory, but that is meaningful not just physically, but also culturally, especially in a non-Western context as this is where the majority of urban growth is occurring. This may require a process of integration, articulation, joining and inclusion, of bringing together urban, social and economic fragments – of defragmenting the city.

One of the shared concerns that unites many chapters in this book is a common interest for accuracy in the use of concepts and a quest for the right term to express an idea. That is not, as Derudder and Witlox note, 'a trivial matter of semantics'. Maijala and Sairinen show how the lack of definition of a concept introduced as a planning policy – i.e. 'urban consolidation' in Finnish planning – has lead to otherwise unexpected negative outcomes. If certain urban phenomena, which is identified as negative, is conceptualised as urban fragmentation, it would be desirable to propose a set of urban policies that could address this problem, and these policies would have to be defined. The term put forward here is 'urban defragmentation'.

The reason for favouring the term 'defragmentation' over the perhaps most obvious choice 'integration', is related with this same quest for conceptual accuracy. 'Integration' is attached to a string of meanings and connotations that are not necessarily related to the discussion behind the concept of urban fragmentation. In various discourses in the social sciences the concept of integration is associated with quite particular meanings. In some cases, it is understood as the process in which a subjugated entity (e.g. an ethnic minority) *integrates*, or *is integrated*, to the dominant entity (e.g. the majority culture). Integration is also sometimes linked to the idea of homogenisation, and for that reason is resisted in postmodern discourses. As Roland Barthes ([1953] 1968, p. 16) argues, 'language is never innocent: words always carry attached a second memory that mysteriously persists in the midst of new meanings'. So, the term integration does not always work as an instant opposite of fragmentation. Conversely, urban defragmentation cannot be understood as anything else except as its exact opposite. Thus the first step before developing urban defragmentation policies is to define clearly what is understood by urban fragmentation.

There are many obvious signs of urban fragmentation, the clearest being the increasing numbers of gated communities – defensive bastions against the poor, or selling 'desirable' lifestyle aspirations. These exclude the majority, and often exclude real life and the benefits of the urban experience. The privatisation of housing areas is also reflected in many city centres, or in new centralities, through the privatisation of space – the ubiquitous shopping mall being a prime example.

Even 'good practice' can lead to urban fragmentation. Urban regeneration is a much heralded policy, in bringing life and economy back to city centres. Too often this leads to gentrification, steep increases in land and property values, and the banishment of the poor and even middle classes to the peripheries. The recognised need to link these disconnected places with public transport can also end up spawning another form of fragmentation. Unless very heavily subsidised, unlikely in a world dominated by neoliberal economies, mass rapid transit provides an efficient but expensive mode of transport, and thus the preserve of the middle classes. Roads dominate transport for the very rich (through choice) and the poor through lack of choice. In many developing countries, poorly maintained but cheap buses, motor cycles and cars are the dominant mode of transport on usually inadequate and congested roads. Again the rich benefit from toll roads and urban motorways, while the less well off suffer the public roads. And of course the roads, especially the urban motorways add to the physical fragmentation of the environment.

The traditional view of the compact city needs to be seen from a different perspective. There may be a network of compact centralities or cities. The key is not just the nodes, but the network links between them that are equally significant. These must not exclude, but must be affordable to all. This can never happen unless there is significant public investment for the public good. Neoliberal economies favouring private investment for the private good is a paradigm doomed to failure if sustainability is to be the goal. The excuse that China is now responsible for much the same level of greenhouse gas emissions as the USA is risible, when *per capita* a Chinese citizen consumes around one sixth of that of an American citizen. For the poorest countries higher standards of living are justified, but for the wealthiest, reductions in the bloated consumption of resources are needed. It may be that polycentric cities and regions can take forms that are more integrated – that a process of defragmentation can occur, within a physical framework of compact nodes that enables citizens to behave more sustainably. The defragmented city must be one that is not only physically defragmented, but also socially and economically derfagmented as well.

We end on a note of caution. There is a limit to how far urban planning and physical form can go. It can provide a framework to enable, rather than disable, sustainability. But it is lifestyle, and aspirations that may need to change. The discussion about world cities has pointed to the use of vague, undefined terms, and the same is true of one of the key aspects 'quality of life'. This is the underlying aim for all the world's citizens, but it is not an absolute, and pursuance of quality of life at any cost in a planet whose resources are limited is clearly unachievable and unsustainable. In a networked world, the commodification of quality of life is clear, spurred on by globalisation. There are alternatives that may not be so dependent on the endless pursuit of wealth and consumption. In 2006 the New Economics Foundation derived a measure called the 'Happy Planet Index (HPI)' which took into account life satisfaction, life expectancy and ecological footprint (Marks *et al.*, 2006). When HPI is measured it is interesting to note that happiness and high GDP do not correspond: USA is placed at 150 (out of 178 countries) on the HPI (ibid.), yet it has the world's second highest GDP (Smith, 2003). Indeed most of the countries in which cities in this book appear are a long way down the happiness list, and almost the reverse of the world city rankings. Of course, the measurement of 'happiness' is controversial and it could be endlessly argued about; nevertheless it does indicate the impact of including sustainability (ecological footprint) as a part of the picture. This point returns us to the genesis of this book in UPE7. The keynote address was given by Jigmi Thinley, the Bhutanese Minister for Home and Cultural Affairs. He introduced the way that the Royal Government of Bhutan had replaced the measurement of Gross Domestic Product (GDP) with Gross Domestic Happiness

(GDH).[3] It is interesting to note that the criteria for GDH closely match those of sustainable development, and bring in a care for environment, culture and social development in a balanced and sustainable way. While we can provide some ways of planning and design to world cities, and mega-cities, and suggest frameworks and process that might aid achieving sustainable urban form, it is clear we still have a lot to learn.

Notes

1. The factors outlined here that are considered to contibute to sustainable urban forms can be found in, for example, Jenks *et al.* (1996); Williams *et al.* (2000); and (in a UK context) in many government documents to be found at www.communities.gov.uk.
2. See CABE 'Building for Life' website (http://www.buildingforlife.org/). For example, schemes that embody some, or many of the design issues raised, are shown on this website.
3. The full text of the keynote address can be found on the UPE7 website: http://www.arch. ku.ac.th/upebangkok/index.php?menu=down.

References

Barthes, R. (1968) *Writing degree zero*, Hill and Wang, New York [First published 1953].

Jenks, M., Burton, E. and Williams, K. eds (1996) *The compact city: a sustainable urban form*, E & FN Spon, London.

Marks, N., Abdallah, S., Simms, A. and Thompson, S. (2006) *The (Un)Happy Planet Index: An index of human well-being and environmental impact*, New Economics Foundation, London.

Smith, D. (2003) *The State of the World Atlas*, Earthscan, London.

Taylor, P.J. (2005) Leading World Cities: Empirical Evaluations of Urban Nodes in Multiple Networks. *Urban Studies*, 42(9): 1593-1608.

UPE7 (2007) http://www.arch.ku.ac.th/upebangkok/index.php?menu=down Chulaborn Research Institute, Bangkok. Thailand.

Wikipedia (2007) http://en.wikipedia.org/wiki/Global_city

Williams, K., Burton, E. and Jenks, M. eds (2000) *Achieving Sustainable Urban Form*, E & FN Spon, New York.

Index

Printed and bound by CPI Group (UK) Ltd, Croydon, CR0 4YY

01/11/2024

01782605-0006